NEURONS AND NETWORKS

NEURONS AND NETWORKS

An Introduction to Neuroscience

JOHN E. DOWLING

THE BELKNAP PRESS OF

HARVARD UNIVERSITY PRESS

CAMBRIDGE, MASSACHUSETTS LONDON, ENGLAND 1992

Library of Congress Cataloging in Publication Data

Dowling, John E.
  Neurons and networks : an introduction to neuroscience / John E. Dowling.
    p.   cm.
Includes bibliographical references and index.
ISBN 0-674-60820-8 (alk. paper)
1. Neurology.   I. Title.
[DNLM: 1. Nerve Net.   2. Nervous System.   3. Neurons.   WL 102.5   D747n]
QP355.2.D68   1992
591.1'88—dc20
DNLM/DLC
for Library of Congress   91-23583
                                                                              CIP

Designed by Gwen Frankfeldt

FOR ALEXANDRA

Contents

PART TWO Networks: Integrative Neuroscience

Preface

"To know the brain . . . is equivalent to ascertaining the material course of thought and will, to discovering the intimate history of life in its perpetual duel with external forces."

—Santiago Ramón y Cajal, *Recollections of My Life* (1937), describing why the study of the nervous system attracted him "irresistibly" in the 1880s

"It is essential to understand our brains in some detail if we are to assess correctly our place in this vast and complicated universe we see all around us."

—Francis Crick, *What Mad Pursuit* (1988)

The brain—how it works and how it controls our actions and behavior—has long fascinated scientists and philosophers, yet substantial progress in understanding brain mechanisms began only about a century ago. Since then the pace of brain research has accelerated greatly, and in the past twenty-five years neuroscience has become one of the most active branches of all the natural sciences.

A Young Field Coming of Age

As a separate but integrated area of inquiry, neuroscience is a relatively new field. It is derived in large part from the merging of three quite separate disciplines: neuroanatomy, which focused on the structure of neural tissue and nerve cells; neurophysiology, which investigated how neural tissue and nerve cells function; and neurochemistry, which was concerned mainly with the kinds of substances found in brain tissue. Prior to the 1960s these fields were centered mainly in separate departments in medical schools, although the study of nervous systems, particularly their comparative aspects, could be found in departments of zoology and biology in colleges and universities. But there was minimal communication among researchers in the separate disciplines; each moved steadily ahead but without much interaction.

The beginnings of an integrated field of neuroscience came about in the early 1950s, when two new techniques were used to study neural tissue: electron microscopy (in neuroanatomy) and intracellular electrical recording (in neurophysiology). Electron microscopy enabled us to see for the first time what is inside nerve cells, while intracellular recording allowed us to record electrical responses generated in single nerve cells. These two techniques led almost simultaneously in the early 1950s to the discovery of small vesicles within nerve terminals and to the realization that small, discrete electrical responses occur in a nerve cell when it receives input from another nerve cell. These observations would lead to important advances in our understanding of the brain.

It had been known for nearly a century that the brain is made up of billions of individual nerve cells that communicate at specialized junctions called synapses, and it had long been supposed that understanding the nature of these synaptic interactions is the key to learning how the brain works. But progress in understanding synaptic mechanisms was slow. Early in the twentieth century investigators first suggested that nerve cells communicate at synapses by chemical means. They proposed that nerve cells release chemicals that excite or inhibit the receptive cells. Not until mid-century, however, was this theory universally accepted—but the details of the synaptic process remained largely unknown.

The discovery of small vesicles in nerve terminals adjacent to synaptic junctions suggested that the chemicals released at synapses are stored in these vesicles. Moreover, it was realized that the small electrical responses in the postsynaptic cell were related to the released chemicals: the contents of one synaptic vesicle caused a quantal electrical event. These findings immediately linked neuroanatomy and neurophysiology. The anatomical discoveries provided a plausible explanation for the physiological findings, while the physiological findings indicated the function of the newly visualized anatomical structures.

What came next was an explosion of activity that cemented the bond between neuroanatomy and neurophysiology. For example, once the sites of synaptic interaction between nerve cells could be recognized, neuroanatomists began to work out the wiring patterns among nerve cells. Soon it was possible to correlate anatomical specializations with specific synaptic interactions, namely those that excited adjacent nerve cells and those that inhibited them. The term *neurobiology* was coined, and new, interdisciplinary departments of neurobiology were formed that embraced the contributions of both the neuroanatomist and the neurophysiologist.

Neurochemistry was also advancing at this time, particularly in the search for and identification of the substances released at synaptic sites. But close links between chemistry and anatomy–physiology were not forged until the 1970s, when researchers realized that two kinds of substances are used to communicate information in the brain: neurotransmitters, which lead directly to the excitatory or inhibitory electrical responses of nerve cells, and neuromodulators, which exert their effects on nerve cells by modifying the cell's biochemistry. Neurotransmitters mediate fast information processing in the brain, whereas neuromodulators appear responsible for mediating slower, longer-lasting changes in the brain. Memory and learning, for example, are thought to be initiated by the action of neuromodulatory substances.

After the integration of neurochemistry with anatomy and physiology, the term *neuroscience* came to the fore—the term now generally used for the study of the brain. In neuroscience laboratories, anatomical, physiological, and chemical experiments are done side by side, often by the same investigator. At the same time neuroscientists are being joined by computer scientists, psychologists, and molecular biologists as they turn toward questions of brain function. New terms such as *cognitive neuroscience*, *computational neuroscience*, and *molecular neuroscience* signal that even more interdisciplinary work is taking hold in the field. Where we will be in the next century is anyone's guess, but I am confident that along the way we will witness exciting advances.

Why a New Book?

The origins of this book lie in an introductory course in neuroscience, Biology 25, that I have taught at Harvard for the past twenty years. The course is designed for sophomores who have had introductory biology and chemistry in college and high school physics. It is intended to introduce students to neuroscience relatively early in their college careers, so that they can explore further in the field if they so desire. Many have, and each year a number of Biology 25 graduates write an undergraduate thesis on a topic in neuroscience.

Over the years, I have used in the course virtually every textbook available. None has been entirely satisfactory, and so I decided to write my own. Textbooks in neuroscience (and most other branches of biology) have become more and more encyclopedic and, in my view, less and less useful for the beginning student or for others interested in the principles of a field. I have been selective in the topics presented in this book, but I cover those topics in more depth than is generally the case in an introductory text. I also emphasize results and concepts that have stood the test of time rather than more recent work that is likely to change as new research is carried out.

In writing this book, I often chose examples from the visual system, not only because it is my own area of expertise but also because it is perhaps better understood than any other system in the brain. No matter which part of the brain is under discussion, however, whenever possible I describe the experimental observations that led to a discovery and then explain or speculate on the significance of the findings. Most of the drawings are based on real recordings and experiments, but in many cases I have simplified or stylized the figures for ease of interpretation. I have tried to present the major ideas in neuroscience in a clear and simple way with a minimum of technical terms, and with the hope the book conveys the excitement of this rapidly expanding field.

The book is divided into two sections. The first, on cellular neuroscience, focuses on nerve cells, their structure and function. Particular attention is paid to the mechanisms by which neurons receive, carry, and transmit information. The second half of the book, on integrative neuroscience, uses examples of the processing of information by nervous systems in vertebrate and invertebrate animals to describe how behavior emerges from this neural processing. The concluding chapters explore the development of the brain and higher brain function in humans. A short appendix on the concepts of electrical charge and circuits is provided for those whose physics is rusty.

Earlier versions of the manuscript were used successfully in Biology 25 in 1990 and 1991, and I appreciate the many comments, suggestions, and corrections made by both students and teaching fellows. A number of colleagues, including Evan Balaban, Anne Bekoff, Dave Bodznick, Ron Calabrese, Tom Carew, John Donoghue, Dan Jay, Bob Josephson, Ann Kelly, Eve Marder, Elwin Marg, and Steve Zottoli have read parts or all of the book and I am grateful to all of them for their suggestions. I owe special thanks to Ron Hoy and his colleagues at Cornell University for helpful advice as I was revising the final draft.

Barbara Barnett beautifully rendered the figures from original drawings made by Patricia Sheppard. Pat also prepared the half-tone prints and, with Stephanie Levinson, who typed innumerable drafts of the manuscript, dealt with the endless administrative details of the project. They were indispensable. Last, but by no means least, Howard Boyer and Kate Schmit of Harvard University Press expertly edited the book and improved it immeasurably. Although the primary audience for the book is the beginning student of neuroscience, Howard, Kate, and I have aimed to make the book accessible to general readers curious about this exciting field.

PART ONE

NEURONS

Cellular Neuroscience

Over the past forty years many neuroscientists have focused their investigations on the cells of the nervous system. The structure of neurons and synapses has been elucidated in great detail. We now understand how individual nerve and receptor cells generate, carry, and transmit electrical and chemical signals, and we have identified many substances that are used by neurons to communicate information. Today, molecular biological approaches are revealing the ways in which molecules carry out the various tasks of neural activity.

Probing the Brain

NEUROSCIENTISTS use a variety of special methods to learn about the structure and function of the nervous system. Specialized anatomical methods reveal, for example, what individual nerve cells, or *neurons,* look like, where and how far they extend their processes, and how they connect together. In essence, they have helped us map out how the nervous system's cells are wired together. Physiological techniques record the tiny electrical signals that nerve cells use to carry and process information; they elucidate how nerve cells communicate. Most recently, molecular biological methods have been applied to neural tissue. These new approaches are unveiling the structure and function of the molecules involved in neural mechanisms.

I focus first on the special anatomical and physiological methods that have provided much of the information we have about the brain; at the end of the chapter I will review some of the newer molecular biological techniques and their potential for helping us to understand the neural mechanisms. Progress in understanding the brain has depended in large measure on the development of new procedures and on innovations in technology. A survey of the techniques used to study the brain, therefore, provides an overview of the history of modern brain research.

Anatomical Approaches: Visualizing Neurons

Using the retina as test material, I present first a series of examples of what conventional light and electron microscopic techniques can reveal about neural tissue. Figure 1.1 shows a section of the retina of a small mammal (the ground squirrel) photographed through a light microscope. The retina was preserved (fixed), cut in thin sections, and then stained with a dye that reacts with biological structures depending on their pH. Cell nuclei contain acidic material (nucleic acids) and thus stain particularly well.

Although the retinas of all vertebrates are built on a common plan, the ground squirrel retina contains more neurons per unit area than most retinas so it nicely illustrates neural tissue organization. We observe, first of all, that the retina is not homogeneous but is layered: some layers contain cell *bodies* and other layers are made up of cell *processes,* branch-like extensions from the cell bodies. The layers containing the cell bodies are called *nuclear* layers; those made up of cell processes are termed *neuropil* or *plexiform* layers. It is in the neuropil layers where most sites of interaction (synapses) between nerve cells are located.

In the retina there are three nuclear layers and two plexiform layers, as Figure 1.1 shows. Layering of cell bodies and processes is seen in the cerebellum, cerebral cortex, and certain other brain structures, but in many places in the brain another cellular organization is characteristically observed. Cell bodies group together in clusters called *nuclei.* The nerve cells of one nucleus connect with the nerve cells of another nucleus via long, thin processes called *axons,* which are usually wrapped with a membranous sheath of myelin. (The myelin sheath and other cellular features of neurons are discussed in detail in Chapter 2.) Axonal processes typically run together in bundles or nerve tracts, which stain prominently because of the myelin sheaths. Even in unstained material it is possible to distinguish axonal tracts because of the glistening white appearance of the myelin. In sections of fresh brain, *white matter,* the bundles of myelinated axons, can be differentiated from *gray matter,* which consists of cell bodies and nonmyelinated cell processes.

The electron microscope provides much higher resolution than the light microscope, as can be seen in Figures 1.2A and 1.2B, which show portions of the inner and outer nuclear layers and the outer plexiform layer of the ground squirrel retina at two different magnifications. Even though these are low-magnification electron micrographs, they reveal cytoplasmic structures in individual cells and even hint at relationships among cells. A striking feature of neural tissue in comparison with other tissues is the virtual absence of extracellular space; that is, the spaces between the neurons are filled by supporting cells, called *glia.* Extracellular space in the brain is mainly limited to narrow (~ 20 nm) clefts between neurons, between glial cells, or between neurons and glial cells.

Higher-power electron micrographs reveal the small size of

1.1   Light micrograph of the ground squirrel retina showing nuclear layers and plexiform layers. *ONL,* outer nuclear layer; *OPL,* outer plexiform layer; *INL,* inner nuclear layer; *IPL,* inner plexiform layer; *GCL,* ganglion cell layer (another nuclear layer). At the top of the micrograph are the photoreceptors, *Ph.* The boxes enclose the approximate areas shown in the electron micrographs of Figures 1.2 and 1.3.

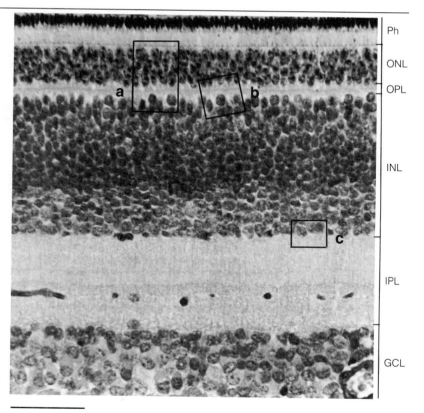

50 μm

many neural processes and the complexity of neuropil structure. In Figure 1.3, presenting a portion of the inner nuclear and inner plexiform layers of the ground squirrel retina, we see many processes that are less than 1 μm in diameter, details that are invisible in an ordinary light microscope. For a view of the junctions or synapses between nerve cells, still higher magnification is required; Figure 1.4 illustrates a chemical synapse in the inner plexiform layer of a frog retina.

Certain characteristic features identify chemical synapses throughout the brain. Within the process passing on information, the *presynaptic* process, many *synaptic vesicles* are clustered near

1.2A   Electron micrograph of a portion of the outer and inner nuclear and outer plexiform layers of the ground squirrel retina. The approximate area covered corresponds to box *a* in Figure 1.1.

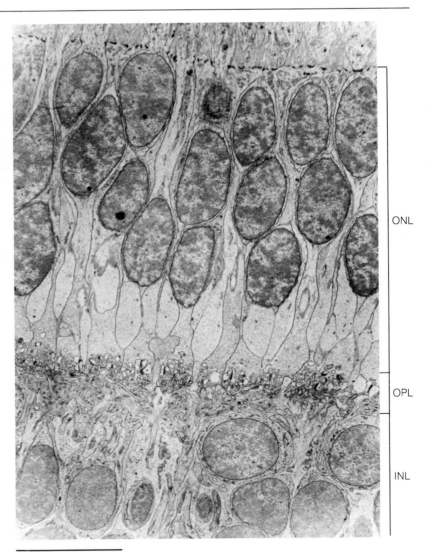

ONL

OPL

INL

10 μm

1.2B A higher-magnification electron micrograph of an area similar to that shown in box *b* in Figure 1.1. Note that glial cell cytoplasm *(g)* fills the spaces between the neurons *(n)*.

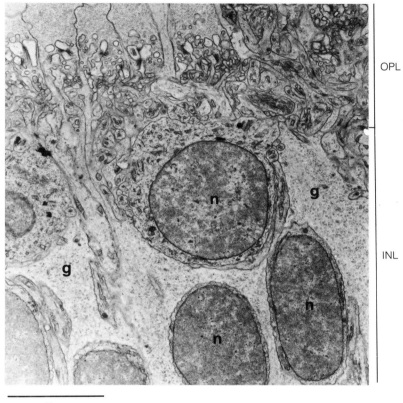

5 μm

the cell membrane. These vesicles contain substances that are released into the gap and interact with molecules on the *postsynaptic* process, thereby leading to excitation, inhibition, or modulation of the postsynaptic neuron. Fine filamentous material fills the gap between the two contacting processes (see Figure 1.4). Often, electron-dense material (material that appears dark in the electron microscope) is observed on or near the membrane of the postsynaptic process, indicating the presence of abundant protein or other biological molecules at these sites.

To summarize, conventional light and electron microscopic methods reveal the cellular organization of neural tissue, details

1.3   Electron micrograph at the border of the inner nuclear and inner plexiform layers. The area included in this micrograph is shown by box *c* in Figure 1.1.

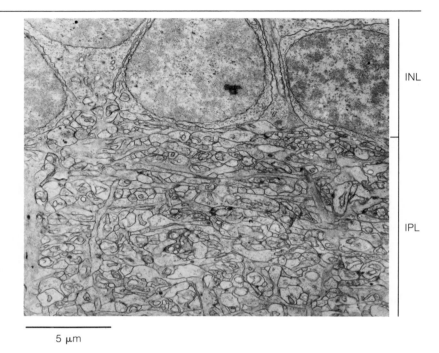

INL

IPL

5 μm

1.4   A high-magnification electron micrograph showing a synapse in the inner plexiform layer of the frog retina. The drawing on the left illustrates the essential features of the synapse. *Pre*, presynaptic terminal; *post*, postsynaptic process; *sv*, synaptic vesicles; *m*, mitochondrion.

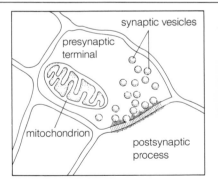

synaptic vesicles

presynaptic terminal

mitochondrion

postsynaptic process

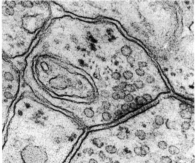

0.2 μm

of individual cells and intracellular organelles, and the structure of synapses between nerve cells. But many of the most important questions are left unanswered. What is the arrangement of the processes of an individual nerve cell? How far do the processes of a neuron extend? Which processes connect with which processes? In other words, how is the nervous system wired together?

It might seem at first glance that the answers could be gleaned simply by cutting serial thin sections of neural tissue and following the paths of the different structures. A simple calculation proves that reconstructing nerve cells in such a manner is not practical. For example, if electron micrographs were taken at a magnification of 10,000× (a modest magnification), we would need a hundred million electron micrographs to reconstruct just one cubic millimeter of tissue! Not only is this an unwieldy number of electron micrographs, but, as we shall see, the processes of many nerve cells extend well over a millimeter in length. Obviously, special methods are required to visualize individual nerve cells.

Observing Individual Nerve Cells

THE GOLGI METHOD

Single nerve cells were first glimpsed in the middle of the nineteenth century. Investigators such as Otto Dieters carefully dissected individual cells from neural tissue hardened with fixative. These early experiments showed that neurons have abundant processes, some of which are quite long (see Figure 1.5), but the method was tedious, difficult, and only occasionally successful. Today, single nerve cells are seldom dissected out, and when they are, the purpose is usually to isolate large neurons for biochemical analyses.

Over a hundred years ago, a new anatomical technique revolutionized the study of nerve cells and the nervous system. The Golgi technique laid the foundation for our present understanding of neural mechanisms and is still used widely today. Discovered by Camillo Golgi in Italy in the early 1870s, the method was most successfully developed by the great Spanish neurohistologist, Santiago Ramón y Cajal. Neural tissue is fixed in a solution containing a heavy metal, such as potassium dichromate, and then soaked in a silver solution for days or even months. Silver impregnates (precipitates in) some of the neurons, and in those cells the staining is usually complete—all the cell's processes are revealed (Figure 1.6).

In well-impregnated Golgi material, only about 2 percent of the

1.5   Drawings of neurons dissected out from hardened neural tissue, made by Otto Dieters in the 1860s.

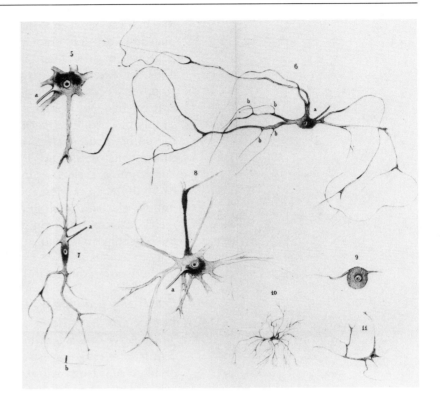

cells stain. The low percentage is not a flaw of the method but the key to its success. Because so few of the cells take up the silver, individual cells can be visualized in thick (100 μm) sections, relatively unobscured by other stained cells. Photomicrographs of Golgi-impregnated cells seldom do justice to the wealth of detail revealed by this method because the light microscope offers only a limited depth of focus. Drawings show the fine details better, but three-dimensional aspects of cell structure are lost (Figure 1.7).

We do not know why just 2 percent of the cells stain and for this reason the method was highly suspect in the early days of its use. Ramón y Cajal, a great believer in the method, devoted virtually his entire career to studying neural cells stained with this technique, and single-handedly he described the cells in almost every part of the brain. Recognized today as the father of neu-

1.6   A Purkinje cell in the cerebellum stained by the Golgi method. A drawing of a Golgi-stained Purkinje cell is in Figure 2.1B.

roanatomy, Ramón y Cajal (pictured in Figure 1.8) shared the 1906 Nobel Prize for Physiology with Golgi.

OTHER STAINING TECHNIQUES

The Golgi technique became widely accepted when a very different staining technique, employing methylene blue, was introduced in the 1890s by Paul Ehrlich, a physician and chemist. When methylene blue was applied by Ehrlich to pieces of living neural tissue, individual nerve cells were revealed that looked virtually identical

1.7 A drawing by Ramón y Cajal of neurons in the frog retina stained by the Golgi method. The drawing is a composite based on stained cells observed in many preparations.

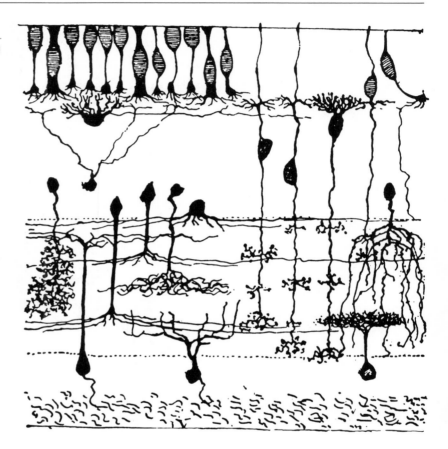

to cells stained by the Golgi method. Ehrlich's method does not show fine details of neuronal cell structure as well as the Golgi method, but it did put to rest an early objection to the Golgi technique—that the Golgi method stains only 2 percent of cells because only dead or abnormal cells take up the silver. Following the groundwork laid with the Golgi method, several other staining techniques have enabled investigators to trace or study whole cells or parts of nerve cells. Prominent among these special stains are those that helped investigators trace axons long distances, from one part of the brain to another. For instance, the Weigert method stains specifically the myelin sheath, whereas the Marchi and Nauta staining methods are used to target degenerating axons and

1.8 Ramón y Cajal in his laboratory in Spain in about 1890.

synaptic terminals of dying neurons, which tend to stain or take up silver differently from the way normal axons do. These latter methods are used to explore axonal pathways: neurons are destroyed in one region of the brain and the stain reveals the degenerating extensions of the same cells elsewhere in the brain.

More recently, techniques have been developed to stain neurons of a given chemical composition. Nerve cells can be identified on the basis of the specific neuroactive substances released at their synaptic junctions. Neuroactive substances in a neuron can be

1.9   The use of fluorescent tags to target nerve cells. The retinal cells in (**A**) contain the neuroactive substance dopamine. After exposure of the tissue to formaldehyde vapors, the dopamine fluoresces. The arrow points to a process that extends from the inner plexiform layer *(IPL)* to the outer plexiform layer *(OPL)*. The cell in (**B**) contains a neuropeptide, substance P. The tissue, another retinal section, was exposed to an antibody that reacts with substance P, and the antibody was linked with a molecule that fluoresces.

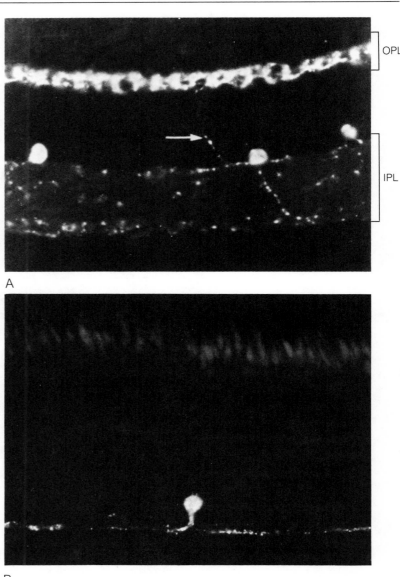

OPL

IPL

A

B

20 μm

induced to fluoresce, or they (or the enzymes that synthesize or degrade them) may be tagged by a specific antibody, a protein that binds selectively to other molecules. By linking the antibody to a fluorescing or dark-staining molecule, researchers can now locate the sites of interaction and hence the cells containing a certain transmitter substance. Some examples are shown in Figure 1.9.

Neurons also generally have efficient mechanisms for reabsorbing back into the cells the neuroactive substances released at their terminals. This is an important mechanism: by ridding the synaptic cleft of the neuroactive agent, the cell can terminate synaptic action. Investigators take advantage of this phenomenon by inducing neurons to take up radioactively labeled neuroactive substances or altered neuroactive substances that will fluoresce. The cells taking up these substances are then identified by autoradiography or by fluorescence microscopy.

Virtually all staining techniques developed initially for light microscopy can be modified for electron microscopy, but often with even better results because of the superior resolving power of the electron microscope. So, for example, Golgi-impregnated cells can be studied in the electron microscope, as can the synapses made by a cell containing a specific transmitter substance. Hence, chemical-specific pathways in the brain can be mapped. It is even possible to pinpoint one or more neuroactive agents to a certain type of synaptic vesicle. Figure 1.10 presents an electron micrograph of an axon terminal exposed to an antibody that binds to a specific neuroactive substance—in this case a small peptide. The antibody was tagged with an electron-dense molecule so that it would stand out in an electron micrograph. The especially dark spots in the figure, confined to the large vesicles in the nerve terminal, indicate the presence of the antibody, and hence the neuroactive peptide, in these vesicles.

Physiological Approaches: Recording Electrical Activity

Nerve cells carry information by electrical signals, and so the first tactic for decoding this information is to record accurately the small electrical potentials generated by neurons. Until the late 1920s, neurophysiologists had no way to record the activity of single neurons. They could measure the sum of the electrical activity of many cells from the brain's surface, from large nerve

1.10   An electron micrograph of a nerve terminal exposed to an antibody (linked to an electron-dense marker) that reacts with a specific neuropeptide. The dense staining in the terminal is confined to the large vesicles.

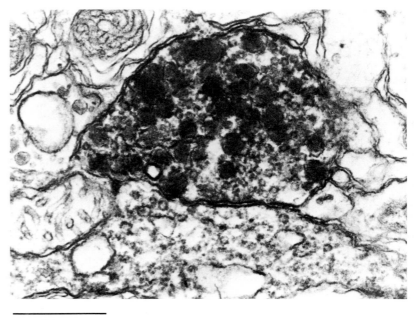

0.5 μm

bundles, or from sense organs such as the eye, but these *field recordings* were not very revealing about the mechanisms underlying the generation of electrical potentials in individual nerve cells or about information processing in the brain. The challenge for the greater part of this century has been to uncover these mechanisms through single-cell recordings.

EEGs and Other Field Potentials

The electroencephalogram, or EEG, is a summed (field) potential recorded from the brain's surface. When a low-resistance electrode is placed on the skull's surface, very small electrical potentials reflecting the summed activity of the cells underneath the electrode can be detected with the help of sufficient amplification (see Figure 1.11). The signals are small, partly because they are recorded some distance from where they were generated. Also, because the electrical signals produced by the individual cells are not synchronous

1.11 Field potentials recorded from the surface of a human skull. The drawing at the top illustrates the recording technique. (**A**) A typical recording from a living subject *(left)*; after death *(right)*, electrical activity in the brain stops. (**B**) The potentials become more synchronous and of greater amplitude when the subject relaxes and closes his or her eyes (bracketed portions of record). (**C**) Potentials evoked at the back of the skull at the onset and cessation of a light flash.

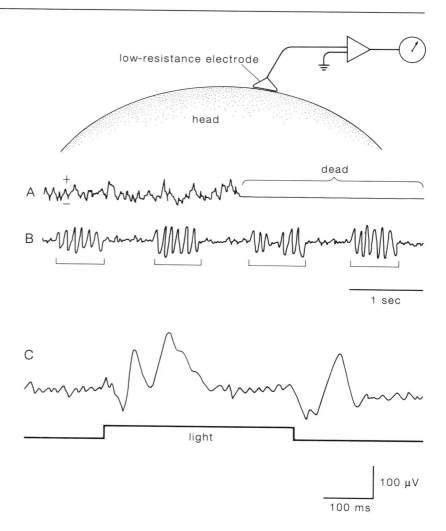

and are both positive and negative, the potentials tend to cancel one another.

In humans, the activity recorded from the skull can be made more synchronous by having the subject relax with eyes closed. The signals recorded become larger and several distinct waves can be distinguished according to their frequency. These waves are called Berger waves after their discoverer, Hans Berger, an Austrian psychiatrist. That the potentials are neural in origin is suggested by several lines of evidence. Their size and frequency composition differ depending on where the recording electrodes touch the skull; they often change significantly when sensory stimuli are presented to the subject; and they disappear when the subject dies.

These potentials do not provide much information about underlying brain mechanisms, but since they can be readily recorded in a noninvasive manner they are useful in a clinical context. For example, the EEG can detect brain abnormalities—or whether the brain is functioning at all. Indeed, a patient is declared legally brain-dead when no EEG can be recorded over a period of time, usually 24 hours.

EEG recordings can also be used to identify which parts of the brain of humans and animals are active under various conditions. Distinct and reproducible potentials are recorded from different parts of the skull when different stimuli are presented to a subject. Such potentials, called *evoked potentials*, reflect the activation of a specific sensory system in the brain. So, for example, when a light stimulus is presented to the eye, we detect a characteristic evoked potential from the back of the head. This potential is not recorded in response to a light stimulus if the electrode is moved to the side of the head. On the other hand, an auditory stimulus will evoke a characteristic potential in this area. EEG recordings thus provide evidence that the primary visual areas are located in the occipital lobes of the cerebral cortex, which are found under the back of the skull, and that the primary auditory areas are in the temporal lobes of the cortex, under the sides of the skull.

As already noted, field recordings are not ordinarily useful for finding out how individual units in the brain respond. There are some exceptions. A field recording from a peripheral nerve bundle can provide an approximate indication of the response of an individual nerve cell axon because all axons in peripheral nerve tracts are similar and their electrical responses are nearly identical. If the axon trunk is stimulated with a current pulse so that all the axons are activated simultaneously, the summed potential roughly resembles the activity of individual axons. The potential recorded is smaller than the response of any individual axon, but the response's waveform approximates that of an individual axon (see Figure 1.12).

1.12 A field recording from a nerve bundle. When all of the axons are stimulated simultaneously with a current pulse *(left)*, and a recording is made along the axon *(right)*, the summed potential *(below)* resembles the activity of individual axons.

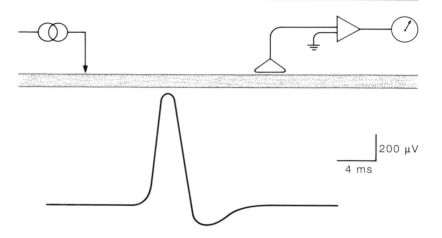

200 μV

4 ms

Single-Cell Recordings

In the 1920s, the first recordings from single nerve cells were made. The activity in individual axons dissected from nerve trunks was measured with wick or wire electrodes. In most nerve trunks, though, the axons are tightly bound together and so this technique is not used routinely today. An exception is the optic nerve of the *Limulus* (horseshoe crab, shown in Figure 9.7) eye, which is easily dissected. Figure 1.13 shows typical recordings from a single axon from a horseshoe crab eye. The axons generate transient, all-or-nothing action potentials whose frequency depends on stimulus intensity. On the left is shown the relative intensity of the light used to stimulate the eye, and below the recordings the duration of the stimulus is indicated. With a weak light stimulus (relative intensity of 1), few action potentials are generated. With light intensities 100 and 10,000 times more intense, the number of action potentials generated is very much greater. Each action potential has the approximate form of the potential shown in Figure 1.12. The time scale of the recordings shown in Figure 1.13 is much more compressed, however, so individual potentials appear as vertical lines.

1.13 Recordings from a single axon dissected from the optic nerve of the horseshoe crab. The response of the axon depends on the intensity of the light delivered to the eye. With a weak stimulus (relative intensity of 1), only a few action potentials were generated in the axon; with a bright stimulus (relative intensity of 10,000), the axon fired action potentials at a rapid rate for as long as the light was on. The duration of the stimulus is indicated at the bottom.

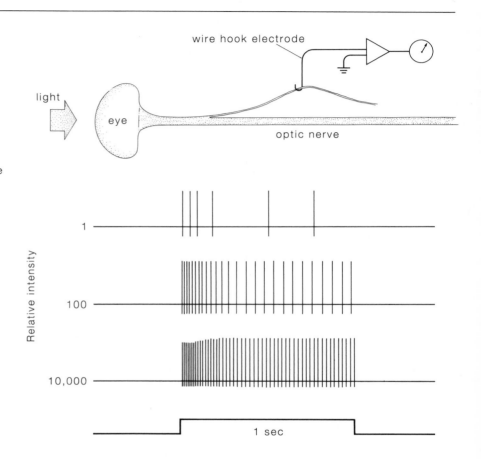

EXTRACELLULAR RECORDINGS

Today, neurobiologists use two principal techniques to record the activity of single nerve cells; one method records responses extracellularly, the other intracellularly. Extracellular recordings of single cells, introduced in the late 1930s, provided for the first time information concerning the electrical responses of a wide variety of neuronal cell types. Usually, metal wires that are electrolytically sharpened to a tip size of 1–5 μm are used as electrodes. The wire is insulated, except at the tip, and connected to an amplifier and a recording device such as an oscilloscope. The electrode is positioned close to a single nerve cell or axon in the brain with the aid of a micromanipulator, a device that permits small, precise

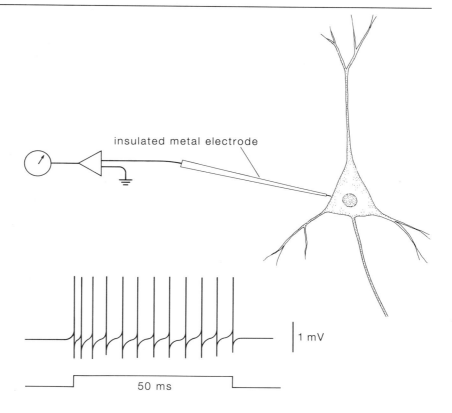

1.14 How an extracellular recording is made from a neuron *(top);* a typical extracellular recording *(bottom).*

insulated metal electrode

1 mV

50 ms

movements of the electrode, and potentials generated by the cell or axon are recorded. An extracellular recording from a neuron is shown in Figure 1.14.

Wire electrodes readily record the relatively large transient electrical signals (called *action potentials,* which are the subject of Chapter 4) that nerve cells use to transmit information along axons. They are not good for recording small or sustained potentials generated by neurons because the metal wire accumulates electrical charge (that is, it polarizes). The buildup results in spontaneous electrical changes that obscure the biological signals. Another disadvantage of this type of recording is that, because it is extracellular, remote from the site of potential generation, the size of the potential recorded is only a fraction of its true amplitude. With this method it is extremely difficult to detect small signals.

INTRACELLULAR RECORDINGS

In the early 1950s, the first intracellular recordings from neurons were made. This technique is possible because of a remarkable property of glass tubing. Upon heating, glass tubing softens and when pulled it becomes thinner. Regardless of how thin the tubing becomes, though, it remains a tube; it does not close. Thus, if thin (capillary) glass tubing is heated and pulled out rapidly, very fine tipped pipettes are made. Depending on the temperature of the glass and how forcefully it is pulled, sharp pipettes with tip openings of 0.5 μm or less can be fashioned. The pipettes are filled with a conducting (salt) solution, roughly similar to the solution found in cells, and inserted into a single cell by micromanipulation.

In essence, the pipette opens a pathway that is electrically continuous with the inside of the cell. An electrode, connected to an amplifier and recording device, is inserted into the pipette; as if it were inserted into the cell itself, it registers potential changes inside the cell relative to the outside (that is, potential changes across the cell membrane). An intracellular recording is shown in Figure 1.15. The electrical resistance of these fine pipettes is often high—up to several hundred million ohms. This causes some technical difficulties for the experimenter, but they usually can be overcome.

The advantages of intracellular recording are many. Large, transient action potentials and small, slow potentials can be recorded, as well as steady resting potentials. Furthermore, the voltages recorded are an accurate reflection of the potentials generated across the cell membrane. It is also possible to include a dye or an electron-dense substance in the pipette that may be squirted into the cell during or following a recording. The recorded cell can then be identified by light or electron microscopy and physiological responses correlated with a specific type of neuron or synaptic junction. Finally, ions can be passed into nerve cells with intracellular pipettes, thereby changing the membrane potential. Such pulses of current are helpful in determining the properties of nerve cell membranes and the mechanisms underlying potential generation by neurons. These topics will be developed more fully in Chapters 3 and 4.

There are also disadvantages to intracellular recording that restrict its use. Intracellular pipettes sometimes damage cells, the duration of recording is usually limited, and small neurons often

1.15 How an intracellular recording is made from a neuron *(top)* and a typical intracellular recording *(below)*. At time 0, when the pipette penetrates the cell *(arrow)*, a steady resting potential of about −70 mV is recorded. Superimposed on the steady resting potential are small and large potential changes indicative of cell activity.

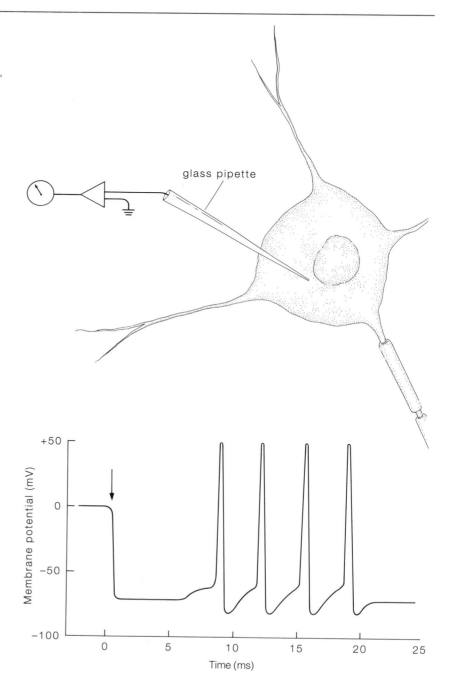

cannot be penetrated. It is particularly difficult to record from small neurons in many parts of the mammalian brain; even today, most of our information on cell responses in the mammalian brain is based on extracellular recordings.

PATCH-CLAMP RECORDINGS

In the late 1970s a new recording method was introduced by Erwin Neher and Bert Sakmann, who were awarded the Nobel Prize in 1991 for their work. The technique enables neuroscientists to record the activity of single membrane channels. Channels are protein molecules through which charged ions cross the cell membrane, resulting usually in voltage changes across the membrane. Called patch clamping, this method, like intracellular recording, depends on another interesting property of glass, namely that it can bind extraordinarily tightly to cell membranes. A patch-clamp pipette is fashioned by breaking the tip of an ordinary intracellular pipette and fire-polishing the broken tip, which leaves a blunt but smooth end of 1–3 μm diameter.

When pressed against a cell membrane, the tip seals onto the membrane so tightly that when the pipette is withdrawn, a patch of membrane remains covering the tip. It is then usually possible to measure the flow of ions across the patch of membrane through single protein channels. Channels in nerve cell membranes open and close in an all-or-none fashion, so in patch-clamp records one observes step-like changes in current flow across the membrane patch. With one channel in a patch, the current levels fluctuate between two states; that is, the current is at one level when the channel is open and at another when the channel is closed. If there are several channels in the patch of membrane, the current levels fluctuate over several levels. (The method and some sample recordings are illustrated in Figure 1.16.)

It is also possible to use patch electrodes to record activity in intact cells. The patch pipette is sealed onto the cell membrane and the membrane enclosed by the tip of the pipette is ruptured. The patch pipette now has entry to the cell's interior, as a conventional intracellular pipette does. But the tip of a patch pipette is much larger than that of an intracellular pipette; it will permit the passage of fairly large molecules into the cell. With patch electrodes, therefore, investigators can perfuse a cell's interior; that is, introduce drugs or other substances directly into the cell.

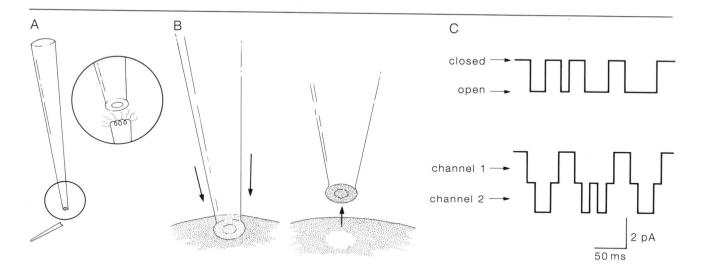

1.16 Patch-clamp pipettes are made by heat polishing the tips of broken glass pipettes (**A**). The pipette tip seals onto a neuronal membrane so tightly that it tears a patch of the membrane from the cell (**B**), and currents across the membrane can then be measured. In (**C**), two recordings of channel events are shown: current flow across a single channel *(top)*, and flow across a membrane patch containing two channels *(bottom)*.

Molecular Biological Approaches: Mapping Molecules

New techniques of molecular biology have revolutionized biochemistry and the study of cellular mechanisms. These methods are already having an impact in neuroscience, and we are only at the beginning of their use in the analysis of neural mechanisms. Throughout this book, specific examples of the new technology will be described; in the limited scope of this introductory chapter I will merely outline some of the principles involved and the kind of information that can be obtained when they are applied.

Recombinant DNA Technology

Many of the techniques derive in large part from recombinant DNA technology, which has permitted the detailed analysis and alteration of both genes and proteins. This technology developed as a result of the tremendous progress made over the past forty years in understanding genetic mechanisms and how genetic information is transformed into protein structure. Everyone today is familiar with the central dogma of molecular biology:

$$\text{DNA} \xrightarrow[\text{(transcription)}]{} \text{RNA} \xrightarrow[\text{(translation)}]{} \text{Proteins}$$

DNA specifies RNA through a process called transcription; RNA then specifies proteins via translation. Put in other words, the DNA of a cell codes for the proteins made by that cell, but DNA is not the direct template for protein synthesis. Rather, RNA molecules are made from the DNA, and the proteins are made from RNA templates on ribosomes. There are several kinds of RNA; the one that carries the information specifying how to make a particular protein is called messenger RNA (mRNA).

Information is coded by DNA in a linear sequence; thus DNA molecules are very long. A stretch of DNA that codes for a protein is called a gene, and in mammals there are about 100,000 genes. The brain utilizes more genes than any other tissue, perhaps as many as 50,000, and of these 30,000 may be unique to neural tissue.

Four slightly different subunits make up the DNA chains and code the genetic information. Each subunit consists of three molecular groups—a base, a sugar, and a phosphate group. Only the bases are different in the subunits and thus they code the genetic information. A sequence of three bases specifies one amino acid in a protein, and since proteins can contain several hundred amino acids, genes can contain over a thousand bases. Indeed, genes usually contain several thousand bases because in most organisms they contain, in addition to regions of DNA that code for proteins (exons), regions that are noncoding (introns). An important aspect of DNA structure is that the bases, of which there are four kinds, pair off in complementary fashion; the base adenine always pairs with thymine, and guanine pairs with cytosine. Thus one DNA chain specifies a second chain. In fact, DNA ordinarily consists of two chains coiled around a common axis in the shape of the double helix (Figure 1.17).

RNA is also a long polymer very similar to DNA. It too consists of four subunits whose bases are complementary to the DNA bases (indeed, 3 of the 4 bases are similar in RNA and DNA; however, thymine is replaced by uracil in RNA). RNA is usually single-stranded and is formed from DNA. Thus, DNA determines the structure of RNA, and the amino acid sequence of a protein is in turn determined by messenger RNA.

Enzymes have been discovered that synthesize DNA and RNA molecules, and so have enzymes that transcribe DNA into RNA and even RNA back into DNA. Furthermore, there are enzymes

1.17 A schematic diagram of DNA *(top)* and how DNA specifies RNA and RNA codes for protein structure *(bottom)*. The ribbon-like structures are the sugar and phosphate backbones of the DNA molecule; the bases on one strand extend into the middle of the molecule and pair in a complementary fashion with bases on the other strand. The base thymine always pairs with adenine, cytosine with guanine. The two chains of the DNA molecule are coiled about each other, forming a double helix. RNA is a very similar molecule but it is single-stranded, and thymine is replaced by uracil. RNA is specified by DNA, and protein by RNA. A sequence of three bases in the RNA molecule specifies one amino acid, of which four are shown here *(AA₁–AA₄)*.

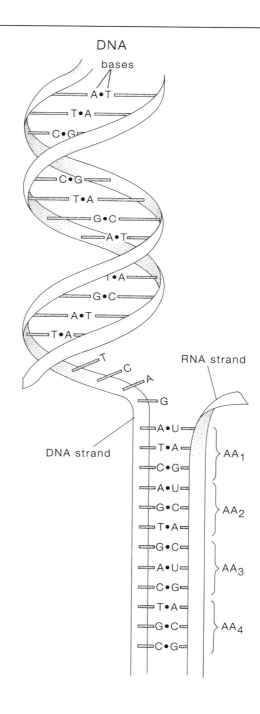

that cut DNA molecules in specific ways, and still other enzymes that will join DNA fragments back together again. These enzymes are used in the laboratory to cut DNA molecules into small pieces, to identify the sequence of bases in a fragment, to synthesize new fragments, and to make new DNA molecules. Hence the name *recombinant DNA:* imagine taking apart a jigsaw puzzle, making a few adjustments to some pieces, and reassembling a somewhat different picture. The new DNA molecules can then be injected or inserted into cells, where they will often incorporate into the cell's own DNA and be translated eventually into protein. If the cells multiply rapidly, as bacteria do, an abundant supply of the recombinant DNA and the new protein will be made.

The structure of proteins can be readily deduced from the genes that code for them. In addition, genes can be altered, one or a few bases at a time, and these changes will alter one or a few amino acids in the protein that is produced, or "expressed." The effects of the alterations on the function of the protein can then be tested. Today in neuroscience, we are particularly interested in membrane channels, which can be studied easily in the frog egg or oocyte. When mRNA coding for a membrane channel, for example, is injected into the oocyte, the channel protein usually is expressed and inserted into the egg membrane. The physiological properties of that channel can then be studied using patch-clamp methods.

Monoclonal Antibodies and Genetic Mutants

In addition to recombinant DNA technology, other molecular biological techniques are also finding important uses in neuroscience. Monoclonal antibodies are an example. Antibodies are proteins made by one kind of white blood cell (lymphocytes) that interact with specific molecules. They bind to their target molecules and can be used to identify the target. Usually lymphocytes will make a variety of antibodies, each lymphocyte producing a somewhat different antibody. But now it is possible to fuse antibody-producing lymphocytes with lymphocyte tumor cells (which multiply indefinitely) and raise large numbers of fused cells that derive from one lymphocyte. In other words, all the cells are identical—they are clones—and they all make the exact same antibody. Monoclonal antibodies are used to identify types of neurons, the neuroactive substances released at synapses, enzymes that synthesize and degrade molecules in neurons, and other structures.

The use of genetic mutants to help analyze neural structure, function, and development is another area of research that is yielding much important information. And with the recombinant DNA techniques presently available, once a mutant gene has been identified and isolated, its product can be expressed and analyzed. The nature of the change resulting from a mutation can then be determined. It is also possible to use recombinant DNA techniques to induce specific mutations in single genes and explore how they affect neural cell structure or function.

Summary

With techniques presently available, it is possible to study neuronal mechanisms down to the level of a single cell, even to a particular molecule. This chapter has presented a sampling from the major approaches to the study of the brain. With silver-impregnation and staining methods, for example, the anatomy of individual neurons can be explored and their axonal projections followed via light microscopy. The superior resolving power of electron microscopy permits the identification of synaptic junctions between neurons and the mapping of neural circuitry. The workings of this circuitry have been investigated through various physiological techniques. Extracellular recordings describe the responses transmitted along nerve cell axons and provide clues concerning the information processing within the brain. Intracellular recordings reveal the details of potentials generated by nerve cells and, when combined with intracellular staining, allow correlations to be made between structure and function. Patch clamping of cell membranes elucidates the underlying molecular events that give rise to the generation of electrical potentials by neurons, the means of neural communication.

Molecular biological techniques permit the mapping of gene and protein structure and, with recombinant DNA techniques, these structures can be modified. The new approaches offer an understanding of the function of particular molecules in neural function, even of intramolecular mechanisms. Monoclonal antibodies are used to identify specific molecules in neural tissue, and genetic mutants can help analyze the normal structure, function, and development of nerve cells and circuits.

Further Reading

BOOKS

Adelman, G., ed. 1987. *Encyclopedia of Neuroscience*. Boston: Birkhauser. (Short reference articles on all aspects of neuroscience. A good place to start.)

The Brain. 1979. A Scientific American Book. San Francisco: W. H. Freeman and Company. (A comprehensive collection of articles on brain mechanisms.)

Ramón y Cajal, S. 1937. *Recollections of My Life*. Trans. E. H. Craigie and J. Cano. Philadelphia: University of Pennsylvania. (A fascinating autobiography.)

Sakmann, B., and E. Neher. 1983. *Single-Channel Recording*. New York: Plenum Press. (The details of patch-clamp recording by the originators of the technique and others.)

Shepherd, G. M. 1988. *Neurobiology*. New York: Oxford University Press. (A basic textbook with a section on molecular neurobiological approaches.)

ARTICLES

Benzer, S. 1973. Genetic dissection of behavior. *Scientific American* 229(6):24–37.

Heimer, L. 1971. Pathways in the brain. *Scientific American* 225(1):48–60.

Milstein, C. 1980. Monoclonal antibodies. *Scientific American* 243:66–74.

Stretton, A. O. W., and E. A. Kravitz. 1968. Neuronal geometry: Determination with a technique of intracellular dye injection. *Science* 162:132–134.

Yellen, G. 1984. Channels from genes: The oocyte as an expression system. *Trends in Neuroscience* 1:457–458.

Cells and Synapses

THE BRAIN, like all organs of the body, is made up of discrete cellular elements. But it is the *interactions* between nerve cells that underlie much of the brain's accomplishments, that allow organisms to behave, learn things, remember things, abstract things, and create things from mathematical theories to symphonies.

The cellular mechanisms used by nerve cells are for the most part similar to those used by cells in other systems. Most nerve cells communicate with one another chemically, much like cells of the endocrine system. Neurons carry information by means of electrical signals, but cells in all tissues generate steady potential differences between inside and out. The brain, though, is enormously more complex than other organs, especially in higher vertebrates. It has more cells, more intricate cell structure, and more various cellular interactions. In the human brain there are between 10^{11} and 10^{12} nerve cells, and each cell can make as many as 10^3–10^4 connections with other cells. A single neuron may possess an enormous number of processes, and some processes (axons) may extend several feet from the cell body.

Neurons and Glia

Two types of cells make up the brain: neurons and glia. *Neurons* are involved in the processing of information in the brain; they receive signals from sensory organs or other neurons; and they integrate the information and transmit it. They are responsible for the most interesting aspects of neural function.

Neurons differ from other cells in the body in at least two important ways. Once neurons have differentiated, they never divide again. Furthermore, new neurons are not produced following brain injury; in most organs, cells are replaced after a trauma. Recovery of brain function, in contrast, results only from the use of existing cells. The brain of a one-year-old human contains about as many cells as the brain ever will have, and throughout life neurons are continually lost. The number of neurons lost per day is surprisingly high—perhaps 200,000 or so in humans*—but because the brain contains so many neurons, most of us can get through life without losing so many that we become mentally debilitated. Some people are not so lucky, however. Alzheimer's disease is marked by excessive brain cell loss, for reasons that are not clear; sufferers may show severe loss of mental abilities in their forties or fifties. If one lives long enough, Alzheimer-like symptoms are almost always seen, but we do not know why some individuals maintain keen mental abilities much longer than others.

Neurons also differ from cells of other tissues in their requirement for oxygen. Deprived of oxygen, nerve cells almost always die within a few minutes. Neurons cannot build an oxygen debt, as can muscle cells or other cells of the body; that is, they cannot survive anaerobically. This constraint has enormous medical implications. When oxygen flow is shut off to tissues as a result of a heart attack or suffocation, the brain dies first. Only if oxygen is restored to the brain in a few minutes will the brain survive. Today, with technological life-support systems, it is not uncommon for an individual to be permanently brain-dead following a period of oxygen deprivation while other organs survive the insult and recover completely.

Glial cells, or *neuroglia* (literally, "nerve glue"), are supporting cells whose function is still not entirely understood. They do not appear to participate directly in information processing in the brain, although their membrane potential changes when the brain is functioning—and they may even respond to substances released by the neurons. More numerous than the neurons, they fill up the extracellular space between the neurons, thus providing a structural framework for neurons, especially during brain development. Glial cells also form the insulating myelin sheath around nerve cell axons and regulate the concentration in the extracellular space of various constituents, such as ions and neuroactive substances released from neurons.

---

* This estimate is derived as follows: There is typically a loss of 5–10 percent of brain tissue with age. Assuming a brain loss of 7 percent over a life span of 100 years, and 10^{11} neurons (100 billion) to begin with, approximately 200,000 neurons are lost per day.

Two categories of glial cells are present in the vertebrate brain: *macroglia,* the major glial cells, and *microglia.* Macroglia are of two classes: *astrocytes,* which are star-shaped cells with many processes and often with numerous filaments, and *oligodendrocytes,* which have fewer processes and filaments and which mainly help to make the myelin sheath. In the peripheral nervous system (so named because it is outside of the brain and spinal cord), the myelin sheath is made by glial cells called *Schwann cells.* Microglia are small glial cells that serve, among other things, as macrophages—cells that phagocytose (eat) dead cells or other debris. I will say more about glial cells when I describe the formation of the myelin sheath, but the rest of this chapter will focus on neurons.

Neuronal Structure

Neurons are varied in form. Most have numerous and often long processes that allow the cells to contact one another in complex and intricate ways. Processes account for more than 90 percent of the volume of many nerve cells; thus much of the brain is made up of cell processes. Also, only a small percentage of the synaptic contacts are on cell bodies, so the overwhelming majority of interactions in the brain occur between neuronal processes.

Each part of the brain has neurons with unique shapes, presumably related to the function of the cells and of that part of the brain. Figure 2.1, for example, shows two neurons from the cerebral cortex and one from the cerebellum, all of very different shape. Although the total number of neurons in the different brain areas is tremendous, the cells can be grouped first of all in two broad classes, which Ramón y Cajal termed *Golgi type I* and *Golgi type II* cells. Furthermore, in each brain region we can classify the neurons into distinct types, of which there are usually a reasonable number. Each cell of a certain type looks similar to the other cells of that type and generally has a similar function. The retina, for example, has five types of neurons in addition to the photoreceptors, and the cerebellum also has five neuronal cell types. In the cerebral cortex two major types of cells are observed.

Anatomists further divide the cell types into subtypes, of which there may be many. Over 20 structural subtypes of one cell class in the vertebrate retina have been described, and physiological

2.1 Neurons from the cerebral cortex (**A** and **C**) and cerebellum (**B**), stained by the Golgi technique and drawn by Ramón y Cajal. Shown here are a pyramidal (**A**) and a Purkinje cell (**B**), which are Golgi type I neurons (long-axon cells), and a stellate cell (**C**), a Golgi type II neuron (short-axon cell).

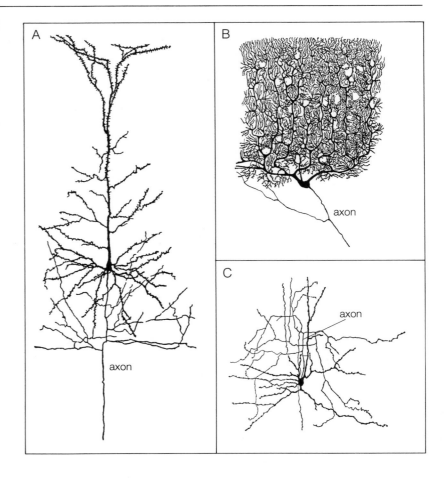

experiments indicate variations in the response properties that the subtypes display. Correlations between anatomical and physiological subtypes of cells have been made in many instances. Many subtypes of cells have specific neuroactive substances they presumably employ to communicate information. Classification of neurons on the basis of their structure, function, and chemistry remains an active area of research. By discovering the role of specific cells in each part of the brain, it is hoped we will gain insights into brain mechanisms and function.

Golgi Type I Neurons

Golgi type I cells are long-axon neurons that carry information from one part of the brain to another (that is, from one nucleus or nuclear layer to another) or from the brain or spinal cord to effector organs such as muscles. They are larger than Golgi type II neurons and more is known about them. Consider, for instance, a motor neuron in the base of the spinal cord that innervates muscles in the foot, as illustrated in Figure 2.2A. In an adult this cell's axon extends about 3 feet (1 m), yet its cell body is not more than about 100 μm; so the ratio of axon length (1 m or 10^6 μm) to cell body diameter (10^2 μm) is about 10^4 or 10,000 to one. If we were to draw the cell body as 6 inches in diameter, and include the entire axon, the axon would be about 5,000 feet long (0.5 feet × 10,000), close to a mile in length!

The cell body of a motor neuron is typically angular and has numerous processes extending outward. Two types of processes are distinguished. *Dendrites* are like the branches of a tree; they are usually bushy and become thinner at each branching point. Primary, secondary, and tertiary dendrites can be distinguished, and the finest dendritic processes often have small lollypop-shaped spines (but not on spinal cord neurons). Figure 2.2B shows spines on Purkinje cell dendrites. On the spines may be found much of the synaptic input to many Golgi type I neurons (Figure 2.2C).

The other type of process extending from the cell body is the *axon*. Each neuron has just one axon, which as it comes off the cell body is thinner than the primary dendrites. It remains roughly the same diameter its entire length and usually branches few times, if at all, until near its termination. The axon terminals form synapses on other neurons or effectors, and at these points the information carried by the axon is passed on. Virtually all of the input to Golgi type I neurons is onto the dendrites or cell body, whereas output occurs at the axon terminals. Dendrites transmit information *toward* the cell body, axons transmit messages *away* from it.

The cell body and main dendrites of neurons contain cytoplasmic organelles typical of all cells, including mitochondria, ribosomes, endoplasmic reticulum, and microtubules. Early studies on the cytology of nerve cells suggested that neurons contain special structures, but electron microscopy has disproved this. Neurons can have an abundance of a certain type of organelle; an

2.2 **(A)** Schematic drawing of a motor neuron, Golgi type I. **(B)** Spines on Purkinje cell dendrites revealed by Golgi staining. **(C)** An electron micrograph of a Purkinje cell dendrite showing its lollypop-shaped spines *(asterisks)*. The boxed area in **B** is equivalent to the area shown in **C**. Because the shaft of the spine is thin, continuity between the spine tip and dendrite is not always seen. Arrows indicate synapses onto the spines. Note the mitochondria *(M)* and other organelles in the dendrite.

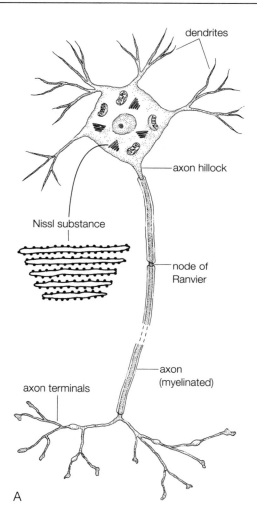

example is *Nissl substance,* dense, basophilic-staining material seen in many neurons. When first observed in the nineteenth century by neuroanatomists, Nissl substance was thought to be unique to nerve cells; however, electron microscopy showed it to consist of stacks of endoplasmic reticulum membranes densely covered

B

5 μm

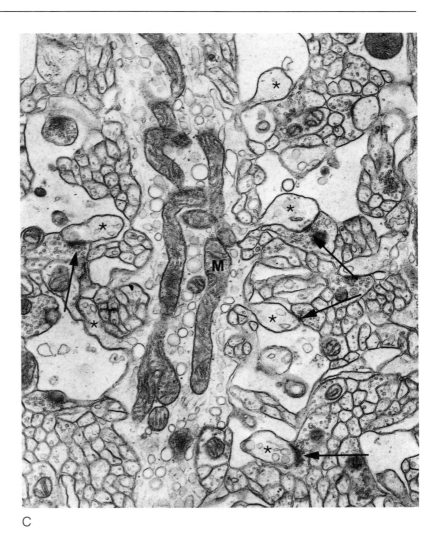

C

with ribosomes, particles on which protein synthesis occurs. Studded endoplasmic reticulum, as such material is often called, is observed in cells of all tissues but in the cell bodies of neurons it is especially dense (see Figure 2.2A).

Many nerve cell processes contain numerous microtubules of 25 nm diameter or filaments of 10 nm diameter. Observed prominently in nerve cells by early cytologists, these structures were named *neurotubules* and *neurofilaments*. We now know that they too are not limited to cells of the nervous system but are part of the cytoplasmic structure (cytoskeleton) common to all cells. It may be that some of the proteins that make up the neurofilaments in neurons are unique to the brain, but we do not know whether this is so, nor what it may mean if it is so.

Axonal Transport

Adjacent to the exit point of the axon from the cell body is a specialized cone-shaped region called the *axon hillock*. No ribosomes are in the axon hillock; indeed, no ribosomes are observed from the axon hillock down the entire length of the axon. Lacking ribosomes, the axon is incapable of synthesizing proteins. How, then, are axons and axon terminals maintained? Two mechanisms contribute to the maintenance of the axon. First, materials are transported from the cell body down the axon (*axonal transport*); second, glial cells are believed to help maintain axons.

Axonal transport was discovered in the 1940s by Paul Weiss, then at the University of Chicago. He and his colleagues tied a ligature around a nerve trunk in a living animal and observed within a few days a swelling of the nerve on the side of the ligature closer to the cell body (see Figure 2.3). They surmised that material flowing down the axons was being dammed by the ligature, causing the swelling. By moving the ligature along the nerve and watching the development of the swelling, they estimated that

2.3   When a ligature is tied around an axon, a prominent swelling of the axon occurs on the side toward the cell body, indicating a damming of the transport of materials from the body to the axon terminal (orthograde transport). There is also some damming of materials on the other side of the ligature, indicating a flow of material in the opposite direction as well (retrograde transport).

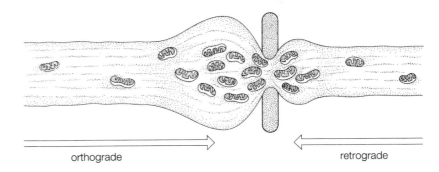

orthograde   retrograde materials were moving down the axons at a rate of about 1 mm per day.

In the early 1960s Bernard Droz and his colleagues in France studied axonal flow more directly. They injected radioactive amino acids or other small molecules into areas of the brain known to be rich in cell bodies and then observed the displacement of radioactive materials along the axons. Amino acids and other small molecules are taken up by the cell body and incorporated into proteins and other substances. If the small molecules are radioactively labeled, the synthesized substances are also tagged. They can then be traced by autoradiography, a technique for detecting radioactivity on a photographic film or plate. These experiments confirmed that many substances move down the axon at a rate of 1 mm or so per day, but they also showed that other materials travel down the axons at much faster rates, up to 100–400 mm per day. The slower wave of material consists mainly of cytoskeletal components and soluble proteins, whereas the substances traveling at the faster rates include small lipid vesicles, cell organelles, certain proteins, and some sugars.

This experimental technique not only yields information concerning the mechanisms of transport within nerve cells but also allows us to trace axons in the brain. The radioactive materials in effect map out the route taken by and the termination point of the cell's axon. They may also be released from the axon terminals of a labeled cell and taken up by postsynaptic cells, revealing the synaptic pathways beyond a single neuron (see Figure 16.8).

Electron microscopy of axons tied with a ligature shows that cytoplasmic organelles accumulate not only on the cell-body side of the constriction, but on the other side of the ligature as well. This suggests that substances flow in both directions along axons. Substances injected in the vicinity of the axon terminals of a cell may be taken up by the terminals and transported to the cell body. Transport from the cell body to terminals is called *orthograde* transport, and transport from the terminals to the cell body is *retrograde* transport. Retrograde transport of materials can be followed to discover the cells of origin of synaptic terminals.

Both slow and fast axonal transport are active processes, requiring energy and the presence of microtubules in the axonal cytoplasm. Metabolic poisons stop axonal transport, as do substances or other manipulations that disrupt the microtubules. Using high-contrast television imaging, investigators can observe substances moving along microtubules in living axons. Two proteins, kinesin and dynein, are crucial for the transport process; they move the materials that are being transported along the microtubules. Kinesin is involved in orthograde transport, dynein in retrograde transport.

That the axon of a neuron depends largely on substances (especially protein) synthesized in the cell body explains the dense Nissl substance observed in many neurons, especially those with long axons. When an axon is severed from a neuron (and the neuron survives) the Nissl substance usually diminishes in amount, presumably because the need for newly synthesized protein decreases. If the axon regrows, Nissl substance reappears, roughly in proportion to the length of the regenerated axon.

The Myelin Sheath

All axons are encased in glial cells. In the peripheral nervous system, the glial cells sheathing the axons are the Schwann cells, and in the central nervous system they are the oligodendrocytes. Two arrangements of axonal sheathing exist and are shown in Figure 2.4. Small axons are simply surrounded by the glial cells, and a single glial cell may surround several small axons. Around larger axons, the glial cells wrap themselves around the axon and form the myelin sheath. In the peripheral nervous system, one Schwann cell forms myelin around only one axon, although many Schwann cells are required to ensheathe the entire axon. One Schwann cell covers at most 1–2 mm of axon; for an axon 1 m long, as many as 1,000 Schwann cells may encase the axon. In the central nervous system, one oligodendrocyte forms myelin around many axons, perhaps as many as fifty. But again, many oligodendrocytes are required to ensheathe an entire axon. Axons with the myelin sheath are called myelinated axons; those without the sheath are unmyelinated axons. Most axons in the vertebrate brain are myelinated.

The myelin sheath is best understood by considering its formation in the developing peripheral nervous system (see Figure 2.4). Initially, a Schwann cell surrounds but does not ensheathe the axon, much as glial cells surround small unmyelinated axons in the adult nervous system. With time, the glial cell wraps its mem-

2.4   Axonal sheathing by glial cells. (**A**) The axons at top are unmyelinated; the sequence at bottom shows the process of myelination. (**B**) An electron micrograph of a myelinated axon on the left and unmyelinated axons on the right.

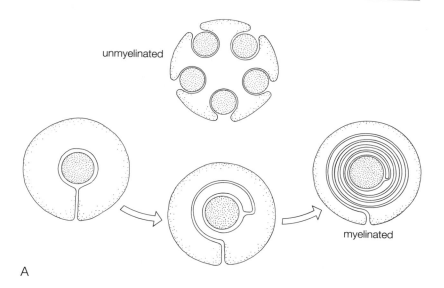

A

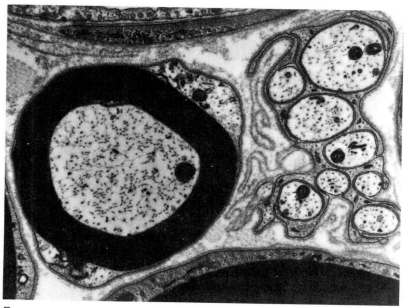

B

5 μm

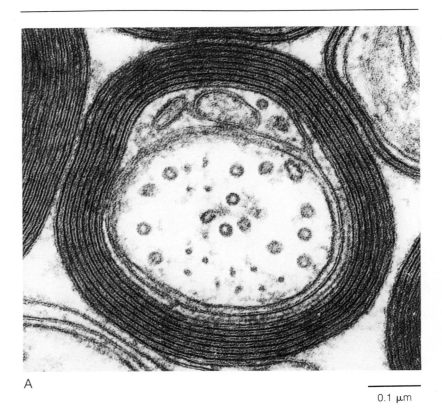

A

0.1 μm

brane around the axon. The number of wraps depends roughly on the axon's diameter; that is, larger axons have thicker myelin sheaths. The glial cell's cytoplasm is squeezed out, leaving behind layers of glial cell membrane, from which much of the protein is lost. What remains is a highly compact layer of glial cell membrane, with a high lipid content, surrounding the axon.

Figure 2.5 shows the myelin sheath in greater detail. In the electron microscope, cell membranes appear as a double leaflet, with the inner (cytoplasmic) side somewhat more electron-dense. Myelin consists of alternating thick and thin lamellae, "sheets" of apposed leaflets of the membrane. The distance between two thick or two thin lamellae is approximately 14 nm, twice the thickness of a single cell membrane.

2.5 (**A**) An electron micrograph of a myelinated axon in the spinal cord. (**B**) Myelin as it would be seen in a high-magnification electron micrograph: the thicker lamellae consist of apposed inner leaflets of membrane; the thin lamellae, apposed outer leaflets. (**C**) A blow-up of the myelin sheath shown in (**A**). The thick and thin lamellae are easily seen.

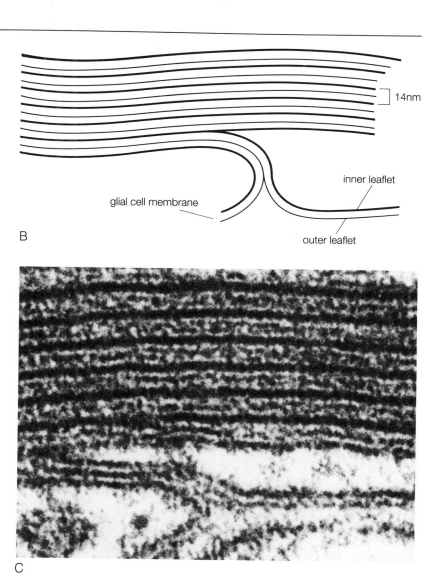

14nm

glial cell membrane

inner leaflet

outer leaflet

B

C

Along an axon, the myelin sheath is not continuous. Indeed, it is interrupted at regular intervals by myelin-free patches (see Figure 2.2A). These myelin-free areas are called nodes of Ranvier, after their discoverer, Louis Antoine Ranvier. The distance between nodes varies, depending on axon diameter. Larger axons have

2.6 A node of Ranvier. The asterisks indicate pools of glial cell cytoplasm remaining between the myelin membrane layers adjacent to the node; arrows mark the location of specialized junctions between the glial cell and the axonal membrane.

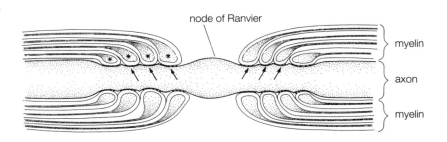

larger internodal distances, but an average internodal distance is 1 mm. One glial cell contributes myelin to just one internodal region.

The structure of a node of Ranvier is shown in Figure 2.6. The myelin-free area is about 1–4 μm long. Two features of the myelin sheath adjacent to the nodal regions are noteworthy. First, the myelin is not compact immediately adjacent to the node—glial cell cytoplasm remains between the layers of membrane (asterisks in Figure 2.6). Second, specialized junctions lie between glial cell and axonal membranes adjacent to the node (arrows). It has been proposed that both of these structural features suggest a supportive or nutritive role for the glial cells in the nodal regions of the axon. The important physiological function of the nodes of Ranvier in the transmission of action potentials down the axon will be discussed in Chapter 4.

As our description suggests, the glial sheath and nerve cell axon are closely related, but we do not know how closely. We have known for some time that nerve cell axons in the peripheral nervous system of mammals will regenerate, but not axons in the mammalian central nervous system. Thus, it is possible to reattach human limbs surgically after they have been severed and to expect full function and sensation to return to the limb because the nerves regenerate. In contrast, spinal cord injuries that result in the severing or loss of central nervous system axons cause permanent paralysis of the body and limbs served by those axons.

In the past few years, Albert Aguayo and his colleagues in Montreal have shown that axons in the central nervous system will regenerate if the damaged neurons or axons are brought in contact with glial (Schwann) cells of the peripheral nervous system. It is therefore not a difference between peripheral and central nervous system neurons that allows axonal regeneration in one but not in the other: the explanation lies with associated glial cells. Peripheral nervous system glial cells may release a substance, or factor, that promotes the regeneration of axons, whereas central nervous system glial cells either do not have this factor or do not release it. Another possibility (for which there is some evidence) is that central nervous system glial cells may release a factor that inhibits the regeneration of axons. The importance of these findings for treatment of brain injury, especially spinal cord damage, is obvious. Much research is under way to identify what it is about glial cells that may induce or inhibit axonal regeneration.

Golgi Type II Neurons

Golgi type II neurons are characterized by having short or, sometimes, no axons. They usually have smaller cell bodies than do Golgi type I neurons, and they do not go from one part of the nervous system to another. Rather, their processes are generally confined to a single nucleus or neuropil layer. Golgi type II neurons are involved in local interactions between nerve cells and are often called association neurons. Figure 2.7 shows two Golgi type II neurons, one with a short axon and one with no axon.

Electron microscopy has revealed that the dendrites and axon terminals of Golgi type II neurons are often both pre- and postsynaptic. That is, the dendrites are not exclusively receiving input, nor are the axons exclusively providing output, as is true of Golgi type I neurons. Rather, both dendrites and axon terminals can receive and transmit information. The dendrites in Figure 2.8, for example, both receive and make synaptic contacts. Often the synapses are arranged in a serial fashion. One consequence of this feature of Golgi type II cells is that local interactions can be mediated by a single dendrite or part of an axon terminal, without involving the entire neuron. Local-circuit interactions of this type are thought to be important in mediating subtle neuronal interactions.

Ramón y Cajal long ago pointed out that the brains of more highly developed animals contain more Golgi type II cells than Golgi type I cells. He noted also that the main difference between

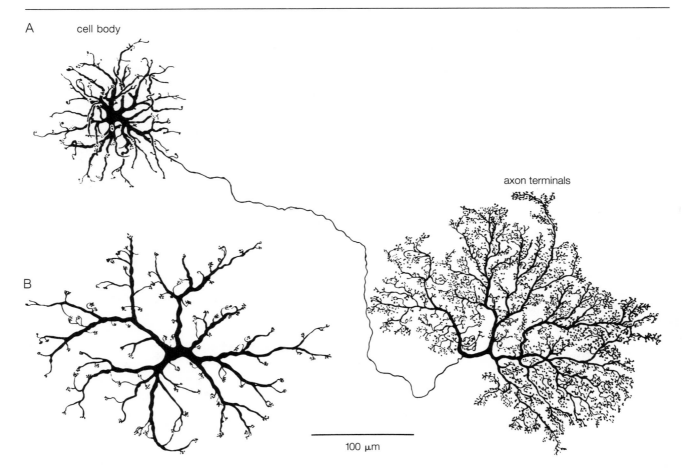

A cell body

B

axon terminals

100 μm

2.7 Two Golgi type II neurons found in the cat retina: (**A**) a short-axon horizontal cell and (**B**) an axonless horizontal cell. Cells stained by the Golgi method.

the mouse and the primate cerebral cortex was in the relative number of Golgi type II neurons. Others have found the same difference in other parts of the brain and in other comparisons.

Figure 2.9 shows in a highly schematic way the cellular organization of a nucleus in the vertebrate brain. In a typical brain nucleus, there are present both Golgi type I and II neurons; information leaves or arrives at the nucleus via the axons of Golgi type I cells. The processes of the Golgi type II neurons are confined to the nucleus, and the dendrites and axon terminals of these cells

2.8 Electron micrograph showing synapses *(arrows)* made between the dendrites of Golgi type II neurons. Note the serial arrangement of synapses on the right and the reciprocal synaptic interaction between the two dendrites on the left.

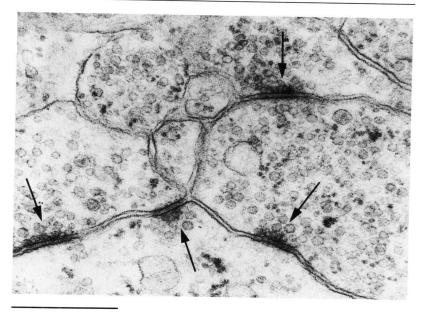

0.5 µm

2.9 The cellular organization of a brain nucleus. Inputs to and outputs from the nucleus are carried by the axons of Golgi type I neurons. The Golgi type II neurons and their processes are confined to the nucleus.

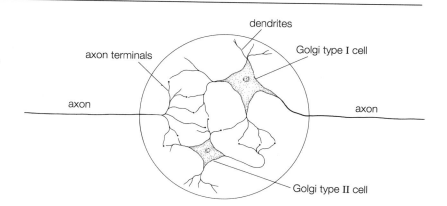

receive and make synaptic contacts. On Golgi type I cells, synaptic contacts are generally found only on the dendrites and cell body.

Synapses

The functional contacts or synapses between neurons are of paramount importance for understanding brain mechanisms. Here neurons are excited, inhibited, or modulated. Modifications of synapses are believed to underlie phenomena such as memory and learning. On any one cell hundreds to thousands of contacts are made. The champion in this regard may be the Purkinje cell of the cerebellum (see Figure 2.1B), which may have as many as 100,000 synapses on its dendritic tree! Most synaptic contacts in the brain are chemical; that is, a substance is released from the presynaptic side of the terminal, diffuses across a narrow (20–40 nm) cleft of extracellular space, and interacts with specific receptor sites on the postsynaptic side of the contact. Electrical synapses are also known but are less common. Two interacting elements come into close apposition, allowing membrane channels to link the two cells, and ions and small molecules to cross from one cell to the other.

Chemical Synapses

The basic features of chemical synapses are shown in the electron micrograph of Figure 2.10. On the presynaptic side of the junction, synaptic vesicles store the neuroactive substances to be released by the terminal. Adjacent to the site of release, synaptic vesicles are typically clustered close to the presynaptic membrane. Electron-dense material associated with the presynaptic membrane may indicate the presence of structures involved in the binding of the synaptic vesicles to the membrane. In the extracellular space between the pre- and postsynaptic membranes, filaments are often observed; the function of these filaments is not yet understood—they may anchor the two sides of the junction to each other. On the postsynaptic side of the junction also, prominent electron-dense material is associated with the membrane. It is likely related to the events initiated in the postsynaptic cell during synaptic activity.

In many parts of the brain, two anatomical types of chemical synapses are distinguished: type I and type II synapses (see Figure

2.10 In this chemical synapse, the presynaptic terminal contains numerous synaptic vesicles. These are of two types: large vesicles containing an electron-dense core and small, clear vesicles. A cluster of the clear vesicles marks the site of transmitter release *(arrow)*. Electron-dense material occurs near the membrane on both sides of the release site.

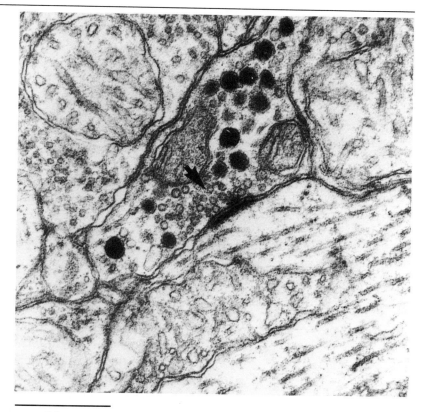

0.5 μm

2.11), first described by George Gray of University College, London. The type I synapse, an excitatory synapse, is found mainly on dendrites. It is characterized by spherical synaptic vesicles of 40 nm diameter, a widened synaptic cleft, and particularly prominent electron-dense material at the postsynaptic membrane. The type II synapse, an inhibitory junction, is found mainly on cell bodies. The vesicles in the presynaptic terminal at these junctions are varied in shape, but they are generally flatter than the vesicles in type I synapses and have dimensions of 25 nm by 50 nm. The synaptic cleft is not widened and the specialized structures in type

2.11 Type I synapses are excitatory; type II synapses inhibitory.

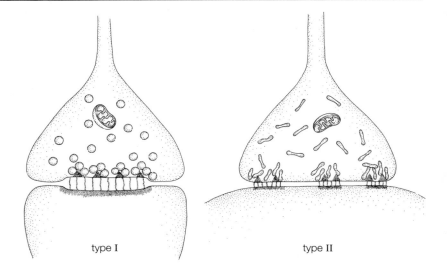

type I   type II

II pre- and postsynaptic membranes that appear dark in electron micrographs are clustered in groups, not evenly distributed as they are in type I synapses.

Perhaps the most striking difference between type I and II synapses is in the morphology of the synaptic vesicles. Although the shapes we see in micrographs may be influenced by the fixatives used, the difference is consistent in many parts of the brain and probably relates to the different substances transmitted at various kinds of synapses. At excitatory (type I) synapses, round vesicles are observed, and the transmitters released at such junctions in the vertebrate brain are usually an acidic amino acid (glutamate) or acetylcholine. At inhibitory (type II) synapses, the vesicles are typically flattened and they tend to store and release γ-aminobutyric acid (GABA) or glycine.

Other substances released at synaptic sites modify or modulate neural activity rather than initiate it (see Chapter 6), and the vesicles that store them have distinct features. In terminals containing monoamines (such as catecholamines or indoleamines), the vesicles are small, 50 nm in diameter, and have an electron-dense interior or core. Terminals that release small peptides are characterized by large spherical synaptic vesicles (120 nm in diameter), also with an electron-dense core. It is possible, therefore, to determine whether a certain synaptic junction is excitatory, inhibitory, or modulatory according to the morphology of the synaptic vesicles in the presynaptic terminal. Moreover, often we can surmise the substances released by the terminals under observation. A single terminal may release more than one neuroactive substance, in which case they may contain more than one type of synaptic vesicle (see Figure 2.10).

The most common synapses in the brain are those between axon terminals and dendrites; these are called *axodendritic* synapses and are mainly excitatory. Synapses between axon terminals and cell bodies are called *axosomatic* (*soma* means "body") and are generally inhibitory junctions.

Other synaptic arrangements are also observed, particularly on Golgi type II cells. For example, synapses between dendrites, *dendrodendritic* junctions, are known, as are *axoaxonic* synapses. Axoaxonic junctions are often inhibitory, and this interaction is termed *presynaptic inhibition*. Dendrodendritic interactions can be quite complex. There are *serial* synaptic arrangements, where a series of synapses link dendrites within a small area, and *reciprocal* synapses, where a dendrite receives a synapse from and makes a synapse back to the same process (refer to Figure 2.8). In at least one instance where the physiology of a reciprocal synapse has been explored, one of the junctions is excitatory in nature and the other is inhibitory. In electron micrographs, the excitatory junction is seen to have round synaptic vesicles and the inhibitory junction, flattened vesicles.

Neuromuscular and Sensory Cell Junctions

I will just briefly mention two other chemical synaptic contacts here; they will be discussed in some detail in Chapters 6, 7, and 14. The *neuromuscular junction* is the synapse between nerve and muscle. The elongated terminal of an axon typically runs along the muscle fiber in a depression called the *synaptic gutter*. Figure 2.12 shows a section of an innervating terminal in the gutter. Within the terminal are numerous synaptic vesicles that cluster opposite infoldings in the muscle membrane. In the vicinity of the vesicle clusters, electron-dense material is again noted along the

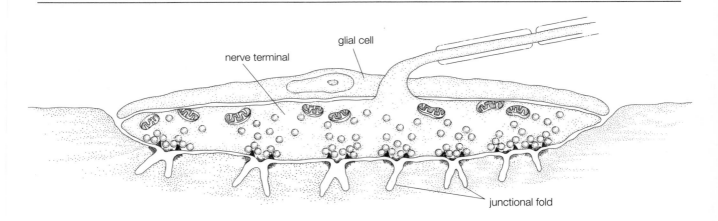

2.12 At a neuromuscular junction, the nerve terminal runs in a depression in the muscle fiber. Synaptic vesicles cluster in the nerve terminal opposite the junctional folds in the muscle cell membrane. Presumably transmitter is released at these zones.

pre- and postsynaptic membranes. The presumption is that these areas are active zones, where transmitter is released and where the muscle fiber responds to the released substances.

The *ribbon synapse* is characteristic of many sensory cells, retinal bipolar cells, and invertebrate neurons. Photoreceptors, many auditory hair cells, and electroreceptor cells make this kind of junction, and synapses throughout the nervous systems of many invertebrates have characteristics typical of ribbon synapses. In all cases, an electron-dense ribbon, bar, or other structure is observed in the presynaptic terminal. Synaptic vesicles cluster around this bar, and it is believed that release of transmitter occurs in its vicinity. Often, two or three postsynaptic processes are observed at a ribbon synapse, and usually electron-dense material is present on all of the postsynaptic elements. The evidence is that at many ribbon synapses, more than one postsynaptic process receives input from the presynaptic cell. Figure 2.13 shows a ribbon synapse typical of those made by bipolar cells in the retina, where two postsynaptic processes are present.

Electrical Synapses

At electrical synapses there is direct continuity between the interiors of the contacting cells. Ions and small molecules (up to a

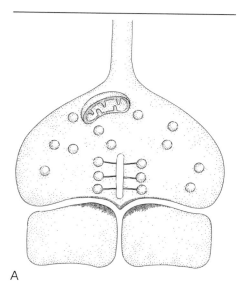

A

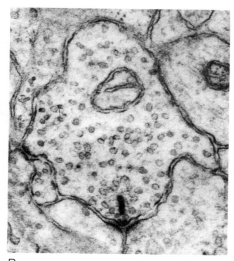

B

———————
0.5 μm

2.13 (A) A ribbon synapse typical of a reti-nal bipolar cell. Two postsynaptic elements are present at these junctions, and presum-ably both receive input when the synapse is active. (B) An electron micrograph of a bi-polar cell terminal from the chicken retina.

molecular weight of 1,200 daltons) pass through connecting chan-nels from one cell to the next. Electrical changes in one cell are transmitted virtually instantaneously to the other cell by ions pass-ing from one cell to the other via the channels. At most electrical synapses, ions can pass in either direction across the junction; thus, information flow is bidirectional. At some electrical junc-tions, however, ion flow is limited to one direction. These junctions are termed *rectifying* junctions and, like chemical synapses, are characterized by one-way-only transmission of information.

The striking structural feature of electrical junctions is the close apposition of the synapsing elements. At low magnification, it may appear that the membranes of the two cells or processes are touch-ing each other, but higher-magnification electron micrographs in-dicate a small gap of 1–1.5 nm between the two membranes (see Figure 12.14). Electrical junctions are referred to as *gap junctions* because of this small separation. The gap, though, is not contin-uous; it is interrupted by structures (proteins) extending from both membranes and linking together to form the channels between the two cells. Freeze-fracture electron microscopy (more about this technique will be presented in Chapter 3) has revealed many details about the structure of gap junctions and about the connecting protein channels, particularly their number and distribution. A freeze-fracture micrograph showing dense packing of membrane channels at a gap junction is shown in Figure 2.14.

Summary

The brain, like all tissues of the body, is made up of cells. Two kinds of cells are found in the brain: neurons, which process and transmit information, and glial cells, which play a supporting and perhaps a nutritive role. Neurons are varied in form, but in general they have many processes which branch out to facilitate complex interactions between cells. Two classes of neurons are distin-guished: Golgi type I neurons are large cells that transmit infor-mation from one part of the nervous system to another; the smaller Golgi type II neurons assist in processing and integrating infor-mation locally. There are many types and subtypes of Golgi type I and II cells; each part of the brain has its particular cell types.

Dendrites are relatively short, bushy processes, whereas axons are usually longer, thinner elements that end in a profusion of

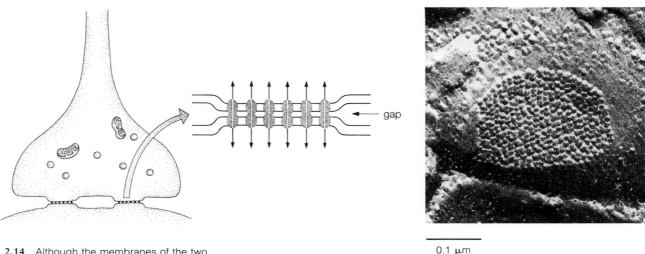

gap

0.1 μm

2.14 Although the membranes of the two cells making an electrical, or gap, junction are closely apposed, high-magnification electron micrographs show that a small gap always remains between the cells. The gap is interrupted by protein channels that link the two cells and allow the passage of ions and small molecules between the cells. These membrane channels, as seen in the freeze-fracture electron micrograph, can be very tightly packed.

terminals. In Golgi type I neurons the dendrites only receive input and the axons only transmit information. The dendrites and axon terminals of Golgi type II cells, by contrast, can both receive and transmit information.

The axon contains no protein-synthesizing particles or ribosomes, so it depends for its maintenance on materials made in the cell body. Substances are actively transported up and down the axon to facilitate this process. Glial cells also help maintain axons, and they form a special insulating membrane, the myelin sheath, around the larger axons.

Interactions between nerve cells occur at synapses. Most synaptic interactions in the brain are chemical, but electrical junctions are also known. Chemical synapses are characterized morphologically by storage (synaptic) vesicles in the presynaptic terminal and specialized structures in or near the pre- and postsynaptic membranes. Excitatory, type I synapses contain round vesicles; inhibitory, type II synapses contain flattened vesicles. The size and shape of the vesicle depend on the substance it stores. In electrical synapses, the two cell membranes are in close apposition and protein channels form links between the contacting neurons. A variety of synaptic arrangements are possible: axodendritic, axosomatic, dendrodendritic, and axoaxonic contacts.

Further Reading

BOOKS

Peters, A., S. L. Palay, and H. F. Webster. 1990. *The Fine Structure of the Nervous System.* 3d ed. Oxford: Oxford University Press. (A richly illustrated atlas of neurons and glial cells.)

Ramón y Cajal, S. 1911. *Histologie du Système Nerveux de l'Homme et des Vertébrés,* 2 vols. Paris: Maloine. Reprinted in 1955 by the Institut Ramón y Cajal, Madrid. (Even if you can't read French, these classic volumes are worth the effort of finding them for the illustrations alone.)

ARTICLES

Bunge, R. P. 1968. Glial cells and the central myelin sheath. *Physiological Review* 48:197–251.

David, S., and A. J. Aguayo. 1981. Axonal regeneration after crush injury of rat central nervous system innervating peripheral nerve grafts. *Journal of Neurocytology* 14:1–12.

Kuffler, S. W., and Nicholls, J. G. 1966. The physiology of neuroglial cells. *Ergebnisse der Physiologie* 57:1–90.

Heuser, J. E., and T. S. Reese. 1977. Structure of the synapse. In *Cellular Biology of Neurons, Part 1,* ed. E. R. Kandel, pp. 261–294. *Handbook of Physiology,* Sec. 1, *The Nervous System,* Vol. 1. Bethesda, MD: American Physiological Society.

Palay, S. L., and V. Chan-Palay. 1977. General morphology of neurons and neuroglia. In *Cellular Biology of Neurons, Part 1,* ed. E. R. Kandel, pp. 5–37. *Handbook of Physiology,* Sec. 1, *The Nervous System,* Vol. 1. Bethesda, MD: American Physiological Society.

Staehelin, L. A., and B. E. Hull. 1978. Junctions between living cells. *Scientific American* 238(5):140–152.

Vale, R. D., T. S. Reese, and M. P. Sheetz. 1985. Identification of a novel force generating protein (kinesin) involved in microtubule-based motility. *Cell* 41:39–50.

Vallee, R. B., and G. S. Bloom. 1991. Mechanisms of fast and slow axonal transport. *Annual Review of Neuroscience* 14:59–92.

Membranes and the Resting Potential

NEURONS CARRY and encode information by means of electrical signals generated across the outer membrane of the cell. That is, voltage changes occur across the cell membrane when the neuron is excited or inhibited (when it receives a signal from another neuron) and when information is transmitted along the cell's axon (when it sends a message to its terminals). Underlying these electrical signals is a maintained resting voltage (potential) that exists across the outer membranes of all cells. In this chapter I describe the kinds of electrical signals generated by neurons, the properties of neuronal cell membranes, and the means by which resting potentials are established across cell membranes.

Figure 3.1 indicates in general terms how the nervous system functions. This simple scheme is the central dogma of neuroscience. Sensory stimuli from the environment impinge on receptors, which respond by producing *receptor potentials*. Receptor potentials lead to the generation of *action potentials* (also called *spikes* or *impulses*), which carry information substantial distances in the brain (>1 mm). As information passes from one neuron to another at synaptic junctions, *synaptic potentials* are evoked in postsynaptic neurons and lead to the generation of action potentials in the postsynaptic cells. Receptor potentials can also lead directly to the generation of synaptic potentials in adjacent neurons, and in instances where distances between input and output synapses are not great, synaptic potentials themselves can result in the activation of nearby synapses and to the generation of synaptic potentials in adjacent neurons. Effectors, such as muscles, are activated by synaptic potentials, resulting eventually in behavior.

Properties of Neural Potentials

Receptor and synaptic potentials are alike in many ways. Because both may lead to the production of action potentials in neurons,

3.1   The central dogma of neuroscience.

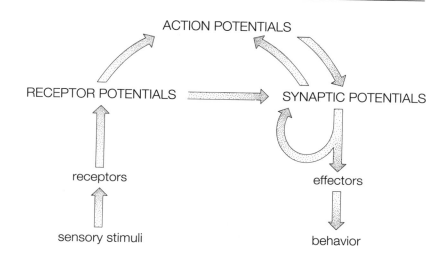

they are frequently termed *generator potentials*. Receptor and synaptic potentials are, however, quite distinct in their behavior from action potentials, as indicated in Table 3.1. In succeeding chapters, I explain the mechanisms underlying the properties of all three potentials, but here I wish to emphasize important differences in how they code information. Receptor and synaptic potentials code information by response amplitude: weak stimuli evoke small potentials; strong stimuli evoke large potentials. Action potentials code information by response frequency: weak stimuli cause neurons to generate few action potentials per unit time, whereas strong stimuli cause neurons to fire many action potentials per unit time. Stated differently, receptor and synaptic potentials code information by amplitude modulation (an AM system), and action potentials code information by frequency modulation (an FM system).

Figure 3.2 shows intracellular recordings from a second-order cell in the horseshoe crab eye. The recordings illustrate both the receptor potentials evoked by light in the receptor cells as well as the action potentials generated in the postsynaptic cell. In this case the receptor cells (first-order cells) are connected by electrical (gap) junctions to second-order cells called *eccentric cells*. The potentials generated in the receptor cells pass directly into the eccentric cell, where the action potentials are generated. Note that with brighter

Table 3.1   Differences between generator (receptor and synaptic) potentials and action potentials

Generator potentials	Action potentials
1. Graded response Amplitudes are graded with stimulus intensity; the stronger the stimulus, the larger the potential	*1. All-or-none response* All potentials are the same size: approximately 0.1 volt (100 millivolts), regardless of stimulus strength
2. Sustained response Potentials may last for as long as the stimulus lasts	*2. Transient response* Potentials are of constant duration, about 0.0015 sec (1.5 ms)
3. No discrete threshold for response Tiny potentials may be evoked with the smallest possible stimulus; that is, 1 quantum of light or transmitter from a single synaptic vesicle will elicit a small receptor or synaptic potential	*3. Threshold for response* Action potentials require a substantial change of membrane potential (up to 15 mV) for their generation
4. Responses summate Potentials sum when two stimuli are presented close together	*4. Responses are refractory* A refractory period of about 1.5 ms occurs after every action potential during which time another action potential cannot be generated, regardless of stimulus strength
5. Response is local Potentials spread passively from the site of generation; they are largest where they are produced and become progressively smaller away from that point	*5. Response propagates* Potentials are actively regenerated along an axon; a potential recorded at the end of an axon is the same size as one recorded at the beginning of an axon

stimuli the amplitude of the underlying receptor potential grew larger and the frequency of the action potentials' firing increased correspondingly. In contrast, the amplitude of the action potentials, measured from baseline to peak, did not change appreciably. The receptor potential lasted as long as the light stimulus, although initially it was larger than the maintained, plateau potential. Decay

3.2   The relationship between receptor and action potential generation is indicated by these intracellular recordings from the horseshoe crab eye. (**A**) Because the receptor cell is connected by gap junctions to the eccentric cell, the receptor potentials pass directly into the eccentric cell, from which intracellular recordings may be made. (**B**) As these recordings show, with dim light stimuli a small receptor potential is evoked in the receptor cell, and the eccentric cell generates only a few action potentials. With brighter light stimuli, larger receptor potentials are produced and more action potentials are generated in the eccentric cell.

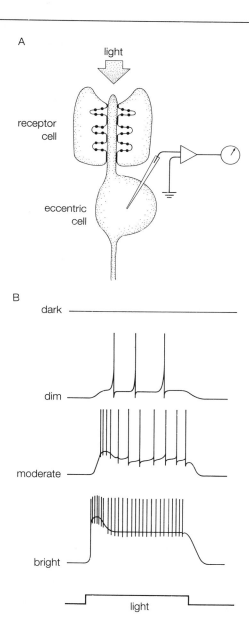

of response amplitude with time is characteristic of receptor potentials and is called *adaptation* (see Chapter 7).

Membranes: Their Structure and Electrical Properties

All electrical potentials in biological systems are generated across cell membranes. They depend on the properties and constituents of cell membranes for their generation; thus, knowing about membranes is fundamental to an understanding of the electrical potentials generated by neurons.

All membranes consist of a lipid bilayer in which proteins are imbedded. Phospholipids, the predominant lipids found in membranes, are made up of a glycerol backbone to which are attached two long fatty-acid chains and a shorter phosphate-linked alcohol molecule. This structure is shown in Figure 3.3. A key feature of phospholipids is that the glycerol and the phosphate-linked alcohol are highly hydrophilic (have a strong affinity for water), whereas the fatty-acid chains are highly hydrophobic (weak affinity for water). In simple terms, one end of the molecule seeks water and the other end avoids it. The consequence is that these lipids spontaneously form bilayers in aqueous solutions with the polar (hydrophilic) heads of the molecules on the outside and the hydrocarbon (hydrophobic) tails on the inside. Because the interior of

3.3   The chemical structure of a membrane phospholipid *(left)* and the arrangement of the lipid bilayer *(right)*. The interior of the bilayer is hydrophobic because the hydrocarbon tails of the lipid repel water. The head groups, on the other hand, are hydrophilic because the polar nature of a number of the atoms in this part of the molecule attracts water.

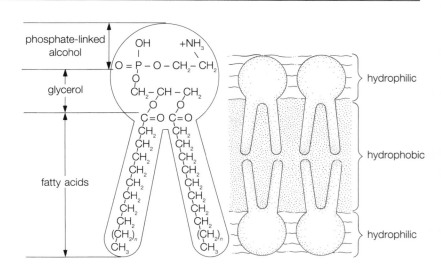

3.4 The overall organization of cell membranes. Most proteins are imbedded in the lipid bilayer; others are found just under the inner leaflet of the bilayer, in association with the cell's cytoskeleton. Sugar groups are often attached to proteins on the membrane surface.

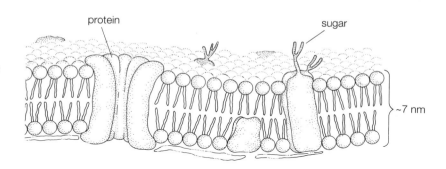

protein

sugar

~7 nm

the bilayer excludes water, the lipid part of a membrane is highly impermeable to ions and other water-soluble substances.

The movement of ions and small polar molecules across membranes is regulated by the membrane proteins, which form aqueous pores or channels through the lipid bilayer or pump ions or molecules across the membrane. Most membrane proteins occupy the entire thickness of the membrane, but some are found on just one side of the bilayer. The arrangement is shown schematically in Figure 3.4. In many membranes, proteins can move laterally in the membrane; indeed, the viscosity of membranes may be as fluid as that of a light motor oil, facilitating the mobility of the proteins. Minor lipid components in membranes, such as cholesterol, can reduce membrane fluidity. Membrane proteins can also be restricted in their movements by attachments to proteins and cytoskeleton components located just underneath the membrane in the cell's cytoplasm.

Freeze-Fracture Studies

Visualization of membrane proteins and their distribution within membranes has been made possible by the technique of freeze-fracture electron microscopy. Tissue is frozen in liquid nitrogen or liquid helium and, when solid, it is fractured with a blow from a sharp blade. The fracture plane follows paths of least resistance through the tissue; one such path is through the interior of membrane bilayers, where there is no ice. The membrane splits apart, exposing intramembranous particles—the membrane proteins. A replica of the fractured surface is made by depositing a thin layer of metal on the frozen tissue; then the replica is examined in an electron microscope.

Figure 3.5 depicts the freeze-fracture technique and shows a freeze-fracture electron micrograph, in which one can see the random distribution and heterogeneous nature of the particles in a typical membrane. The arrow points to a small gap junction. The inset illustrates at higher magnification the regular distribution and dense packing of the particles found at the gap junction; the presumption is that these particles are the protein channels that connect the two contacting cells. Membranes with a low protein content, such as myelin membrane, have few particles; membranes with a high protein content have many particles.

As in the micrograph of Figure 3.5, after the bilayer splits most proteins remain with the leaflet on the membrane's cytoplasmic side. This may be because the proteins and cytoskeleton elements inside the cell form attachments with the membrane proteins.

Reconstituted Membranes

Phospholipids, extracted from natural sources or made synthetically, spontaneously form bilayers in aqueous media. When examined by conventional electron microscopy, these reconstituted membranes look like natural membranes: they have the railroad-track structure common to all cell membranes and are between 6–9 nm thick.

By studying reconstituted membranes we gain further understanding of natural ones. Reconstituted bilayers have a high electrical resistance and are impermeable to ions. Their capacitance, however, is similar to that of natural membranes (see Appendix). The addition of membrane proteins to the reconstituted bilayers can restore the properties of natural membranes. For example, adding membrane proteins that form channels in the membrane reduces the electrical resistance of the membrane; that is, ions can now cross the membrane.

Membrane Channels

Five basic types of ion channels are found in nerve cell membranes. *Leak channels* account for the natural permeability of membranes to ions. *Voltage-sensitive channels* vary in permeability depending on membrane voltage; they are found mainly in axons. *Ligand-sensitive channels* respond to specific chemical agents and open or

3.5   The freeze-fracture method *(top)* and freeze-fracture micrographs of a photoreceptor inner segment at low and high magnification (inset). The arrow in the micrograph points to a collection of particles that are believed to be gap-junctional channels; the inset micrograph illustrates this cluster and nearby membrane particles at higher magnification.

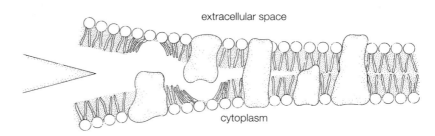

extracellular space

cytoplasm

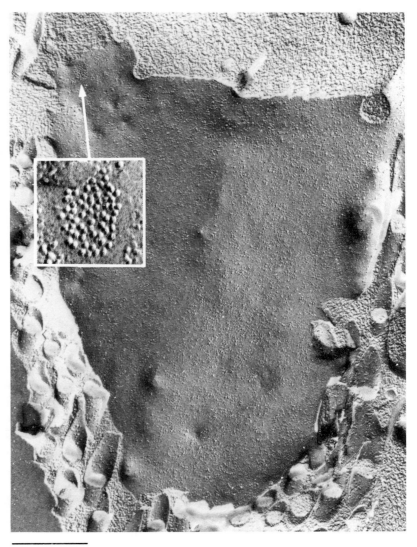

1 μm

close in the presence of that agent; they are found in dendritic and cell body membrane at postsynaptic sites. *Mechanosensitive channels* respond to deformation of the channel or the membrane surrounding the channel; they are found in certain receptor cells. And, finally, *gap-junctional* channels allow ions to move between cells, particularly at electrical synapses. Most channels show a high degree of ion specificity. They will allow only certain cations or anions to cross the membrane and they may even discriminate between quite similar ions such as Na^+ and K^+.

Measuring Membrane Resistance and Capacitance

Since the electrical potentials generated by neurons depend on the movement of ions across membranes through protein channels, neuroscientists frequently investigate neural mechanisms by measuring the electrical resistance of nerve cell membranes. This is done by injecting a known number of charged ions into a cell and recording the potential (voltage) established across the membrane. If the membrane is freely permeable to the injected ions, the ions quickly equilibrate on the two sides of the membrane and little potential difference develops across the membrane. Conversely, if the membrane has little permeability to the injected ions, a voltage develops across the membrane in accord with Ohm's law:

$$V = IR$$

where V is the voltage, I is the current (the number of ions injected into the cell per unit time), and R is the resistance of the cell membrane (the ease with which ions cross the membrane).

Figure 3.6 presents a schematic view of how resistance measurements are made. Ions are passed into a cell through one microelectrode while the voltage developed across the membrane is recorded with another microelectrode. When a positive pulse is applied to the current pipette, positive ions such as K^+ move into the cell (I is positive), and the inside of the cell becomes positive with respect to the outside. When a negative pulse is applied to the current pipette, negative ions such as Cl^- are forced into the cell, and the inside of the cell becomes negative relative to the outside. Records of the voltage developed across the membrane following the injection of positive and negative ions into a cell are also shown in Figure 3.6. (Solid lines are used for the positive

3.6 The measurement of membrane resistance and capacitance. In this experimental arrangement, ions (either K⁺ or Cl⁻) are injected into the cell from the current *(I)* pipette. Membrane voltage *(V)* is measured via the other pipette. On the right are shown records of the current *(I)* injected into the cell and the resulting voltage *(V)* changes that occur across the cell membrane. The discrepancy in time between the injection of current and the increase in voltage (shaded area) is due to membrane capacitance *(C)*. See the Appendix for further discussion of capacitance.

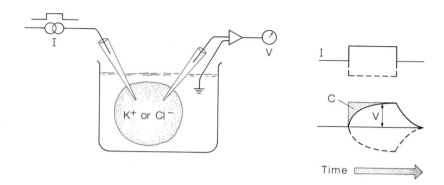

current pulse, dashed lines for the negative.) The final voltage *(V)* following the injection of current *(I)* reflects the resistance *(R)* of the membrane.

Note that it takes some time for the voltage to reach its final value. The delay is a result of the *capacitance* of the membrane. That is, as ions are injected into the cell, the effects of their charge are initially offset by interactions with charges present on the other side of the membrane; thus, the injected charge is initially buffered. Neuroscientists talk of this delay as "charging of the membrane." The *time constant* (τ) for this process depends on the membrane's capacitance *(C)* and resistance *(R)*: $\tau = RC$. Membrane capacitance initially depresses the voltage developed across the membrane relative to the amount of current flowing into the cell, and this can have consequences for neural signaling. For example, membrane capacitance affects the speed that action potentials travel down axons; I will return to this point in Chapter 4.

The Equivalent Circuit of Membranes

From an electrical point of view, biological membranes can be represented as resistors and capacitors wired in parallel; such circuits behave electrically like membranes. Figure 3.7 presents the equivalent electrical circuit of a membrane. As noted earlier, reconstituted lipid bilayers have the same capacitance as natural membranes; therefore the capacitance of biological membranes reflects mainly the influence of the lipid bilayer. In contrast, the

3.7 An equivalent circuit of a biological membrane. The capacitance of the membrane is determined by the lipid bilayer, whereas the resistance of the membrane is regulated by the membrane proteins (the channels).

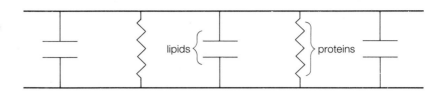

membrane's resistance relates mainly to the protein channels that regulate membrane permeability.

Finally, a word about the units used by neuroscientists to describe the resistance and capacitance of cell membranes. Membrane area is important for both properties, but it affects the capacitance and resistance of a cell in opposite ways. The larger a cell is, the more membrane it has and the greater is its total capacitance. A larger cell with more membrane, however, has a lower total resistance; that is, it has more ion channels than does a small cell. To take into account membrane area, the units used for membrane resistance are ohms *times* membrane area and for capacitance, farads *divided* by membrane area. These units are called *specific resistance* and *specific capacitance*. A typical nerve cell membrane has a specific capacitance of one microfarad (10^{-6} F) per centimeter squared (1 μF/cm^2), whereas the specific resistance of a nerve cell membrane is 10^3–10^5 ohm · cm^2. An artificial membrane, consisting only of a phospholipid bilayer, has a similar specific capacitance but a higher specific resistance of 10^7–10^8 ohm · cm^2.

Resting Potentials

Neurons at rest typically maintain a voltage across the membrane of between 60 and 70 millivolts. The inside of the cell is negative relative to the outside, so the membrane potential is conventionally written as −60 to −70 mV. In essence, the membrane potential reflects an unequal distribution of charge on the two sides of the cell membrane; that is, an excess of negative charge is inside the cell and an excess of positive charge is outside the cell. The charge is carried by ions, and the unequal distribution of charge results from (1) a difference in the concentrations of various ions inside

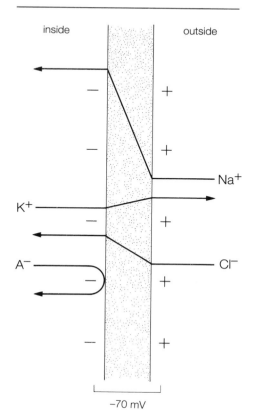

3.8 A schematic rendering of the location of the principal ions inside and outside a nerve cell and the relative permeability of the cell's membrane to the ions. K^+ and organic anions (A^-) are the major ions inside nerve cells; Na^+ and Cl^- predominate outside nerve cells. The membrane is more permeable to K^+ than to Cl^- or Na^+ and virtually impermeable to A^-.

and out and (2) a difference in the permeability of the membrane to different ions.

Figure 3.8 depicts the situation at rest. Inside cells, the major cation is potassium (K^+), outside it is sodium (Na^+). The major anions inside are small organic molecules (A^-) such as aspartate, acetate, pyruvate, or isothionate, while outside chloride (Cl^-) is the major anion. As I introduce the concepts involved in understanding resting and action potentials, I will most often use data from the squid giant axon and sea water (see Table 3.2 on page 72). The concentrations of ions in sea water approximates those in the blood of the squid and are therefore used here for the concentrations on the outside of the cell membrane.

There is relatively little Na^+ inside the squid axon (50 millimoles, or mM) relative to the outside (460 mM), whereas there is much more K^+ inside (400 mM) than outside (10 mM). Cl^- is like Na^+, there being considerably more Cl^- outside (540 mM) than inside (40 mM). The concentration of organic anions is high inside (400 mM) and virtually nonexistent outside the axon. Ratios of concentrations between inside and out can be given for Na^+, K^+, and Cl^-, but not for the organic anions (A^-).

The permeability of the nerve cell membrane in the resting state is also different for the various ions. This is depicted schematically in Figure 3.8 by the steepness of the barrier the ions must traverse to cross the membrane. The relative permeabilities of the squid giant axon membrane at rest to K^+, Na^+, and Cl^- are also given in Table 3.2. The membrane is most permeable to K^+, but sparingly permeable to Na^+. Permeability to Cl^- is intermediate between that of K^+ and Na^+, while the membrane is essentially impermeable to the organic anions (A^-). It is important to remember that the membrane is not *freely* permeable to K^+. If that were so, nerve cell membranes would have little electrical resistance, which is not the case. Of the major ions inside and outside of cells, however, the membrane is considerably more permeable to K^+ than to other ions.

With different concentrations on either side of the membrane, ions try to move down a concentration gradient, from an area of high concentration to one of lower concentration. The ions Na^+ and A^- cannot move across the membrane to any appreciable extent because of the low permeability of the membrane to these ions. The ions K^+ and Cl^- can cross the membrane. As they do so, an excess of negative charge builds up on the inside of the membrane, while positive charge builds up on the outside, and this imbalance results in the resting potential.

A Simplified Model

To understand better the principles underlying resting potentials, consider the simplified situation involving just K^+ and A^- on the inside and outside of a cell (Figure 3.9). Just as in real cells, we make the concentration of K^+ and A^- high inside our hypothetical cell and low outside, and permeability of the membrane relatively high to K^+ and nonexistent to A^-. (In real life, the situation depicted in Figure 3.9 could not exist because of the osmotic difference between the inside and outside of the cell. To equalize ion concentrations between inside and out, water would rapidly move into the cell and eventually make the cell explode.) In our hypothetical situation, there is diffusion pressure on K^+ and A^- to cross the membrane, to go from an area of high concentration to one of lower concentration. K^+ can pass through the membrane, but A^- cannot. Every K^+ ion that exits the cell leaves an extra negative charge inside the cell and adds an extra positive charge

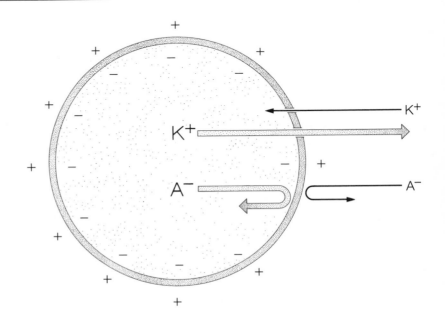

3.9 A simplified model of how a resting potential is established. The membrane is permeable to K^+ but not to A^-. With a higher concentration of K^+ and A^- inside than outside, K^+ and A^- try to leave the cell. K^+ can but A^- cannot, resulting in excess negative charge inside the cell and excess positive charge outside. A potential difference, a resting potential, is thus established.

to the outside. The resulting charge separation establishes a voltage across the membrane; that is, a resting potential.

Can K^+ equilibrate completely? Can it reach equal concentrations on both sides of the membrane? The answer is no, because as K^+ leaves the cell, making the inside negative, an electrostatic force pulls the positive K^+ ions back in. In other words, the system reaches an equilibrium in which the diffusion pressure on K^+ to move out is balanced by the electrostatic attraction of the negative charge pulling the positive ions back in.

The potential established across the membrane depends on the relative concentration of ions inside and out. The greater the concentration difference between inside and out, the greater is the diffusion pressure on an ion to move across the membrane, and the larger is the potential developed across the membrane. Resting potentials are thus often described as diffusion potentials.

The Nernst Equation and Equilibrium Potentials

The membrane potential at equilibrium for any two ion concentrations across a membrane (assuming permeability of the membrane to that ion) can be calculated with an equation derived in the late nineteenth century from basic thermodynamic principles by Walter Nernst, a physical chemist. It is called the *Nernst equation:*

$$E = \frac{RT}{nF} \log_e \frac{[C]_{out}}{[C]_{in}}$$

where E is the equilibrium potential; R is the universal gas constant; T, the absolute temperature; n, the charge on the ions involved; F, the Faraday constant (electrical charge per gram-equivalent of ion); $[C]_{out}$ is the outside concentration of the ion and $[C]_{in}$ is the inside ion concentration.

R, n, and F are all constants. When T is set at 18°C and the natural logarithm is converted to the base-10 system, the equation reduces to

$$E = (58 \text{ mV}) \log_{10} \frac{[C]_{out}}{[C]_{in}}$$

If the concentration of K^+ inside decreases, or if the concentration outside increases, the equilibrium potential will decrease because the concentration gradient across the membrane decreases.

Hence, less voltage will develop across the membrane. Conversely, increasing inside K^+ or decreasing outside K^+ will raise membrane voltage, because the concentration gradient for K^+ is higher.

For K^+ concentrations inside and outside the squid giant axon (400 mM and 10 mM, respectively), the K^+ equilibrium potential (E_K) equals -92 mV:

$$E_K = (58 \text{ mV})\log_{10} \frac{10}{400} = (58 \text{ mV})\log_{10} 0.025$$

$$= (58 \text{ mV})(-1.6) = -92 \text{ mV}$$

Consider also the situation where the membrane is totally impermeable to K^+. In this case, no potential will develop across the membrane because no K^+ can leave the cell, and there will be no charge separation across the membrane. Thus, every positive ion inside the cell is balanced by a negative ion. The same holds true for the outside, even though the total number of ions inside is different from the total number outside. What happens if the membrane is equally permeable to K^+ and A^-? Again no potential develops, because both ions would move from inside to out, thereby maintaining electrical neutrality on both sides of the membrane.

Is it possible to prevent K^+ from crossing the cell membrane if the membrane is permeable only to K^+ and there is a concentration difference between inside and out? The answer is yes. It can be done by making the inside of the cell negative. If the inside of the cell is maintained at the equilibrium potential (-92 mV), no K^+ will leave the cell; that is, the negative charge inside the cell attracting the positively charged K^+ ions will exactly balance the diffusion pressure on K^+ to leave the cell. If the inside of the cell is made even more negative than the equilibrium potential (> -92 mV), K^+ ions will move into the cell from the outside; that is, they will move against the concentration gradient! The electrostatic force on K^+ to move into the cell exceeds the diffusion pressure on K^+ to move out of the cell in such a case.

Equilibrium potentials for Na^+ and Cl^- can also be calculated by using the Nernst equation. For Na^+, whose concentration is much higher outside than inside a cell, the diffusion pressure on Na^+ is to enter the cell, which would make the inside of the cell positive and the outside negative (Figure 3.10). If we assume that

3.10 An idealized scheme showing how Na$^+$ could establish a membrane potential. If Na$^+$ and A$^-$ are concentrated outside the cell and the membrane is permeable to Na$^+$ but not to A$^-$, Na$^+$ will enter the cell and make the inside of the cell positive relative to the outside.

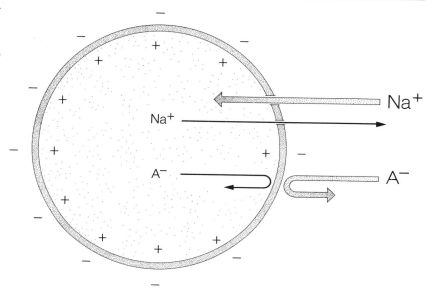

the membrane is permeable only to Na$^+$, the equilibrium potential calculated for Na$^+$ concentrations found inside and outside of the squid giant axon is +55 mV. For Cl$^-$, which has a higher concentration outside the squid giant axon than inside, the diffusion pressure is inward, as it is for Na$^+$. But Cl$^-$ is negatively charged, so its equilibrium potential is negative; at the Cl$^-$ concentrations found across the squid giant axon membrane (Table 3.2), the equilibrium potential is −67 mV.

The Goldman Equation and the Determination of Resting Potentials

If the membrane of an axon were equally permeable to K$^+$, Na$^+$, and Cl$^-$ (assuming the same ion concentrations used above), the resting potential established across the membrane would be approximately the average of the equilibrium potentials for these ions, −35 mV. But axons have resting potentials about twice that value, −70 mV. Why? Because, as pointed out earlier, nerve cell membranes at rest are much more permeable to K$^+$ than to Cl$^-$ or Na$^+$, and therefore the resting potential of most nerve cells is determined mainly by K$^+$. This means that K$^+$ is crucial for the

Table 3.2   Ion concentrations and membrane properties of the squid giant axon

	Concentration in axoplasm	Concentration in sea water	$\dfrac{[C]_{out}}{[C]_{in}}$	Relative permeability of membrane	Equilibrium potential
Na^+	50 mM	460 mM	9/1	1/30	+55 mV
K^+	400 mM	10 mM	1/40	1	−92 mV
Cl^-	40 mM	540 mM	14/1	1/10	−67 mV
A^-	400 mM	—	—	∞	—

establishment of the resting potential, and that alterations of K^+ concentrations either inside or out profoundly affect it. The resting potential is not at the equilibrium potential for K^+ (−92 mV) because the membrane is not totally impermeable to Na^+ or Cl^-. In other words, the difference between the resting potential and the K^+ equilibrium potential is due to the contribution of the other ions to the resting potential.

We can calculate the resting membrane voltage (V_m) by taking into account all relevant ion concentrations *and* the permeabilities of the ions. To do so, we use an equation similar to the Nernst equation formulated by David Goldman, an American biophysicist, in 1943:

$$V_m = \frac{RT}{F} \log_e \frac{P_K[K]_{out} + P_{Na}[Na]_{out} + P_{Cl}[Cl]_{in}}{P_K[K]_{in} + P_{Na}[Na]_{in} + P_{Cl}[Cl]_{out}}$$

The absolute permeability coefficients (P_K, P_{Na}, and P_{Cl}) are difficult to measure, so a simplification of the Goldman formulation is often used. This simplification employs the ratios of Na^+ and Cl^- permeabilities relative to K^+ permeability (termed b and c) and it has the form:

$$V_m = (58 \text{ mV}) \log_{10} \frac{[K]_{out} + b[Na]_{out} + c[Cl]_{in}}{[K]_{in} + b[Na]_{in} + c[Cl]_{out}}$$

So, for example, using ion concentrations and permeability ratios given in Table 3.2:

$$V_m = (58 \text{ mV}) \log_{10} \frac{10 + (0.03)460 + (0.1)40}{400 + (0.03)50 + (0.1)540} = -70 \text{ mV}$$

This result is in excellent agreement with the experimental observations.

Variations and Exceptions

The picture developed so far suggests that resting potentials depend strictly on differing ion concentrations inside and outside of a nerve cell and on differing membrane permeabilities to the various ions. Strong evidence in favor of this view has come from experiments with the squid giant axon, whose inside and outside ion concentrations are reversed. The squid giant axon may be as large as 1 mm in diameter, and its size enables neuroscientists to squeeze out the cytoplasm with a small roller, much as toothpaste is squeezed from a tube (Figure 3.11). Not only does this technique allow for chemical analyses of the axoplasmic contents (such as those presented in Table 3.2), but it also permits the replacement of the axoplasm with artificial solutions. When the axons are reinflated with simple salt solutions containing NaCl and KCl, investigators find that the subsequent membrane potentials are appropriate with regard to sign and magnitude of the concentrations of ions on the membrane's two sides. Thus, if K^+ concentration is high on the inside of the axon and low on the outside, the potential across the membrane is negative. If the situation is re-

3.11   (A) A squid giant axon with a recording electrode inside. This axon is about 0.9 mm in diameter. (B) The axoplasm of a giant axon can be squeezed out by means of a small roller.

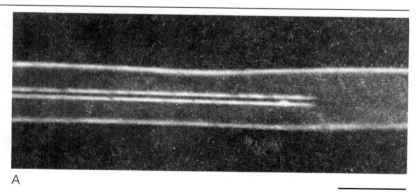

A

1 mm

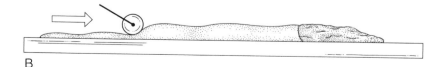

B

Table 3.3 Membrane potential of the squid giant axon under experimental conditions

Salts in solution	Concentration inside cell	Concentration outside cell	Membrane potential
KCl	600 mM	10 mM	−55 mV
NaCl	10 mM	600 mM	
KCl	10 mM	600 mM	+55 mV
NaCl	600 mM	10 mM	

versed (outside K^+ concentration high, inside K^+ low), a membrane potential of the same magnitude but opposite polarity is recorded (see Table 3.3).

Do all animals have the same concentrations of ions inside and out as does the squid, a marine organism? No, but the *ratios* of ion concentrations inside and outside most nerve cells are remarkably constant. A frog, for example, has only about one-third the amount of K^+ in its blood as the squid has (150 mM versus 400 mM), but the inside concentration of K^+ in the nerve cells of the frog is also only about one-third the concentration of K^+ in the squid axon (3.5 mM versus 10 mM). The ratio of K^+ concentrations between inside and out is about 40 to 1 in each case. Since it is the ratio of concentrations that determines equilibrium potentials (the Nernst equation), the resting potentials of frog and squid neurons are virtually the same.

In some cells, though, we find noteworthy differences from the situation just described. For example, the membranes of many muscle cells are more permeable to Cl^- than is the squid giant axon membrane, and the permeability of the membrane to K^+ is less than that of the squid axon membrane. In these cells, Cl^- is more important for establishing the resting potential than is K^+. Furthermore, the concentration of chloride in cells can vary substantially; thus, the equilibrium potential for Cl^- can be significantly different from that of the squid giant axon. In some plant cells, inside concentrations of K^+ are quite low, and so virtually all of the resting potential is determined by Cl^-. These examples are the exception rather than the rule; most neurons in most organisms have resting potentials determined mainly by K^+.

How Many Ions Cross the Membrane?

Obvious questions now confront us. If, in establishing the resting potential, K^+ ions leave the cell, doesn't this mean that the K^+ concentration inside the cell decreases? Does this affect the potential? In other words, how many ions must cross the cell membrane to establish a potential of 70 mV? We can calculate the number of ions involved. The charge (Q) on 1 cm^2 of membrane is equal to the product of the specific capacitance times the voltage, or $Q = CV$. If the charge is carried by a univalent ion (such as K^+), the number of moles of that ion per cm^2 is equal to the specific capacitance times the voltage divided by the Faraday constant:

$$\text{moles/cm}^2 = \frac{CV}{F}$$

Specific capacitance is 1 μF/cm^2, voltage is 70 mV (0.07 V), and the Faraday constant is 96,500 coulomb/mole. Multiplying and dividing through shows that approximately 7×10^{-13} moles of charge must cross each cm^2 of membrane to establish a 70 mV resting potential. Is this a lot or a little? In squid, one square centimeter of axonal membrane encloses about 3×10^{-5} moles of K^+, so the loss of 7×10^{-13} moles is insignificant—only 0.000002 percent of the K^+ ions inside the cell, an amount so small that it is not measurable. In other words, the ion concentrations across the cell membrane do not change appreciably when the resting potential is established.

Maintenance of Ion Concentrations: Ion Pumps

Even though the Na^+ permeability of the membrane at rest is low, some Na^+ does cross the membrane and with time accumulates inside the cell. The addition of positive charge inside a cell also allows more K^+ to leave. Eventually the concentration differences between inside and out would be abolished, so removing Na^+ from inside a cell is critical.

Cell membranes possess active ion pumps that require an energy source (usually adenosine triphosphate, ATP) and that are readily poisoned by metabolic inhibitors that prevent ATP formation. The most common pump is a Na^+–K^+ exchange pump in which Na^+ is pumped out of the cell while K^+ is pumped in (see Figure 3.12). A phosphate group from ATP is transferred to the pump and yields ADP (adenosine diphosphate); the activated pump now moves a

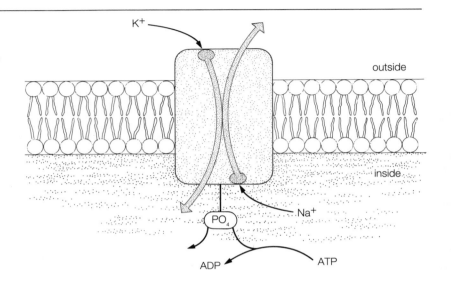

3.12 A Na^+–K^+ exchange pump. Following the addition of a phosphate group to the pump by ATP, a Na^+ ion is pumped out of the cell and a K^+ ion into the cell. The stippled areas represent binding sites of the ions. The mechanism of ion transport by exchange pumps is not well understood, but it appears that the ions move in concert, as indicated by the meeting of the curved lines showing (very schematically) the paths of the ions through the pump.

Na^+ ion from inside to out and a K^+ ion from outside to in. Following this exchange, the phosphate group comes off the pump and the process is ready to begin again.

The pump in Figure 3.12 is a neutral pump, that is, it makes a one-to-one exchange. Most pumps are not neutral; they pump more Na^+ out than K^+ in. The Na^+ pumps found in axons and in most neurons exchange 3 Na^+ ions for 2 K^+ ions, for example. These are called *electrogenic* pumps because they add to the membrane potential: if more Na^+ ions leave the cell than K^+ ions come in, the membrane potential becomes more negative. Ordinarily this uneven exchange makes only a small contribution to the resting membrane potential, but in some cells, particularly in certain invertebrate neurons, electrogenic pumps contribute significantly to the resting membrane potential of a neuron.

Summary

Two types of electrical potentials are generated by neurons. (1) Synaptic and receptor potentials (often called generator potentials) are graded, sustained, and local potentials; they summate and have no specific activation threshold. (2) Action potentials are all-or-none, transient, and propagated signals that are refractory and that require a substantial change of membrane voltage for their generation. Synaptic and receptor potentials trigger the membrane voltage changes that lead to the generation of action potentials, whereas action potentials transmit information long distances in the brain (>1 mm).

The electrical potentials are generated across nerve cell membranes, which, like all biological membranes, consist of a phospholipid bilayer in which proteins are imbedded. The lipid bilayer has a high electrical resistance; it is impermeable to ions. The imbedded proteins form aqueous pores or channels through which certain ions cross the membrane. Biological membranes behave electrically like a circuit consisting of resistors and capacitors in parallel. The phospholipid bilayer is mainly responsible for membrane capacitance, whereas the membrane proteins regulate membrane resistance.

All cells have a resting potential across the cell's membrane. The potential is established by the unequal concentrations of ions inside and outside the cell and by the different permeability of the membrane to the ions. The concentration of potassium (K^+) is high inside and low outside cells, whereas sodium (Na^+) is low inside and high outside. At rest, nerve cell membranes are 30 times more permeable to K^+ than to Na^+, and thus K^+ is most important in establishing the cell's resting potential. Some K^+ ions cross the cell membrane because of the diffusion pressure on K^+ to move from a region of high concentration (inside) to one of low concentration (outside). Their exit leaves unbalanced negative ions on the inside and adds positive charge to the outside of cells; hence a potential difference develops across the cell membrane, the resting potential.

The potential thus established across the cell membrane depends on relative ion concentrations inside and out. The equilibrium potential for any ion, assuming the membrane is permeable to that ion, is given by the Nernst equation, whereas the Goldman equation takes into account the concentrations of *all* relevant ions and the permeability of the membrane to the ions to calculate the resting membrane voltage.

Although various animals may have different absolute concentrations of ions both in and surrounding their cells, the ratios of the ion concentrations are quite constant, as are the relative membrane permeabilities. Some exceptions are known. Muscle cells, for example, are more permeable to Cl^- than are most neurons and thus $Cl-$ is more important for establishing the resting potential in these cells than is K^+.

Relatively few ions need to cross the membrane to establish the resting potential. Nonetheless, the ion concentrations inside and out must be constantly maintained. The ion concentrations inside cells are regulated by active pumps in the cell membrane. For most cells the critical pump is an exchange pump that extrudes Na^+ from the cell's interior while transporting K^+ into the cell.

Further Reading

BOOKS

Kuffler, S. W., J. G. Nicholls, and A. R. Martin. 1984. *From Neuron to Brain*. Sunderland, MA: Sinauer Press. (A textbook that provides an especially good discussion of resting, action, and synaptic potentials.)

Matthews, G. G. 1991. *Cellular Physiology of Nerve and Muscle*. 2d ed. Palo Alto, CA: Blackwell Scientific Publications. (A short, readable description of resting and action potentials.)

ARTICLES

Baker, P. F. 1966. The nerve axon. *Scientific American* 214(3):74–82.

Baker, P. F., A. L. Hodgkin, and T. I. Shaw. 1962. The effects of changes in internal ionic concentrations on the electrical properties of perfused giant axons. *Journal of Physiology* 164:355–374.

Hille, B. 1977. Ionic basis of resting potentials and action potentials. In *Cellular Biology of Neurons, Part 1*, ed. E. R. Kandel, pp. 99–136. *Handbook of Physiology*, Sec. 1, *The Nervous System*, Vol. 1. Bethesda, MD: American Physiological Society.

Thomas, R. C. 1972. Electrogenic sodium pump in nerve and muscle cells. *Physiological Review* 52:563–594.

Action Potentials and Their Propagation

With some understanding of the properties of the nerve cell membrane at rest behind us, we can now explore the mechanisms underlying the electrical signals generated by neurons. In this chapter, I discuss action potentials, the transient, all-or-none voltage impulses that neurons employ to transmit information along axons. A crucial feature of action potentials is that they are *propagated:* action potentials are continuously regenerated along the length of an axon so that their size is as large at the end of an axon as at its beginning. In other words, regardless of the length of the axon, no decrement in the strength of the signal occurs. The advantage of this feature for communicating information in the nervous system is obvious.

The Generation of Action Potentials

If a small amount of electric current is passed into a squid giant axon, the voltage across the membrane changes in accord with the current's polarity (positive or negative) and strength and with the membrane's resistance and capacitance. So long as the change induced in membrane voltage is less than about 15 mV, the response is passive—the voltage change simply reflects the resistive and capacitive properties of the membrane, as we learned in the last chapter (see Figure 3.6). If, however, the membrane voltage is changed by about 15 mV in the positive (depolarizing) direction, additional potential change is often observed across the membrane. Moreover, recovery after the positive current is turned off is slower than we would predict from the membrane properties. (See Figure 4.1.) Negative (hyperpolarizing) currents do not evoke such active responses. No matter how great a change in membrane potential induced in the negative direction, the response remains passive, depending strictly on the membrane's resistance and capacitance.

When the amount of depolarizing current injected into a giant

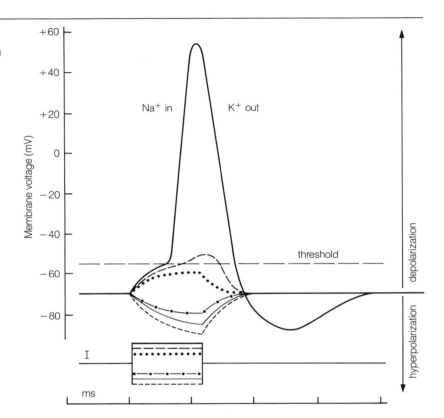

4.1   Effects of current *(I)* on axonal membrane. Hyperpolarizing and weak depolarizing currents evoke only passive responses in the membrane. Stronger depolarizing currents evoke active responses. If the current depolarizes the membrane by more than 15 mV (threshold), an action potential is generated. Voltage rapidly goes to +50 mV and then returns to resting levels within a few milliseconds. The rising phase of the action potential is due to the influx of Na$^+$ ions, recovery to the outflow of K$^+$.

squid axon is greater than 15 mV (when, in other words, membrane voltage is brought from the resting potential of −70 mV to more than −55 mV) the membrane potential undergoes an explosive change. The voltage rapidly reverses across the cell membrane, and the inside of the cell becomes positive relative to the outside. Within about 1 millisecond (ms), the voltage across the membrane rises to about +50 mV and then rapidly falls. Within another millisecond or so, the voltage plummets to about −90 mV and thereafter returns to resting levels of −70 mV. This explosive and rapid change of membrane voltage is the action potential.

What causes the rapid depolarization of the axonal membrane to about +50 mV? In the last chapter we learned that the equilibrium potential for Na$^+$ is +55 mV, which suggests a clue—namely, if the axonal membrane suddenly becomes much more permeable to Na$^+$ than to K$^+$, we would expect the membrane voltage to approach the sodium equilibrium potential (E_{Na}). Indeed, as long ago as 1902, E. Overton showed that Na$^+$ is needed for nerve-muscle excitation. What is responsible for the *repolarization* of the cell membrane? That the membrane potential dips to about -90 mV during the recovery phase, close to the potassium equilibrium potential, suggests that restoration of resting membrane voltage depends on K$^+$ rapidly moving out of the cell.

Voltage-Dependent Permeability Changes

What makes the membrane so permeable to Na$^+$ during an action potential? The answer is membrane voltage. At rest, the membrane is about 30 times less permeable to Na$^+$ than to K$^+$. But, as we depolarize the membrane, it becomes more permeable to Na$^+$. Why this is so is not entirely understood, although we will come back to the question later on in the chapter.

If a membrane's Na$^+$ permeability depends on membrane potential, the explosive change in membrane voltage during action potential generation makes sense (see Figure 4.2). That is, with depolarization, membrane permeability or conductance (g_{Na}) to Na$^+$ increases.* With an increase in Na$^+$ conductance, Na$^+$ enters the axon, causes more depolarization, and increases Na$^+$ conductance even more. More Na$^+$ enters the cell, which leads to more depolarization, and so on. The system acts as a positive feedback, resulting in an accelerating, regenerative response.

When does the response become regenerative? When the Na$^+$ current coming in as a result of membrane depolarization exceeds the rate at which K$^+$ can leave the cell. This occurs when an axon is depolarized by about 15 mV from rest, the threshold for action potential generation. If the membrane voltage is depolarized by less than 15 mV, K$^+$ ions can leave the cell as rapidly as Na$^+$ ions enter the cell. Positive charge does not build up inside the cell, and

---

* Conductance *(g)* is used to describe ion flow across membranes. Conductance is closely related to permeability but not identical to it. Conductance depends not only on the permeability of the membrane to an ion but also on the number and distribution of the permeable ions. For example, if an ion is not present on either side of the membrane, there is no conductance to that ion, even though the membrane may be permeable to it. Conductance, measured in siemens, is inversely related to membrane resistance: $g = 1/R$, where R (as before) is measured in ohms.

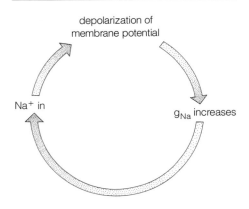

4.2   The effect of depolarization on Na$^+$ conductance (g$_{Na}$). As g$_{Na}$ increases, more Na$^+$ crosses into the axon, further depolarizing the membrane.

regeneration does not occur. Above this level of membrane depolarization, Na$^+$ ions enter the cell more rapidly than K$^+$ ions can leave, resulting in the all-or-none action potential.

During the rising phase of the action potential, Na$^+$ conductance increases about 50-fold, to 20 times the K$^+$ conductance. This is why the membrane potential rises close to the Na$^+$ equilibrium potential of $+55$ mV. In other words, the axonal membrane's resistance during an action potential falls from about 1,000 ohms $\cdot$ cm^2 to roughly 20 ohms $\cdot$ cm^2.

RECOVERY

Following the initiation of an action potential, Na$^+$ permeability remains high for only a short time. Indeed, the Na$^+$ current across the membrane reaches a peak in about 1 millisecond and then subsides. Thereafter, the membrane becomes more permeable to K$^+$, and the membrane voltage heads toward the potassium equilibrium potential. As the membrane voltage repolarizes, the increased K$^+$ conductance decreases and the resting potential eventually returns to about -70 mV. Figure 4.3 shows in a schematic fashion the sequence of changes in membrane permeability that occur during an action potential.

An important point is that Na$^+$ and K$^+$ conductances are dependent on both voltage and time. That is, the conductance changes for Na$^+$ and K$^+$ are caused by changes in membrane voltage, but the timing of the changes differ. Upon depolarization of the membrane, Na$^+$ conductance rapidly increases, but then it shuts down to resting levels after about 1 millisecond. K$^+$ conductance, by contrast, rises slowly upon membrane depolarization and falls only as membrane voltage decreases. Finally, note that Cl$^-$ conductance across the membrane does not change during the action potential; it plays no role in generating action potentials.

Voltage-Clamp Analysis

The action potential is difficult to study quantitatively because ion conductance changes during the action potential are so rapid. For these changes to be analyzed during an action potential, the system must be controlled by voltage clamping the axon. This means that the voltage across the axon is set at a fixed level, and current is passed into the axon to maintain the membrane at the fixed voltage in accord with Ohm's law, $V = IR$. The current required to

4.3 Schematic representations of membrane permeability to Na^+ and K^+; as in Figure 3.8, the steepness of the path of an arrow across a membrane indicates the level of *resistance*. (The lower the resistance, the higher the permeability.) Depolarization increases Na^+ permeability (shown at left as lowered resistance), but as long as K^+ can leave the cell as rapidly as Na^+ enters the cell, no action potentials are generated; below threshold the change in permeability (indicated by double-pointed arrow) depends simply on the magnitude of depolarization. Once Na^+ influx exceeds K^+ efflux, positive charge builds up inside the cell, membrane permeability to Na^+ increases rapidly, and an action potential is rapidly fired *(middle)*. After a millisecond or so, the membrane becomes impermeable to Na^+ again, but now the membrane becomes more permeable to K^+ *(right)*.

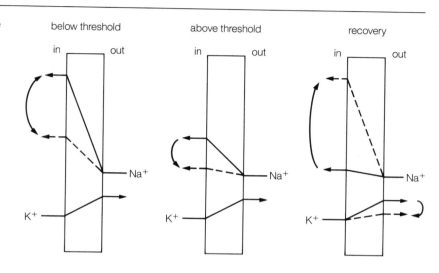

maintain the axon at the fixed voltage, which is easily measured, reflects the current flow across the membrane. From the measurements, the membrane conductance at any given voltage can be calculated.

Voltage clamping of a cell or axon requires a feedback system. It will work so long as the electronics respond faster than the nerve cell membrane, and today this is no problem. As shown in Figure 4.4A, the basis for the voltage clamp is a differential amplifier whose output is determined by the difference between a command voltage (one set by the investigator) and the membrane voltage. The amplifier's output is fed back to the axon to keep the membrane voltage the same as the command voltage. If command and membrane voltages are the same, no output comes from the amplifier; hence, no current is fed back to the axon. But if membrane voltage is different from the command voltage, current is passed into the axon to bring membrane voltage to the same level as the command voltage. When membrane voltage is shifted, time-dependent changes in Na^+ and K^+ conductances occur. A record of the current necessary to maintain membrane voltage at the command level reveals the conductance changes over time at a certain voltage.

Figure 4.4B shows a typical record when the membrane voltage

4.4 **(A)** With a voltage clamp, membrane potential of the axon (V_m) is maintained at the command voltage level by current *(I)* fed back into the axon. **(B)** Immediately after the command voltage *(V)* is shifted from resting voltage (−70 mV) to 0 mV, current *(I)* must be fed into the axon to counter current flows across the membrane. There is an initial capacitive and gating current across the membrane, then a transient inward current followed by a slower outward current that lasts as long as the axon is kept depolarized.

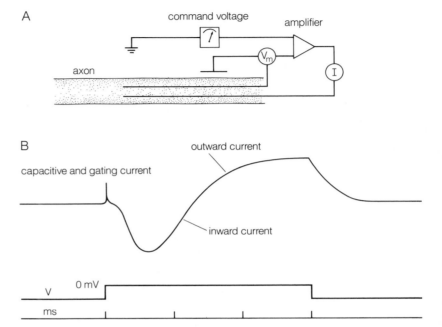

is shifted from rest (−70 mV) to 0 mV. Immediately after the membrane voltage is shifted from −70 mV to 0 mV, a fast, transient deflection in the record reflects a redistribution of charge on the membrane (a capacitive discharge) and a redistribution of charge within the ion channels themselves (the so-called gating current). We will discuss the gating current later. Following this initial deflection, a slower, downward swing in the record indicates current passed into the axon to counter an inward flow of current across the membrane (in other words, to counter the inward Na$^+$ current). The inward flow of current peaks after about 1 millisecond and over the next millisecond the current injected into the axon reverses to counter an outward current flow across the membrane—the K$^+$ current. The outward K$^+$ current persists until the command voltage is returned to rest voltage.

By shifting the command voltage to different levels, we can evaluate the conductance changes at various membrane potentials. Figure 4.5 presents a family of voltage-clamp records when the command voltage was shifted from rest to −20 mV, 0 mV, +50

4.5 A family of voltage-clamp records generated successively by shifting membrane voltage from resting voltage to −20 mV, 0 mV, +50 mV, and +70 mV. Note that the maximum inward Na$^+$ current occurs when the membrane is shifted to 0 mV, and maximum K$^+$ current occurs when the membrane is shifted to +70 mV.

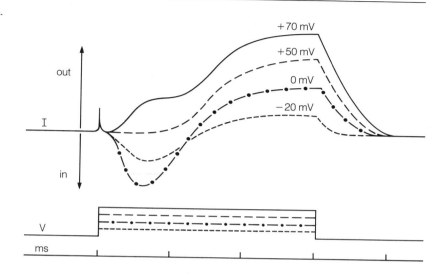

mV, and +70 mV. The initial downward deflections in the records represent the inward Na$^+$ current, the later upward deflections, the outward K$^+$ current. The behavior of the outward K$^+$ current is simpler and easier to understand; the more depolarized the axon, the larger is the K$^+$ current. The inward Na$^+$ current, however, increases in magnitude up to 0 mV, but then it declines as the membrane is depolarized further. When the membrane is shifted to +50 mV, no current flows inward; when shifted to +70 mV, there is an initial outward current flow! What is going on?

There is no inward Na$^+$ current when the inside of the axon is maintained at +50 mV because the voltage across the membrane is at or close to the Na$^+$ equilibrium potential. Hence, the positivity inside the axon counters the diffusion pressure on Na$^+$ to enter the axon. As Na$^+$ channels open, Na$^+$ does not flow across the membrane. When membrane voltage is shifted to +70 mV, the inside of the axon is now so positive that Na$^+$ is driven from the axon against the Na$^+$ concentration gradient; this accounts for the small initial outward current.

From records such as these, certain conclusions can be reached. For example, the inward Na$^+$ current peaks about 1 millisecond after the current begins to flow, and it is at a maximum when the membrane is at 0 mV. The records also reveal the delay before the

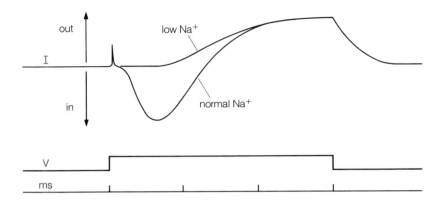

4.6   The effects of lowering outside Na$^+$ (to 50 mM) on axonal membrane currents. The initial inward current is abolished.

outward K$^+$ current begins, and that the K$^+$ current is maintained for as long as the membrane is depolarized.

Further analysis requires the separation of the currents in and out. Put another way, since the two currents overlap in time, eliminating one current permits study of the other current in isolation. One way to do this is to alter levels of Na$^+$ or K$^+$ on one or both sides of the membrane. Such experiments help to confirm the identity of the ions responsible for the two currents. For example, in one experiment (Figure 4.6) the outside Na$^+$ was lowered to 50 mM; thus inside and outside Na$^+$ levels were the same. The initial inward current was abolished, providing evidence that the initial current flow is due to Na$^+$.

Effects of Drugs on Na$^+$ and K$^+$ Currents

A more powerful way to isolate the currents is pharmacologically; that is, add substances to the axon that block one or another of the currents. Fortunately, certain drugs can eliminate either the Na$^+$ or the K$^+$ currents. Tetrodotoxin (TTX), a drug that potently blocks the channels carrying the Na$^+$ current, was originally isolated from the puffer fish, which has high TTX concentrations in its liver and ovaries. Puffer fish is a delicacy of Japanese cuisine, but because TTX is extremely poisonous, chefs preparing this fish must be specially licensed. TTX, by blocking action potential generation, interferes with long-distance (>1–2 mm) neural transmission in the brain and body. In addition to causing profound dis-

4.7 The effects of drugs on the inward Na$^+$ and outward K$^+$ currents. TTX blocks the inward Na$^+$ current, TEA blocks the outward K$^+$ current, and pronase eliminates the time-dependence of the Na$^+$ current (best seen when TEA is also present).

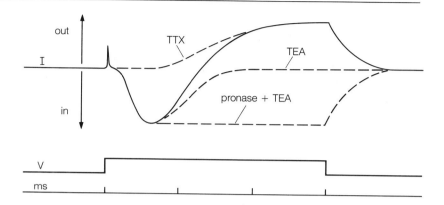

turbances of brain function, TTX renders an individual completely paralyzed and unable to breathe. One of the first reports of TTX poisoning in Western literature is in Captain Cook's diaries from the South Pacific. He describes symptoms of TTX poisoning he suffered after ingesting a small amount of liver from a local fish, undoubtedly a puffer fish. Other toxic substances are also known to block voltage-sensitive Na$^+$ currents. Saxitoxin, for example, is found in high concentrations in the microorganisms that cause "red tide" along many coastlines. Shellfish that have ingested the microorganisms are highly toxic.

Figure 4.7 shows the effect of TTX and other drugs on an axon. Following the addition of an extremely small amount (1 μM) of TTX, the Na$^+$ current is completely abolished but the K$^+$ current is left intact and isolated. The K$^+$ current, too, can be eliminated pharmacologically, most commonly by tetraethylammonium (TEA). In many axons, including the squid axon, TEA must be introduced inside the axon to block the K$^+$ current. After TEA blocks the K$^+$ current, voltage-clamp records indicates only the Na$^+$ current. In preparations blocked with either TTX or TEA it is possible to study the current still operative, K$^+$ or Na$^+$, over a wide range of membrane voltages (Figure 4.8). The Na$^+$ current is maximal at 0 mV membrane voltage and reverses at +55 mV, the equilibrium potential for Na$^+$. The K$^+$ current, on the other hand, increases continuously with increasing membrane depolarization. In other words, the more the membrane voltage is displaced

4.8 Relationship between membrane voltage level and Na⁺ and K⁺ currents (I_{Na} and I_K). I_{Na} is maximum at 0 mV, and it reverses at about +50 mV *(vertical arrow)*. At higher levels of membrane depolarization, I_{Na} is outward. I_K is outward at all levels of membrane voltage.

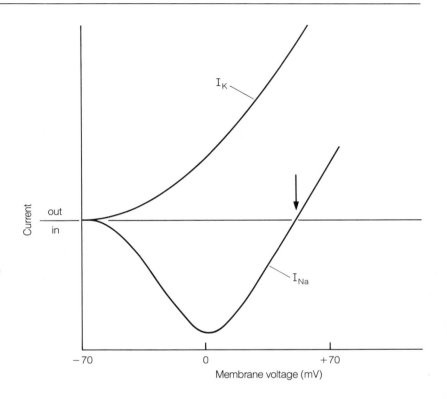

from the K⁺ equilibrium potential, the greater the driving force on K⁺ to leave the cell, so the K⁺ current increases.

Other substances also affect the voltage-sensitive currents in interesting ways. For example, condylactis toxin (CTX) and pronase (an enzyme that partially digests proteins) eliminate the time-dependence of the changes in the Na⁺ current. This is most clearly seen in preparations whose K⁺ current has been blocked by TEA. Following the addition of pronase, the Na⁺ current increases normally upon depolarization of the membrane, but instead of peaking after about 1 millisecond and then decreasing (the usual case), it continues so long as the membrane remains depolarized (see Figure 4.7).

Experiments with CTX and pronase indicate that separate processes control the opening and closing of voltage-sensitive Na⁺

gates. Indeed, substantial experimental evidence indicates that this is so. The turning on of the Na^+ current is called *activation* and the turning off is *inactivation*. Both are voltage-dependent; that is, both activation and inactivation are triggered by membrane depolarization. This is why the Na^+ current is transient even during a sustained current pulse. But, as one would expect, more time is required to initiate inactivation than activation. If activation and inactivation occurred at the same time, there would be no Na^+ current: inactivation would negate activation.

A toxin derived from scorpion venom acts on Na^+ currents in yet another interesting way, by altering the sensitivity of Na^+ channels to voltage. In the presence of this toxin, the channels begin to open at voltages well below resting voltage, and maximum Na^+ conductance occurs at lower levels of depolarization than is normally the case. This means that axons are highly depolarized by this toxin and, as would be expected, normal action potential generation cannot occur.

GATING CURRENTS

The drug studies indicate that separate Na^+ and K^+ channels are present in the axonal membrane. Moreover, they have provided insights into why the Na^+ channels are sensitive to voltage. If the voltage-sensitive currents are blocked by drugs, and if the voltage across the membrane is rapidly changed in the depolarizing direction, the only response the voltage clamp records is an initial transient (see Figure 4.4). As I mentioned earlier, part of this transient response reflects a rapid capacitive discharge of the membrane. When this discharge is compensated for, a deflection in the record remains. This represents a redistribution of charge on the Na^+ channel molecules themselves. The redistribution is believed to relate to the opening of the channels and hence is called the *gating current*. Present evidence indicates that four charges on the Na^+ channels move during the gating process. Thus, as the membrane depolarizes, a change in conformation (that is, a change in the arrangement of the molecules) is induced in the channels and they open. It may be that the voltage change across the membrane drives the charge redistribution on the molecules, thereby changing the conformation of the channel (see Chapter 8).

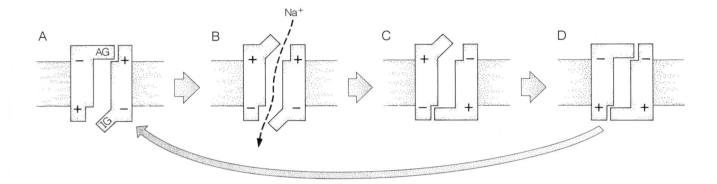

4.9 A model of the voltage-sensitive Na$^+$ channel. **(A)** At rest, the activation gate *(AG)* is closed, the inactivation gate *(IG)* is open. **(B)** Activation, both gates open. **(C)** Inactivation, activation gate open, inactivation gate closed. **(D)** Refractory period, both gates closed.

The Na$^+$ Channel

Figure 4.9 is a simplified model of what voltage-sensitive Na$^+$ channels might look like and how they alter during activation and inactivation. The channel is a transmembrane protein with a molecular weight of about 340,000 daltons. The figure depicts two gates in the channel. The activation gate is closed at rest while the inactivation gate is open. Positive and negative charges are distributed in the channel. Upon depolarization of the membrane, charge in the channel shifts and the activation gate opens. This allows Na$^+$ ions to flow across the membrane, into the cell. The activation gate may close and reopen, but then the inactivation gate closes and shuts off any further flow of Na$^+$ ions. With time the inactivation gate reopens—but slowly. The refractory period is the time during which the inactivation gate is closed ($\sim$1.5 ms). No amount of depolarization can open the channel during the refractory period.

A diagram of the K$^+$ channel present in the squid giant axon would differ from that of the Na$^+$ channel in having no inactivation gate. Upon depolarization the activation gate in the K$^+$ channel opens, but only after a delay of about 1 millisecond. Other K$^+$ channels do inactivate; further discussion of channel structure and function is provided in Chapter 8.

Single-Channel Records

The activity of single Na$^+$ and K$^+$ channels can be recorded by patch clamping an axonal membrane. Figure 4.10A presents typ-

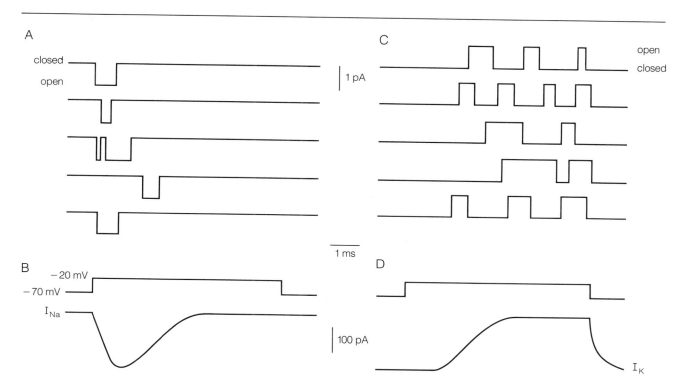

4.10 **(A)** Records of single voltage-sensitive Na+ channels. Individual channels open and close in an all-or-none fashion. **(B)** Sum of the current generated by a single Na+ channel in a patch of membrane depolarized repeatedly. About 200 traces were summed to yield this smooth curve. Note the difference in calibration between **(A)** and **(B)**. **(C)** Records of single K+ channels. Once the channels begin to open, they continue to open and close for as long as the membrane is depolarized. **(D)** Sum of the current generated by a single K+ channel in a patch of membrane that was repeatedly depolarized.

ical single Na+ channel records along with a trace of the summed current flow across an axon patch whose voltage was repetitively depolarized by about 50 mV. The current record in Figure 4.10B, representing the sum of about 200 traces and reflecting the opening and closing of the channels many times, resembles a voltage-clamp record of the Na+ current in that it has a smooth waveform. The individual channels, however, open and close in an all-or-none fashion. When the membrane is depolarized, most channels open almost immediately, but some open a bit later. The individual channels open for various lengths of times, but once their inactivation gates close they stay closed until the refractory period (inactivation) is over.

Single K+ channels in axons also open and close in an all-or-none fashion, but upon depolarization of the membrane, it takes time ($\sim$1 ms) before the first channels begin to open (Figure 4.10C). Furthermore, the channels continue to open and close for as long as the membrane remains depolarized; they do not inactivate. This explains the persistence of the K^+ current during membrane depolarization (Figure 4.10D).

Metabolic Considerations

During an action potential, Na^+ enters an axon and K^+ goes out. The Na^+–K^+ exchange pump takes care of ridding the axon of Na^+ and bringing K^+ back in. One might think intuitively that a large amount of Na^+ enters the axon during an action potential and puts a strain on the Na^+–K^+ exchanger. Yet the amount of Na^+ that comes across the membrane during the generation of an action potential is very small relative to the total number of ions on either side of the membrane. We can calculate the number of Na^+ ions required to generate an action potential in the same way we calculated the number of K^+ ions required to establish the resting potential (see Chapter 3).

It takes about 10^{-12} moles of positive charge per cm^2 to alter membrane potential 120 mV (from -70 mV to $+50$ mV). Relative to the amount of Na^+ present on the inside of the cell (5×10^{-2} M), this is an exceedingly small amount. This fact can be demonstrated experimentally by poisoning the Na^+–K^+ exchange pumps and stimulating an axon repetitively. With a large squid axon (500 μm in diameter), over 100,000 action potentials are generated before the axon fails; that is, before so much Na^+ accumulates inside the axon that the concentration gradient for Na^+ across the membrane is insufficient to allow action potential generation. With tiny mammalian fibers (0.5 μm in diameter), substantially fewer ions are inside, so failure occurs much more rapidly. Even so, such fibers will fire 500 or more times after the Na^+–K^+ exchanger is poisoned.

Ion Conductances during the Action Potential

Much of the work elucidating the basic properties of the action potential and the underlying conductances was carried out in the late 1940s and early 1950s by two English scientists, Alan Hodgkin and Andrew Huxley, working at the Marine Station in Plymouth, England. They experimented on the squid giant axon, first recognized as an axon by another English scientist, J. Z. Young, at the Marine Biological Laboratories in Woods Hole, Massachusetts, in the mid-1930s. Hodgkin spent the summer of 1939 in

Woods Hole learning about this preparation. An American scientist, K. C. Cole, who also worked at the MBL in Woods Hole, first voltage-clamped the squid giant axon in 1948.

Following World War II, Hodgkin and Huxley resumed work on the squid giant axon, and in 1952 they published a set of equations that describe mathematically the behavior of the action potential. The Hodgkin-Huxley equations, which reconstruct the entire course of the action potential almost perfectly, are a remarkably successful example of mathematical model building in neuroscience. For their many contributions, Hodgkin and Huxley were awarded the Nobel Prize in 1963.

From voltage-clamp records and data from experiments similar to those described in this chapter, it is possible to reconstruct the conductance changes during an action potential by using the Hodgkin-Huxley equations. Some theoretical conductances, along with a typical voltage record of an action potential, are shown in Figure 4.11. The Na⁺ conductance rises and falls rapidly, peaking about 1 millisecond after the beginning of the action potential. The K⁺ conductance increases after a latency of about 1 milli-

4.11 Reconstruction of the action potential (V) and the underlying Na⁺ and K⁺ conductances (g_{Na} and g_K).

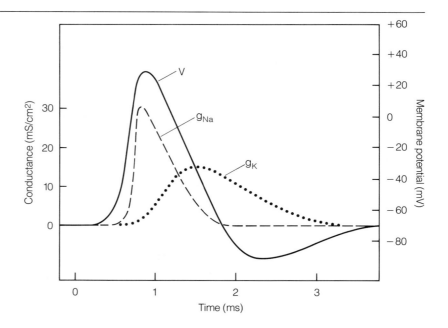

second, peaks after about 2 milliseconds, and then subsides as membrane voltage decreases. Because K^+ conductance is higher 2–3 milliseconds after the initiation of the action potential than it is at resting levels, membrane voltage dips below resting levels, closer to the K^+ equilibrium potential (E_K). This component of the action potential is called the *after-hyperpolarization*.

Ca^{2+} Channels and Other Voltage-Sensitive Channels

The most extensively studied voltage-sensitive channels are those in the axonal membrane, but research over the past several decades has revealed a variety of other voltage-sensitive channels in other nerve cell membranes that also influence neural function. A prominent example is the voltage-dependent Ca^{2+} channel in the presynaptic terminal membrane; it is important for the release of neuroactive substances from nerve terminals, as we shall discuss in Chapter 6. Voltage-sensitive Ca^{2+} channels serve other functions as well. For example, although Ca^{2+} concentrations are low both outside and inside nerve cells, there is a substantial concentration gradient across the nerve cell membrane. Outside concentrations of Ca^{2+} are on the order of a few millimoles, but inside concentrations are in the micromolar range. Thus, if a nerve cell membrane contains enough voltage-sensitive Ca^{2+} channels, Ca^{2+} action potentials may be generated and used by neurons.

All voltage-sensitive channels have the same basic properties as do the voltage-sensitive Na^+ and K^+ channels found in axons. Their permeability depends on membrane voltage, but their ion specificity, time dependence, and sensitivity to drugs differs. Some differences are substantial, while others are small. For example, most voltage-sensitive K^+ channels appear to differ mainly in the timing of the currents they mediate and their sensitivity to pharmacological agents. On the other hand, one voltage-sensitive K^+ channel opens when membrane potential is more negative than -60 mV, and this channel allows only K^+ ions to flow into a nerve cell.

The Propagation of Action Potentials

An action potential recorded at the end of an axon is as large as an action potential recorded at the beginning of an axon. It is, we know, transmitted actively along the axon, but why is propagation needed? Why isn't the large potential change at the site of impulse activation at the beginning of an axon sufficient to have an effect at the end of an axon? To understand why action potential propagation is essential, we must examine the spread of current down an axon.

Current Spread

Figure 4.12 shows an equivalent circuit of a membrane with potential pathways of current flow indicated by dashed lines; records of membrane voltage are included as well. When ions are injected into an axon (left), the current flows along the path of least resistance. The axon's internal resistance (R_i), its axoplasmic resistance, is much lower than the resistance of the membrane (R_m); thus most of the current flows down the axon (to the right). Nevertheless, a small amount of current (some ions) will escape across the membrane and reduce the amount of current flowing down the axon. As the amount of current flowing down the axon decreases, the potential difference (voltage) across the membrane declines. The further down the axon, therefore, the smaller is the voltage difference across the membrane.

How does the voltage produced by an injection of current at one point along an axon change as a function of distance? Two factors are critical: membrane resistance (R_m) and internal resistance (R_i). With a higher R_m, there is less leakage of ions across the membrane, the farther down the axon the current will spread, and the larger the voltage will be along the membrane. With a

4.12 **(A)** In this drawing of the equivalent circuit of a membrane, current, indicated by dashed lines, spreads down an axon following the injection of current into it (indicated by the pipette on left). Most of the current passes down the axon because the internal resistance (R_i) is relatively low. Some current escapes through the membrane even though its resistance (R_m) is relatively high. **(B)** Because current escapes across the membrane, the voltage recorded along the axon (V_m) gradually declines with distance from the source of current.

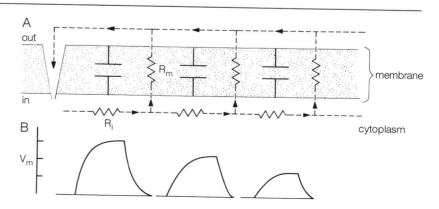

4.13 Exponential decay of membrane voltage *(V)* with axon length. The length constant (λ), the distance from the source at which the voltage has declined by about two-thirds, is 1/*e*. With high R_m and/or low R_i, the length constant is increased; with low R_m and/or high R_i, the length constant is decreased.

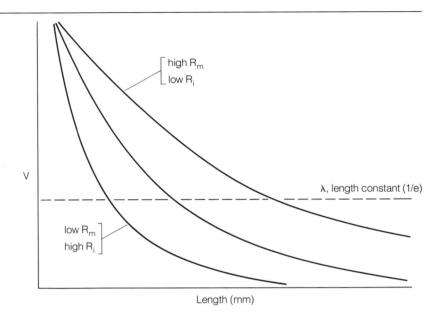

lower R_i, current will also spread farther down the axon. The R_m is determined by the membrane properties—such as the number of channels in the membrane—and whether a myelin sheath covers the axon. The R_i relates mainly to axon diameter: the larger the axon, the lower is the R_i.

Therefore, an interplay between R_m and R_i determines how far current spreads down an axon and how great the voltage will be along the membrane. If voltage is measured as a function of axon length, as shown in Figure 4.13, the decline is exponential. The *length constant* (λ) of an axon is defined as the distance at which the voltage has declined to 1/*e* of its maximum value, or by about two-thirds. This distance varies considerably between axons, depending on their diameter, whether they are myelinated, and the properties of the axonal membrane. Small unmyelinated axons have a length constant of only 10 μm whereas large, myelinated axons can have a length constant of 2 mm. An average length constant is about 1 mm.

Also shown schematically in Figure 4.13 is the alteration in length constant induced by changing R_m, R_i, or both. With increasing R_m, the length constant increases; with increasing R_i, length constant decreases. Regardless of how one manipulates R_m and R_i, however, the voltage across the membrane decays substantially within just a few millimeters of where current is injected into an axon. A voltage of 120 mV occurs at the site of impulse generation; within one length constant, this has decayed to about 40 mV; within two length constants, the voltage across the membrane is down to about 13 mV, and so on. Even if an axon has a length constant of 2 mm, just 4 mm away from the site of action potential generation, the voltage developed across the membrane will be only about 10 percent of the initial voltage. Thus, if signals of significant size are to reach the end of an axon that is longer than a few millimeters, rejuvenation of the signal is essential. And some axons are as long as 1,000 mm!

Impulse Propagation

The need for propagation of the action potential is clear, but how does it occur? Figure 4.14 shows the distribution of charge across an unmyelinated axonal membrane at and around the site of impulse generation, as well as the flow of currents at the site.

4.14 Distribution of charge across an axonal membrane at and around the site of action potential generation. The influx of positive charge into the axon during the generation of an action potential leads to the depolarization of adjacent membrane. Subsequent action potential generation occurs across the membrane in front of the impulse, but not in back of it because of the refractory period. The action potential thus goes in only one direction.

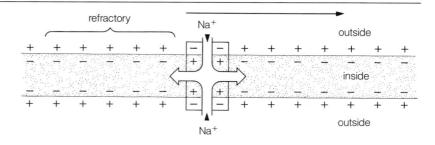

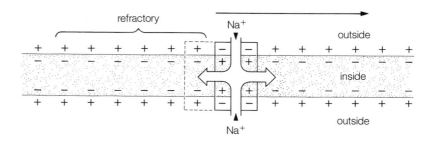

During the generation of an action potential, Na^+ ions flow into the axon. Positive current spreads down the inside of the axon and neutralizes the negative charge on the adjacent membrane, thereby depolarizing it. With depolarization of the membrane, Na^+ permeability increases, leading to further depolarization and eventually to the generation of an action potential across that bit of membrane. As this action potential is generated, the *next* bit of membrane is depolarized, leading to further action potential generation. In this way the action potential is continuously generated down the axon.

Action potentials ordinarily move along an axon in only one direction, but current flowing into the axon at the site of impulse generation moves in both directions along the axon (Figure 4.14). Why doesn't the action potential go in both directions? This doesn't happen because of the refractory state of the Na^+ channels following impulse generation. As noted earlier, during a period of about 1.5 milliseconds after an action potential has been fired, the membrane cannot be induced to generate another impulse regardless of the membrane voltage. It takes time for the inactivation gates to open following the generation of an action potential. A second reason the membrane is refractory immediately after the generation of an action potential is that K^+ conductance is high; more Na^+ must come in to initiate an action potential if K^+ can leave more easily. Thus, the threshold for action potential generation is elevated. For these reasons, new action potentials are generated only in front of the action potential, not behind it. It is important to emphasize that action potentials can, in principle, travel in either direction along an axon depending on the initiation site for impulse generation. But once generated, the impulse travels in only one direction, away from the initiation site.

How far behind an action potential is the membrane refractory? What is the separation between action potentials moving along an axon? This varies according to the axon. For the squid giant axon, action potentials move along the axon at a speed of 20 meters per second, or 20 millimeters per millisecond. Since the refractory period lasts for 1.5 milliseconds, 30 millimeters of axon is in a refractory state behind an action potential. We know that at one length constant away from the site of generation of an action potential the voltage has declined by about two-thirds, to about 40 mV. Two length constants away, the voltage ($\sim$13 mV) is just below threshold for the generation of an action potential. The length constant for a squid giant axon is about 1 mm; thus, the safety factor (30 mm of refractory membrane divided by two length constants, or 2 mm) is about 15. There is no chance that an action potential will be propagated in the wrong direction along a squid giant axon.

Speed of Propagation

The rate of action potential transmission depends partly on the axon's length constant; that is, how far down the axon the membrane is depolarized when an action potential is generated across one part of the membrane. There are two ways to increase the axon's length constant and to improve the speed of impulse transmission. The first is to lower internal resistance (R_i), which is accomplished by making a larger-diameter axon. This strategy is adopted by the squid and other invertebrates who have evolved giant axons. These larger axons are used for mediating escape mechanisms for the animal, and it is obviously advantageous to have information pass along these axons as fast as possible.

The other mechanism that increases the length constant of an axon and speeds action potential transmission is myelination, which is the strategy of vertebrates. (Invertebrates, except for some crabs, do not form myelin.) Myelin significantly increases membrane resistance (R_m), thereby increasing the length constant. However, when resistance increases, the membrane time constant ordinarily increases ($\tau = RC$), which means that it takes longer for the membrane to charge up (see Chapter 3). But myelin also reduces membrane capacitance, decreasing the time constant. In effect, then, the time constant of a myelinated axon is about the same as that of an unmyelinated axon. Myelin reduces membrane capacitance by increasing the thickness of the phospholipid surrounding the axon. This is similar to moving the plates of a capacitor apart, which reduces the ability of a capacitor to store charge (see Appendix). Thus, following depolarization of the membrane, threshold is reached earlier in a myelinated fiber as compared to an unmyelinated one, as illustrated in Figure 4.15. The increase in R_m of the myelinated axon means that the voltage developed across the membrane in response to a set amount of current flowing into the axon is larger, and since the time constant is about the same in a myelinated and unmyelinated axon of the

4.15 The effect of myelination on the properties of axonal membrane. Myelin increases R_m and decreases C_m. If the same amount of current is injected into a myelinated and unmyelinated axon, the voltage developed across the myelinated axon membrane is greater because of the increased R_m. Furthermore, the time constant $(\tau = RC)$ of the membrane is shortened relative to R_m because of the decreased C_m. Thus, an action potential is generated faster in the myelinated fiber.

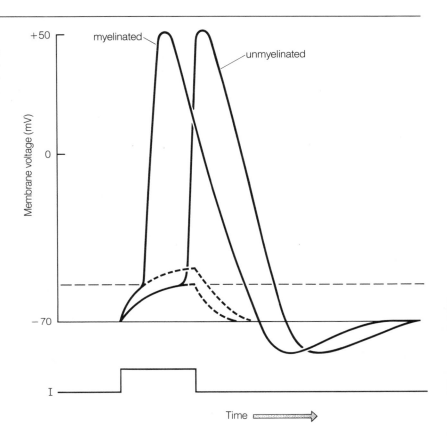

same size, the membrane reaches spike-firing threshold faster in the myelinated axon.

Saltatory Conduction

Myelin provides another important advantage. Action potentials along myelinated fibers jump from one node of Ranvier to the next, rather than being generated all along the length of the axon. In between nodes the spread of potential is passive, and thus action potential generation along myelinated fibers is saltatory, as illustrated in Figure 4.16.

If action potentials are generated only at the nodes of Ranvier, it is essential that the nodes be closer than 2 length constants

4.16 Saltatory conduction down a myelinated axon. Action potentials are generated only at the nodes of Ranvier; between the nodes, there is passive spread of potential.

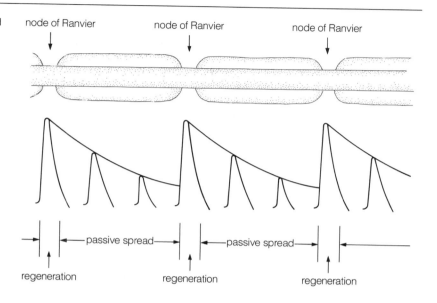

apart; that is, the membrane at a node should be depolarized by more than 15 mV from the current flowing inside the axon after an action potential has been generated at an adjacent node. This means that nodes can be farther apart in larger-diameter fibers and closer together in smaller-diameter fibers. Indeed, the distance between nodes varies from about 0.2 to 2 mm in different fibers, with the average internodal distance being 1 mm.

An important advantage of myelinated axons is that active membrane (membrane containing voltage-sensitive Na^+ and K^+ channels) is restricted to the nodal regions. In this configuration there is less metabolic drain on the fibers. By restricting Na^+ and K^+ flux across the membrane to the nodal regions, the cell has to make and maintain fewer ion channels and to pump fewer ions in and out of the cell. Because myelination also speeds conduction, myelinated fibers can be much finer than unmyelinated ones and perform as effectively and more efficiently. It has been estimated that without myelin the human brain would have to be 10 times larger to do what it does, and we would have to eat 10 times as much to maintain our brain.

Summary

Depolarization of the axonal membrane by about 15 mV (in the squid giant axon, from -70 mV to -55 mV) leads to an explosive change of membrane potential, the action potential. Within 1 millisecond, membrane voltage shifts to $+50$ mV and then rapidly returns to the resting value. The membrane becomes exceedingly permeable to Na^+ (Na^+ enters the axon) during the generation of an action potential; this explains why the peak of the action potential is close to the Na^+ equilibrium potential. Thereafter, the membrane's Na^+ permeability decreases and its K^+ permeability increases (K^+ exits the axon), which leads to the membrane's repolarization.

Voltage clamping of an axon permits analysis of the changes in ionic conductance across the membrane at any membrane voltage. Depolarization of the cell membrane from rest to above-threshold levels reveals an early and transient inward Na^+ current followed by an outward sustained K^+ current. Tetrodotoxin (TTX) specifically blocks the Na^+ current and tetraethylammonium (TEA) blocks the K^+ current, indicating that there are separate voltage-sensitive Na^+ and K^+ channels in the membrane. Other substances that block the transient, time-dependent Na^+ current have shown that activation and inactivation of the channels are independent processes. The Na^+ and K^+ channels are membrane proteins that change configuration when membrane voltage is altered, thus allowing Na^+ or K^+ ions to cross the membrane. Single Na^+ and K^+ channels open and close in an all-or-none fashion; the time the channels remain open is variable.

Action potentials are propagated along the axonal membrane. Generation of an action potential in one bit of membrane causes depolarization of adjacent membrane. This leads to further action potential generation. Action potentials move in only one direction because Na^+ channels are inactive (voltage-insensitive) for a short time, called the refractory period, following their activation. The speed of action potential generation depends on axon diameter, membrane resistance, and membrane capacitance. In invertebrates, large axons allow faster conduction; in vertebrates, axons are myelinated, which increases membrane resistance and lowers membrane capacitance, resulting in faster conduction. Action potentials are generated only at the nodes of Ranvier along myelinated axons; the action potential skips from node to node. Voltage-sensitive Na^+ and K^+ channels are confined to the nodal membrane; thus myelinated axons not only conduct action potentials faster than unmyelinated axons, but they also place less metabolic demand on neurons.

Further Reading

BOOKS

Hille, B. 1991. *Ionic Channels of Excitable Membranes.* 2d ed. Sunderland, MA: Sinauer Associates. (A comprehensive discussion of the physiology and biophysics of membrane channels.)

Hodgkin, A. L. 1964. *The Conduction of the Nervous Impulse.* Liverpool: Liverpool University Press. (Hodgkin's Sherrington Lectures; very readable).

ARTICLES

Armstrong, C. M. 1981. Sodium channels and gating currents. *Physiological Review* 61:644–683.

Catterall, W. A. 1980. Neurotoxins that act on voltage-sensitive sodium channels in excitable membranes. *Annual Review of Pharmacological Toxicology* 20:15–43.

Fuhrman, F. A. 1967. Tetrodotoxin. *Scientific American* 217(2):60–71.

Hodgkin, A. L., and A. F. Huxley. 1952a. Currents carried by sodium and potassium ions through the membrane of the giant axon of *Loligo. Journal of Physiology* 116:499–472. (The first of the classic papers of Hodgkin and Huxley.)

Hodgkin, A. L., and A. F. Huxley. 1952b. A quantitative description of membrane current and its application to conduction and excitation in nerve. *Journal of Physiology* 117:500–544. (The presentation of the Hodgkin-Huxley equations.)

Sigworth, F. J., and E. Neher. 1980. Single Na^+ channel currents observed in cultured rat muscle cells. *Nature* 287:447–449.

Young, J. Z. 1936. The giant nerve fibres and epistellar body of cephalopods. *Quarterly Journal of Microscopic Science* 78:367–386. (Young's classic paper describing giant axons in squid.)

Synaptic Potentials

THE LAST TWO chapters covered virtually all of the principles concerning the electrical properties and activity of neurons that are needed to understand most aspects of neural function. For example, we first discussed the electrical properties of nerve cell membranes and the equivalent electrical circuit of membranes. Resting membrane potentials were next considered and how they are established. In that discussion, the concept of equilibrium potentials was presented, as well as how membrane potentials are calculated.

To understand the generation of action potentials and the conductances of the membrane during an action potential, one must appreciate the notion of variable membrane permeability, presented in Chapter 4, along with voltage- and time-dependent conductance changes. To see why propagation of the action potential is necessary, one must understand the passive spread of currents along axons and the concepts of length and time constants and how they may be varied. Finally, the mechanisms of propagation of the action potential were presented along with discussion of how the rate of propagation is affected by axon diameter and myelination.

Now we can go on and discuss many other aspects of nerve cell activity and consider them in terms of the principles already presented. In this chapter, we examine the properties of synaptic potentials and how they govern action potential generation by a neuron. Then we consider some of the mechanisms underlying synaptic transmission. The analysis of synaptic potentials requires an understanding of variable membrane permeability, equilibrium potentials, and passive spread of currents along nerve cell processes.

Excitatory and Inhibitory Synapses

All neurons are, in a sense, chemoreceptors; they respond to chemicals released at synaptic sites. Thus, although what is covered here deals with the properties of synaptic potentials, much of it also holds for receptor potentials.

Figure 5.1A shows a simplified drawing of a Golgi type I neuron. Excitatory synapses impinge on its dendrites, whereas inhibitory synapses are found on its cell body. Action potentials travel down the innervating axons and depolarize the terminals synapsing on the neuron. This results in the release of neurotransmitter from the terminals. The transmitter diffuses across the synaptic cleft to the postsynaptic membrane, where it causes a small voltage change. A small depolarization, called an EPSP (excitatory postsynaptic potential), occurs at membrane sites postsynaptic to excitatory synapses; at sites postsynaptic to inhibitory synapses, a small hyperpolarization, an IPSP (inhibitory postsynaptic potential), is observed.

At each postsynaptic site such potentials are elicited. Individual EPSPs and IPSPs are too small to have much effect on the neuron. But when many synaptic events are summed, they change the postsynaptic membrane potential substantially; the output of the cell reflects the balance of excitatory and inhibitory input onto the neuron. Excitatory input drives the membrane potential from rest to above action potential threshold levels, whereas inhibitory input lowers membrane potential, often below threshold levels.

The interplay of excitatory and inhibitory input to a neuron is depicted schematically in Figure 5.1B. The resting potential of the neuron is -70 mV, a typical motor neuron resting potential. Invasion of action potentials into presynaptic terminals making excitatory synapses (tics along the abscissa Ex) results in EPSPs that occur virtually simultaneously in the dendritic tree. These small EPSPs summate, so a recording made with a micropipette in the cell body shows an overall depolarization of the neuron of a few millivolts. If additional EPSPs are generated before earlier EPSPs have decayed, the cell depolarizes more, and eventually action potential threshold level is reached in the postsynaptic cell. So long as excitatory input onto the neuron is maintained, the cell rapidly depolarizes after an action potential is generated and further im-

5.1 (**A**) A Golgi type I neuron with excitatory synapses on its dendrites and inhibitory synapses on its body. (**B**) A typical intracellular recording. Although the micropipette is inserted in the cell body, it detects potentials generated in the dendrites, cell body, and axon. When the pipette is inserted into the neuron, a resting potential of −70 mV is recorded. Excitatory input (depicted by tics along the abscissa *Ex*) causes depolarizing, graded (synaptic) potentials, which summate and bring the membrane potential of the cell to action potential threshold (∼−55 mV). Generation of an action potential causes a very rapid reversal of membrane potential, followed by rapid recovery. With continued stimulation of the excitatory input, the cell again depolarizes and additional action potentials are fired. Inhibitory input to the cell (depicted by tics along the abscissa *In*) causes the generation of hyperpolarizing, graded (synaptic) potentials, which negate the effects of the excitatory synaptic potentials and drive membrane potential away from spike-firing threshold. The cell does not fire while the inhibitory input is active, even though the excitatory input continues. When the inhibitory input is terminated, the cell once again depolarizes and fires action potentials.

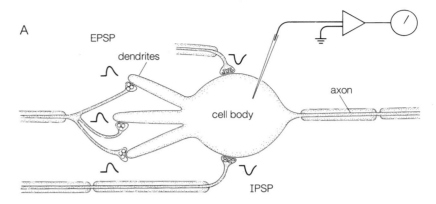

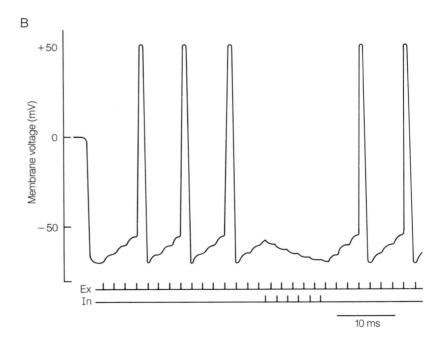

pulse generation occurs. The rate of impulse generation relates to the level of membrane depolarization. (Refer to Figure 3.2.)

Also shown in Figure 5.1 is what happens when inhibitory synapses are activated (tics along abscissa *In*). The IPSPs also summate, hyperpolarizing the cell and opposing the depolarizing action of the EPSPs. If membrane potential falls below the spike-firing threshold, no further action potentials are generated. With cessation of inhibitory input, the postsynaptic neuron once again depolarizes as a result of the excitatory synaptic input, and it resumes firing action potentials.

The Axon Hillock and the Generation of Action Potentials

Action potentials are usually generated at just one locus in a neuron, the axon hillock. This region has the lowest threshold for action potential generation; indeed the threshold for impulse generation rapidly rises away from the axon hillock toward the rest of the cell body and along the dendrites. This is because the voltage-sensitive Na^+ and K^+ channels are highly concentrated in the axon hillock membrane, and few of these channels are in the cell body or in the dendrites. Because most dendritic and cell-body membrane is devoid of voltage-sensitive Na^+ and K^+ channels, the membrane is electrically inexcitable; that is, regardless of the extent of depolarization, no action potentials are generated across the membrane.

Figure 5.2 is a diagram of the current flows induced in a neuron by excitatory synaptic input. Current (mainly Na^+ ions) flows into the cell at the synapses, and from there current flows down the dendrites into the cell body, resulting in overall depolarization of the cell. When the axon hillock region is depolarized by 10–15 mV or so, an action potential is fired and is then propagated down the axon. As shown in Figure 5.2, current flows into the cell at each postsynaptic site. Current flows out of the cell all along the membrane through leak channels, a term used to describe the nongated channels in the membrane that are involved in establishing the cell's resting potential (Chapter 3). Because current flows in at the synapse and flows out elsewhere, synaptic potentials are largest at the site of generation and voltage decays away from that site. Even though the membrane is most depolarized adjacent to the postsynaptic sites, synaptic potentials influence a neuron by

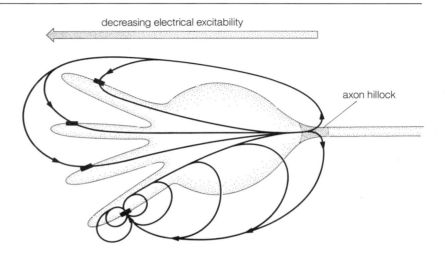

5.2 Diagram of the current flow induced in a neuron by excitatory synaptic input. The current flows into the neuron at synaptic sites (bars) but generates action potentials only at the axon hillock, where the threshold for spike initiation is lowest. Cell body and dendritic membrane has relatively few voltage-sensitive Na^+ and K^+ channels; this membrane is electrically inexcitable.

decreasing electrical excitability

axon hillock

their effect on the axon hillock membrane, where the action potentials are generated.

Why is the neuron organized this way? Why aren't action potentials generated adjacent to the synaptic sites? The reason is simple. If action potentials were generated all along dendrites, impulses would travel along the dendrites in different directions, collisions would occur, refractory periods would prevent some membrane from responding, and so forth. In other words, the output of the cell would not necessarily reflect its synaptic input. The axon hillock region of the cell, though, serves as a summing point for the neuron; it responds to the balance of excitatory and inhibitory input to the neuron and orchestrates a coherent output for the cell.

The dendrites of neurons are usually considerably thicker than are axons, so their length constants are larger. The low internal resistance (R_i) of a stout dendrite facilitates the spread of current from the dendrites into the cell body. Nevertheless, because the spread of potential down the dendrites is passive (decays with distance), synaptic potentials produced on the distal tips of dendrites are attenuated considerably before they reach the axon hillock. A small dendrite on a large neuron, such as a Purkinje cell (see Figure 1.6), may be several hundred micrometers from the cell body. This suggests that unless the dendrite's membrane has

5.3 A drawing of a neuron with patches (stippled areas) of electrically excitable membrane in its dendritic tree. When these patches are sufficiently depolarized, an action potential is generated.

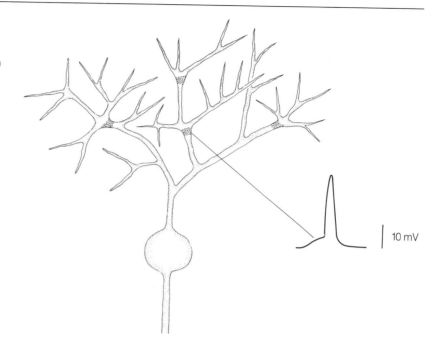

10 mV

a very high resistance, synaptic input onto distal dendrites will have little or no effect on the cell's firing rate.

Some dendritic membranes can generate action potentials. These dendrites have patches of membrane containing voltage-sensitive channels (see Figure 5.3). When sufficiently depolarized, the membrane generates a spike that amplifies the dendrite's summed depolarization. These active responses may be either Na^+ or Ca^{2+} action potentials and are probably not propagated. They are generated in strategic locations in large dendritic trees to ensure, presumably, that a synaptic input has a significant effect on the cell's output.

Analyzing Synaptic Potentials

Several questions immediately come to mind regarding synaptic potentials. What is neurotransmitter doing to the postsynaptic membrane to produce the synaptic potentials? Because current flows across the membrane at postsynaptic sites, an obvious suggestion is that neurotransmitters change membrane permeability and allow one or more ions to enter or exit the postsynaptic cell. An easy way to decide if conductance changes during synaptic activity is to measure membrane resistance. Neuroscientists routinely make such measurements to analyze synaptic mechanisms.

The simplified cell pictured in Figure 5.4A has just a single dendrite with one large excitatory synapse on it. A current pulse is passed into the cell through one micropipette while membrane voltage is measured with a second micropipette. The voltage that develops across the cell's membrane due to the injection of a constant pulse of current into the cell depends on the resistance of the membrane, in accord with Ohm's law, $V = IR$. The higher the resistance, the larger the voltage and vice versa. If intermittent current pulses are passed into the cell, the membrane resistance—

5.4 **(A)** The experimental setup for measuring membrane resistance during synaptic activity. Current pulses *(I)* are passed into a neuron before, during, and after the generation of a synaptic potential. By measuring the voltage *(V)* that develops across the membrane in response to the current pulses, resistance of the membrane can be determined. **(B)** In the record illustrated, the membrane voltage deflection that occurred in response to the current pulse at the peak of the EPSP *(arrow)* was decreased, indicating a decrease in membrane resistance. *Ex* represents the stimulus evoking the EPSP.

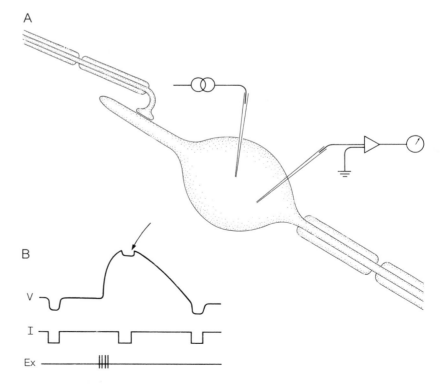

and hence membrane conductance ($g = 1/R$)—can be determined both at rest and during synaptic activation.

During both EPSPs and IPSPs, membrane resistance changes, indicating a conductance change across the membrane. This is shown in Figure 5.4B for an EPSP. The voltage response to the current pulse recorded at the peak of the EPSP (arrow) is much smaller than are the responses to pulses recorded before or after the EPSP. Most often resistance falls, which means conductance increases, during EPSPs and IPSPs; occasionally, though, resistance increases and conductance decreases. These data indicate that neurotransmitter most often opens channels in the membrane, but sometimes channels are closed by transmitter.

What ions move across the membrane at excitatory and inhibitory synapses? Two kinds of experiments are commonly carried out to answer this question. In the first, the membrane potential is altered by injecting current into the cell to a level where activation of the synapse no longer causes a potential change across the membrane. The membrane is brought to a voltage level where there is no net current flow across the membrane. This level is called the *reversal potential* for the response. If just one ion generates the response (the membrane is made more permeable to only one ion), the reversal potential is the equilibrium potential for that ion. If more than one ion is involved, the reversal potential is intermediate between the equilibrium potentials of the permeant ions. The second experimental approach to determine which ions move across the membrane is to vary ion concentrations on one or both sides of the membrane and observe whether the changes in ion concentration affect the synaptic potentials.

Let's begin by describing experiments that seek to determine the reversal potential for a synaptic potential. The same experimental setup as depicted in Figure 5.4 is used, and again neuroscientists commonly make these measurements when studying synaptic mechanisms. To determine a reversal potential, current is continuously injected into the cell to shift the membrane potential by a fixed amount. Synaptic activity is initiated while the membrane is maintained at the fixed level, and a synaptic potential is recorded. It is important to remember that dendritic and cell-body membranes are electrically inexcitable; depolarizing these membranes does not generate action potentials. Hence, the membrane channels in dendrites and cell bodies that respond to the neurotransmitter are not voltage-sensitive. Rather, the permeability of these channels alters in response to transmitter, not voltage, and so they are called *transmitter-* or *ligand-sensitive channels.*

Excitatory Postsynaptic Potentials

Figure 5.5A shows a series of EPSPs elicited at different membrane voltages. Measurements of this type were first made by John Eccles working in Australia in the 1950s, and much of our understanding of EPSPs and IPSPs is due to the work of Eccles and his colleagues. At -70 mV, the cell's resting potential, the EPSP is depolarizing and, in the recording illustrated, the EPSP has an amplitude of about 15 mV. As the membrane is depolarized to -50 and then to -30 mV, the EPSP becomes smaller. At about -10 mV no EPSP is seen. When membrane potential is shifted to values above -10 mV, the EPSP reverses; it becomes hyperpolarizing. The reversal potential for the EPSP is, therefore, at about -10 mV.

None of the ions commonly found inside and outside neurons has an equilibrium potential at -10 mV. What is going on? The results suggest immediately that more than one ion is involved; that is, the membrane is made more permeable to more than one ion. The equilibrium potential for Na^+ is $+55$ mV; for K^+ it is -92 mV. The reversal potential for the EPSP is roughly midway between these equilibrium potentials, suggesting that both ions contribute to the response.

Experiments that alter Na^+ and K^+ concentrations across the neuronal cell membrane show that both ions are involved in the generation of the EPSP. Altering Cl^- concentrations, in contrast, has no effect on the EPSP. The conclusion drawn from these experiments is that the neurotransmitter eliciting an EPSP opens a channel that allows both Na^+ and K^+ to cross the membrane. When membrane potential is more negative than -10 mV (that is, closer to the K^+ equilibrium potential), more Na^+ enters the cell than K^+ leaves and a net depolarizing potential results. When membrane potential is more positive than -10 mV (closer to the Na^+ equilibrium potential), more K^+ leaves the cell than Na^+ enters and hyperpolarization results. When membrane voltage is at -10 mV, equal amounts of Na^+ and K^+ cross the membrane and no net potential change occurs—this is the reversal potential.

It is possible for membrane channels to allow one ion to flow more easily across the membrane than other ions, and so reversal

5.5 Reversal potential measurements for the EPSP (**A**) and IPSP (**B**). The membrane voltage of a neuron is set at a fixed level and a synaptic potential evoked. The reversal (null) potential for the EPSP is about −10 mV; for the IPSP, −80 mV.

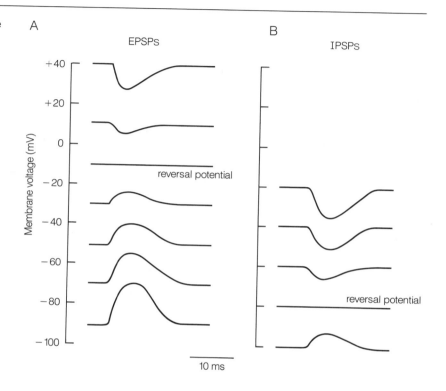

potentials for EPSPs could theoretically be at any value between resting potential (−70 mV) and +55 mV. However, EPSPs generally have reversal potentials between 0 mV and −20 mV, which indicates that these channels become about equally permeable to Na^+ and K^+ ions when they are activated.

Inhibitory Postsynaptic Potentials

What about IPSPs? Figure 5.5B presents the effects of altering membrane voltage on inhibitory potentials. At −60 mV, the IPSP is hyperpolarizing. If the membrane is depolarized, the IPSP becomes larger, which means that the reversal potential for the IPSP must occur when membrane potential is hyperpolarized. The IPSP increases in amplitude as the cell is depolarized because the driving force on the involved ion increases as membrane voltage moves away from its equilibrium potential; that is, the charge inside the cell adds to the diffusion pressure on the ion to cross the membrane and to leave the cell.

When the membrane potential is moved in the hyperpolarizing direction, the IPSP becomes smaller. Often the null point for the IPSP is about -70 mV, and with further membrane hyperpolarization the response becomes depolarizing. The equilibrium potential for Cl^- is about -70 mV in many cells, close to the reversal potential for IPSPs, and alterations in Cl^- concentration either inside or out significantly affect IPSPs. These observations suggest, therefore, that the transmitter released at inhibitory synapses opens channels in the membrane specific for Cl^-. Since Cl^- concentrations are higher outside the cell than inside, Cl^- moves into the cell and hyperpolarizes it.

Other IPSPs reverse at more negative membrane potential levels, suggesting that K^+ channels are also opened by inhibitory transmitters. Since K^+ concentration is higher inside than out, K^+ exits the cell and also hyperpolarizes it. At some inhibitory synapses, the inhibitory transmitter can open both Cl^- and K^+ channels, and the reversal potential for the IPSP lies between E_K and E_{Cl}, or at about -80 mV. The IPSP illustrated in Figure 5.5B has a reversal potential of about -80 mV.

Single-Channel Recordings

The channels in postsynaptic membranes that open in response to neurotransmitters are clearly different from the voltage-gated channels present in axon hillocks and axonal membranes (see Chapter 8). But transmitter- or ligand-gated channels, like voltage-gated channels, open in all-or-none fashion when the appropriate transmitter is present in the synaptic cleft. Recordings of current through a single channel responsive to acetylcholine (ACh), an excitatory neurotransmitter, and a channel activated by γ-aminobutyric acid (GABA), an inhibitory transmitter, are shown in Figure 5.6.

Note that, as is true of voltage-gated channels, the duration of channel opening is variable. Also note that the channels continue to open and close for as long as transmitter is present. Finally, in both cases illustrated in Figure 5.6, there is a net movement of ions across the membrane when the channels are open, but the ions moving across the membrane have a different charge; that is, Na^+ ions are mainly flowing through the ACh channel into the

5.6   Typical records of current through single channels activated by acetylcholine (ACh) or GABA. Both channels open and close in an all-or-none fashion. When ACh channels open, there is a net inward flow of positive current (Na^+); when GABA channels open, there is a net inward flow of negative ions (Cl^-).

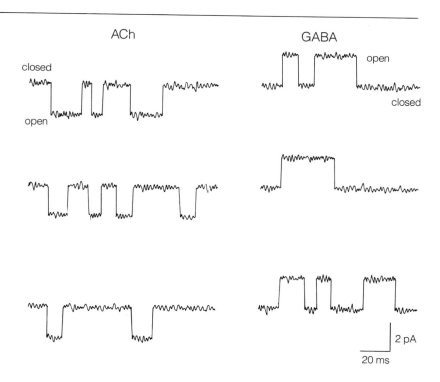

cell, whereas Cl^- ions are entering the cell through the GABA channel. Therefore the current through the ACh channel is positive, and current through the GABA channel is negative.

Presynaptic Inhibition

In Chapter 2, we noted that on occasion a synaptic terminal makes a synapse onto another synaptic terminal. These interactions are often inhibitory, and because the inhibitory signal is mediated presynaptically, it is termed *presynaptic inhibition*.

Figure 5.7A shows schematically an experimental setup for exploring presynaptic inhibition. Terminal *A* impinges on a neuron that is penetrated by a micropipette to record membrane voltage. The axon to terminal *A* is hooked by a wire so it can be electrically stimulated. The axon to terminal *B*, which synapses on terminal *A*, can be similarly stimulated. When terminal *A* alone is stimulated, an EPSP is elicited in the neuron (Figure 5.7B). When *B*

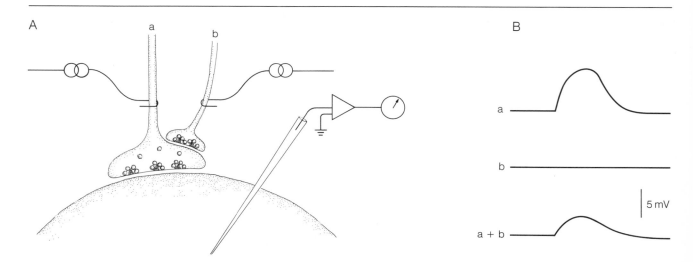

5.7   (A) An experimental set-up for studying presynaptic inhibition and (B) typical records illustrating the phenomenon. When terminal *a* is stimulated alone, a large EPSP is evoked in the postsynaptic neuron. When terminal *b* is stimulated alone, no response occurs in the postsynaptic cell. When *a* and *b* are stimulated together, the response in the postsynaptic neuron is decreased significantly.

alone is stimulated, the neuron does not respond. When both *A* and *B* are stimulated, the resulting EPSP recorded in the neuron is diminished in size; the reduction in EPSP amplitude is called presynaptic inhibition.

Presynaptic inhibition is maximal when terminal *B* is stimulated a few milliseconds before terminal *A*. What is going on? How does activation of terminal *B* decrease the EPSP induced by terminal *A*? Evidence suggests that usually terminal *B* releases a transmitter that opens Cl^- channels in the membrane of terminal *A*. The opening of Cl^- channels in the terminal membrane may or may not cause a membrane potential change, depending on the membrane potential of terminal *A*. If it is close to the equilibrium potential for Cl^- (~ -70 mV), no membrane potential change may occur in terminal *A*. However, regardless of whether the membrane potential changes or not, membrane resistance of terminal *A* decreases significantly because of the membrane's greater permeability. When a spike subsequently invades terminal *A,* the voltage across the membrane is reduced because of the lowered membrane resistance. The current flowing into the terminal as a result of the invading action potential is not altered, but the voltage will be smaller because membrane resistance is lower, in accord with Ohm's law, $V = IR$. Since the amount of transmitter released from

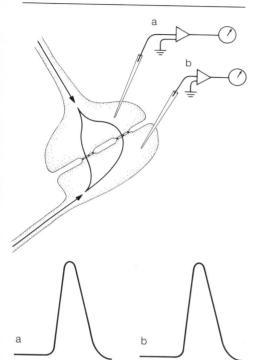

5.8 Current flow across an electrical synapse. Current flows relatively unimpeded across the synapse, usually in both directions. Signals recorded in the contacting elements (*a* and *b*) are virtually indistinguishable.

a synaptic terminal depends on membrane voltage, the EPSP induced in the neuron postsynaptic to terminal *A* is thereby reduced.

Depressed neuronal responsiveness because of decreased membrane resistance is found in other situations. Indeed, an entire neuron can be inhibited in this way; it is called *shunting inhibition*. As in presynaptic inhibition, the transmitter released onto the neuron causes little membrane potential change but exerts its action mainly by reducing membrane resistance. The effects of excitatory synapses on the cell are diminished because of the lowered membrane resistance—that is, the cell's excitability is reduced.

When considering EPSPs, IPSPs, presynaptic inhibition, or shunting inhibition, it is important to keep in mind that both resting potentials and equilibrium potentials will vary to some degree according to which neuron and system we are examining, because different neurons have somewhat different ion concentrations inside and out. A consequence of this variability is that it is not always possible to determine the effect of a transmitter on a cell without careful study. For example, if a neuron has a resting potential of -70 mV and its Cl^- equilibrium potential is -65 mV, a transmitter that opens Cl^- channels in the cell's membrane creates a small depolarizing potential. At first glance, one might think this is an EPSP, reflecting increased Na^+ and K^+ conductance across the membrane, rather than a Cl^--based conductance change similar to an IPSP. Furthermore, the overall effect of this synaptic activity on the cell may be to depress its excitability because of a large drop in membrane resistance, even though a small EPSP-like depolarization of the membrane is recorded.

Electrical Synapses

The physiology of electrical synapses is easy to understand. Current moves relatively unimpeded across gap junctions, so current flowing into a terminal making an electrical synapse on a dendrite or cell body passes directly into the postsynaptic element (see Figure 5.8). Current readily flows across gap junctions because the specific resistance of gap-junctional membrane is very low ($\sim$1–3 ohm $\cdot$ cm^2), much lower than the usual specific resistance of 10^3–10^5 ohm $\cdot$ cm^2. And current, of course, flows down the path of least resistance. The current flows across gap junctions through channels, which, like other membrane channels, open and close in all-or-none fashion. Gap-junctional channels tend to stay open for relatively long periods of time, however, up to 100 ms per opening.

At electrical synapses a potential recorded in the postsynaptic element is virtually indistinguishable from the potential in the presynaptic element (see Figure 5.8). As noted in Chapter 2, gap junctions usually allow current to flow in either direction; they are nonrectifying. Therefore, it is often impossible to decide which is the pre- and which is the postsynaptic side of the junction—both can be either, especially when an electrical synapse lies between two dendrites, two cell bodies, or two synaptic terminals.

What are the advantages and disadvantages of electrical synapses? There is no synaptic delay at electrical synapses, so transmission speed is an obvious advantage. Electrical synapses thus synchronize activity between neurons and also allow for reciprocal innervation between neurons. On the other hand, the signal is not amplified at electrical synapses, signal polarity does not change, and there is usually little in the way of signal modification. The effectiveness of transmission at electrical synapses can be altered by chemical synaptic input to neurons making gap junctions, but gap junctions do not exhibit the variety of signal modifications seen at chemical synapses—the source of many interesting aspects of neural function, some of which are discussed in Chapter 6.

The Neuromuscular Junction

The neuromuscular junction is highly accessible to study, because the large postsynaptic process enables the investigator to insert a micropipette close to the postsynaptic membrane. Many details of synaptic transmission were first elucidated in studies on the neuromuscular junction, and for numerous questions it remains the synapse of choice among many researchers. Furthermore, most of what has been learned from the neuromuscular junction holds for synapses between neurons in the brain.

Figure 5.9 depicts a neuromuscular junction and the pathways of current flow induced in the muscle fiber when the synapse is activated. In brief, depolarizing the innervating terminal leads to the release of a transmitter that alters the muscle membrane's permeability. The increased permeability causes a net inward flow of positive current into the muscle fiber that depolarizes the adjacent muscle membrane. In most muscle cells, voltage-sensitive

5.9 A current flow diagram at the neuromuscular junction. When the innervating nerve terminal is activated, transmitter released from the terminal causes a net inward flow of current into the muscle fiber. The current depolarizes the membrane adjacent to the junction, inducing the generation of action potentials that propagate around the muscle fiber and cause fiber contraction.

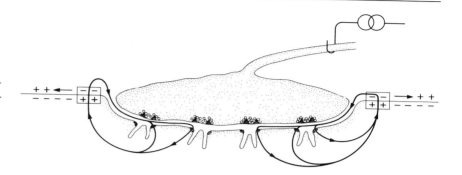

Na$^+$ and K$^+$ channels are in the membrane adjacent to the neuromuscular junction; these allow the generation of action potentials, which propagate down and around the muscle cell, resulting in contraction of the fiber.

A micropipette inserted into the muscle fiber close to the postsynaptic membrane will record small, quantal-like depolarizing events. Bernard Katz and Paul Fatt at University College London discovered this activity in the early 1950s. The potentials are typically 0.5–1 mV in amplitude and have a duration of about 10 ms. They are remarkably constant in any given recording and are observed even when the innervating terminal is not stimulated. Figure 5.10 shows examples of these miniature excitatory postsynaptic potentials or *min EPSPs*.

Without stimulation, min EPSPs have a frequency of 40–50 per second. With depolarization of the innervating axon, the frequency of the min EPSPs increases; with hyperpolarization, the frequency decreases. The min EPSPs summate, so with sufficient depolarization of the innervating axon the muscle membrane will be depolarized to action potential threshold. With action potential generation, the muscle fiber contracts. This usually breaks the micropipette, thereby terminating the experiment.

This technical difficulty can be overcome by bathing the muscle fiber in tetrodotoxin (TTX), which, as we learned in Chapter 4, blocks the voltage-gated Na$^+$ channels that are responsible for action potential generation (and therefore muscle contraction). TTX does not, however, affect transmitter-gated channels, so we can study the EPSPs generated across the muscle membrane with-

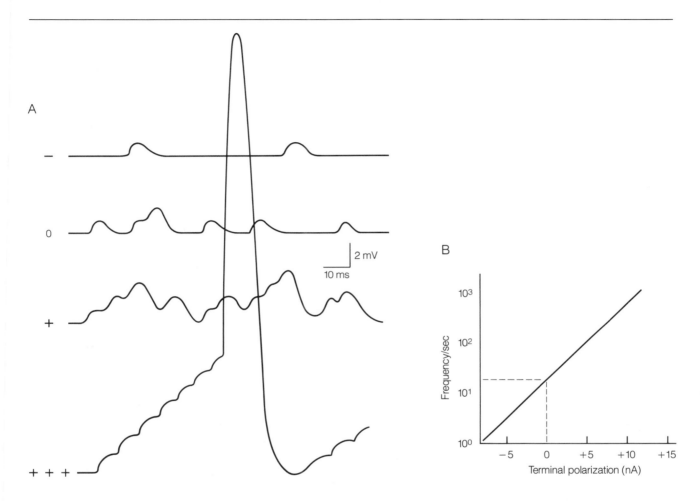

5.10 (**A**) Min EPSPs recorded from a muscle fiber at rest (0), or when the innervating fiber is depolarized (+ or +++), or hyperpolarized (−). (**B**) The relation between frequency of min EPSP generation and terminal polarization. Terminal depolarization increases whereas terminal hyperpolarization decreases min EPSP generation. If the terminal is depolarized sufficiently, enough min EPSPs are produced to bring the muscle fiber to action potential threshold (the bottom record in **A**).

out activating a muscle response. That TTX does not affect postsynaptic potentials provides additional evidence that voltage-gated Na^+ channels are distinct from the transmitter-gated channels in postsynaptic membranes.

With TTX-treated muscle fibers, large EPSPs of up to 20 mV or so can be elicited in the muscle cell. By studying these EPSPs at different levels of membrane potential, investigators have determined the reversal potential for the response, typically between −10 mV and −20 mV. Thus, the muscle EPSP is similar to neu-

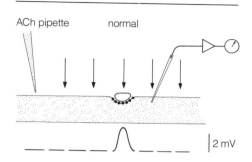

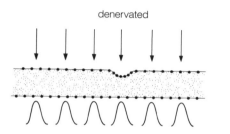

5.11 Sensitivity of the muscle membrane to acetylcholine. In the normal muscle, fibers are sensitive to ACh only at the site of the neuromuscular junction; the ACh channels are restricted to this locus. In the denervated muscle fiber, the cell is sensitive to ACh all over its surface.

ronal EPSPs described earlier; it reflects a change in permeability of the muscle membrane to both Na^+ and K^+.

The picture derived from experiments on the neuromuscular junction is that the muscle EPSP is made of the sum of many min EPSPs and that the fundamental postsynaptic event is the min EPSP. What gives rise to individual min EPSPs? Min EPSPs were discovered just at the time electron microscopy showed synaptic vesicles in presynaptic terminals. The suggestion was made, therefore, that a min EPSP represents the response of the postsynaptic membrane to the transmitter released from a single synaptic vesicle. The quantal nature of the min EPSP can thus be explained by the fact that the synaptic vesicles are quite similar in size and presumably contain about the same amount of transmitter.

The neurotransmitter at the neuromuscular junction in vertebrates is known to be acetylcholine (ACh), so tests of this hypothesis are possible. If small amounts of acetylcholine are squirted onto the muscle membrane, a response similar to a min EPSP is recorded from the fiber. As shown by Stephen Kuffler and his colleagues at the Harvard Medical School, it requires no more than 10,000 molecules of ACh to evoke a min EPSP-like response; this figure matches well the estimate of 6,000–10,000 molecules of ACh stored in a synaptic vesicle.

An important and interesting observation made in the course of carrying out similar experiments was that the sensitivity of the muscle fiber to ACh is confined to the region of the neuromuscular junction. If ACh is squirted onto muscle membrane away from the junction, little or no response is observed, as shown in Figure 5.11. But if the muscle is denervated—that is, if the axon is removed from the muscle a few days before the experiment is carried out— a surprising result is obtained. The muscle is now sensitive to ACh over its entire surface; hence, transmitter-sensitive channels are present all over the membrane. With reinnervation, which requires days to weeks, the membrane's ACh sensitivity is confined once again to the region of the neuromuscular junction.

In response to denervation, the muscle fiber synthesizes new ACh-sensitive channels and inserts them in the membrane. There may also be some spreading out of channels from the region of the neuromuscular junction into nonjunctional membrane. Nevertheless, because the muscle fiber synthesizes new channels in response to denervation, and hence is altered biochemically, we are led to the important idea that neuroactive substances released from synaptic terminals can do more than simply change membrane permeability. In the next chapter, we explore the mechanisms by which substances released from synapses alter the biochemistry of nerve cells.

Summary

The output of a neuron reflects the sum of synaptic input to the cell. Excitatory chemical synapses induce small depolarizing potentials, EPSPs, mainly in the cell's dendrites. Inhibitory chemical synapses cause small hyperpolarizing potentials, IPSPs, in the cell body. Action potentials are generated only at the axon hillock of the cell; the rest of the cell body and dendrites are, for the most part, electrically inexcitable. The spread of potential from synaptic sites to the axon hillock is passive; the axon hillock serves as the summing point for the neuron. It fires action potentials at a rate depending on the balance of excitatory and inhibitory input to the cell.

Both EPSPs and IPSPs result from an increase in membrane conductance. For an EPSP the reversal potential—the membrane potential at which there is no net current flow across the membrane—is about -10 mV, approximately midway between the equilibrium potentials for Na^+ and K^+. Neurotransmitter released at excitatory synapses thus makes the membrane more permeable to both Na^+ and K^+. The reversal potential for many IPSPs is about -70 mV, close to the equilibrium potential for Cl^-. Other IPSPs show a more negative reversal potential (-80 mV), which indicates that inhibitory neurotransmitter can open Cl^- channels or K^+ channels or both. Single neurotransmitter-gated channels open in an all-or-none fashion. They show variable open times, and they continue to open and close for as long as transmitter is present.

Presynaptic inhibition is an example of shunting inhibition. Neurotransmitter released onto a synaptic terminal opens channels in the terminal membrane and significantly decreases membrane resistance in the presynaptic terminal. The voltage developed across the membrane as a result of current flow into the terminal is smaller than is normally the case because of the decreased membrane resistance ($V = IR$). Since the transmitter output of a presynaptic terminal depends on membrane voltage, the amount of transmitter released is depressed, and a smaller EPSP is observed in the postsynaptic neuron.

At electrical synapses, current flows relatively unimpeded from one cell to an adjacent cell through gap junctions. Transmission at electrical synapses is fast, and the potentials recorded on both sides of the junction are virtually identical.

At the neuromuscular junction, small (0.5 mV) potentials called min EPSPs are recorded by micropipettes positioned close to the postsynaptic membrane. These potentials sum to produce the muscle EPSP, which is analogous to neuronal EPSPs. An individual min EPSP appears to result from the release of transmitter (acetylcholine) from a single synaptic vesicle onto the muscle membrane. In an intact muscle cell, the channels responding to acetylcholine are confined to the region of the neuromuscular junction; in the denervated muscle, new channels are made by the cell and are inserted all over the muscle membrane. The denervated muscle is thus responsive to ACh all over its surface.

Further Reading

BOOKS

Eccles, J. C. 1955. *The Physiology of Nerve Cells.* Baltimore: Johns Hopkins Press. (A readable account of Eccles's classic work on synaptic potentials, including the methods involved.)

Eccles, J. C. 1964. *The Physiology of Synapses.* Berlin: Springer-Verlag. (A detailed discussion of the electrophysiology of synapses.)

Edelman, G. M., W. E. Gall, and W. M. Cowan, eds. 1987. *Synaptic Function.* New York: John Wiley and Sons. (An up-to-date collection of articles on synaptic mechanisms.)

The Synapse. 1976. Cold Spring Harbor Symposium on Quantitative Biology, Vol. 40. Cold Spring Harbor, NY: Cold Spring Harbor Laboratory. (A collection of research articles emphasizing synaptic structure and function.)

ARTICLES

Axelsson, J., and S. Thesleff. 1959. A study of supersensitivity in denervated mammalian skeletal muscle. *Journal of Physiology* 147:178–193.

Bennett, M. V. L. 1977. Electrical transmission: A functional analysis and comparison with chemical transmission. In *Cellular Biology of Neurons, Part 1,* ed. E. R. Kandel, pp. 367–416. *Handbook of Physiology,* Sec. 1, *The Nervous System,* Vol. 1. Bethesda, MD: American Physiological Society.

Fatt, P., and B. Katz. 1952. Spontaneous subthreshold activity at motor nerve endings. *Journal of Physiology* 117:109–128. (The first description of min EPSPs.)

Kuffler, S. W., and D. Yoshikami. 1975. The number of transmitter molecules in a quantum: An estimate from iontophoretic application of acetylcholine at the neuromuscular junction. *Journal of Physiology* 251:465–482.

Lester, H. A. 1977. The response to acetylcholine. *Scientific American* 236(2):106–118.

Llinas, R., and M. Sugimori. 1980. Electrophysiological properties of *in vitro* Purkinje cell dendrites in mammalian cerebellar slices. *Journal of Physiology* 305:197–213.

The Chemistry of Synaptic Transmission

THE KEY ROLE of synaptic transmission in neural function has been emphasized several times. Not only are neurons excited and inhibited at synapses, but neural activity is also modulated by synaptic action. In this chapter, we examine the mechanisms by which neurons are modified by synapses. Most synapses in the brain are chemical. A substance released at a synapse diffuses across the synaptic cleft and instigates some change in the post-synaptic membrane. Some substances act directly on postsynaptic membrane channels, altering the permeability of the membrane and allowing ions to flow across it. These substances are called *neurotransmitters* and they mediate the fast excitatory or inhibitory potentials—the EPSPs and IPSPs—that lead to the excitation or inhibition of neurons. EPSPs and IPSPs typically begin within a fraction of a millisecond after transmitter is released from a presynaptic terminal and they seldom last longer than 10–100 milliseconds.

Other substances released at synaptic sites serve to modify neural activity rather than to initiate it. These substances are called *neuromodulators;* their effects usually have a slow onset (seconds) and they can last for minutes, hours, or longer. The keys to synaptic modulation are biochemical mechanisms. In the postsynaptic neuron, enzymes are activated, small second-messenger molecules are synthesized, and many aspects of neural cell function are altered. There is good reason to believe that long-term changes in the brain, which underlie phenomena such as memory and learning, result from neuromodulatory activity.

This chapter reviews the actions of the substances released at synaptic sites and the effects they may have on brain function. I begin by describing the sequence of events during synaptic transmission and how it can be disrupted by drugs or disease. In the second half of the chapter I present an overview of the dozens of substances that have so far been identified as neurotransmitters or neuromodulators.

Mechanisms of Synaptic Transmission

Four steps delineate the progress of synaptic activity from initiation to completion, as illustrated in Figure 6.1. The model here is based in large measure on experiments conducted on the neuro-

6.1   The process of synaptic transmission at an acetylcholine synapse. *(1) Transmitter release.* Upon membrane depolarization, Ca^{2+} enters the terminal and facilitates the binding of synaptic vesicles to the presynaptic membrane. Fusion of vesicles with the presynaptic membrane results in the release of transmitter (ACh) into the synaptic cleft. *(2) Transmitter action.* Once released, ACh diffuses to the postsynaptic membrane and binds to membrane channels that open and allow Na^+ and K^+ to cross the membrane. *(3) Transmitter removal.* Acetylcholinesterase (labeled with asterisks) breaks down ACh to acetate and choline. *(4) Transmitter resynthesis and repackaging.* Choline is transported back into the terminal and joined with acetyl coenzyme A (CoA, an activated form of acetate) to make ACh. ACh is concentrated in vesicles that reform by infolding of the terminal membrane.

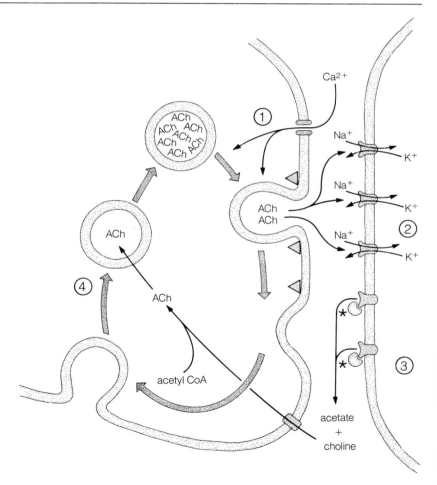

muscular junction. Acetylcholine is released at this site, and it acts here as a neurotransmitter.

1. RELEASE OF TRANSMITTER

Synaptic transmission begins with depolarization of the membrane of the presynaptic terminal. Voltage-sensitive Ca^{2+} channels, present in the presynaptic membrane, open upon depolarization and admit Ca^{2+} into the terminal. The Ca^{2+} facilitates the binding of synaptic vesicles to the presynaptic membrane by a mechanism not well understood. The synaptic vesicles become confluent with the terminal membrane, open, and release their contents into the synaptic cleft.

The electron-dense projections observed by high-resolution electron microscopy along the presynaptic membrane may help bind the synaptic vesicles to the membrane. Ca^{2+}, either directly or through an intermediate activated by Ca^{2+}, promotes the attachment of synaptic vesicles to the dense projections, which leads to fusion of the vesicle to the terminal membrane.

Not everyone agrees that transmitter is released by synaptic vesicles. Some workers maintain that it is released from the cytoplasm of nerve terminals and that the vesicles simply store neuroactive substances. There may be some synapses that operate this way. At some synapses, for example, vesicles are not seen to bind to the presynaptic membrane and Ca^{2+} is not required for transmitter release. In my opinion, however, the evidence is overwhelming that transmitter release is vesicular at the great majority of chemical synapses.

Figure 6.2 provides direct anatomical evidence for the scenario I've just described: vesicles binding to the presynaptic membrane of the neuromuscular junction. The conventional electron micrograph at the top shows clear examples of synaptic vesicle fusion with the presynaptic membrane. The bottom two pictures are freeze-fracture electron micrographs of the presynaptic membrane. The upper micrograph shows the presynaptic membrane 3 milliseconds after the presynaptic terminal had been stimulated. The double row of particles running through the field are thought to be the dense projections found along the presynaptic membrane to which the synaptic vesicles attach. The lower micrograph shows the presynaptic membrane 5 milliseconds after the terminal was

6.2 Electron micrographs showing vesicles binding to the presynaptic membrane at the neuromuscular junction. (**A**) In this conventional electron micrograph, various stages of vesicle fusion with the presynaptic membrane are visible. From left to right is a fortuitous progression, from a vesicle still separate from the membrane to a vesicle almost completely incorporated in the membrane with intermediate stages of fusion in between. (**B** and **C**) Freeze-fracture electron micrographs of the inner leaflet of the presynaptic membrane of the neuromuscular junction 3 and 5 ms after the terminals were stimulated. By 5 ms, synaptic vesicles had begun to fuse with the membrane *(arrows)*.

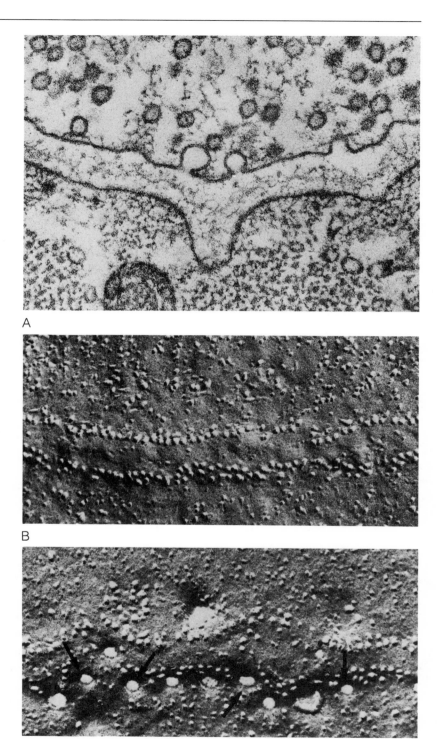

A

B

C

stimulated. Fusion of synaptic vesicles to the membrane adjacent to the row of particles has occurred, and a series of pits (arrows) has been formed.

2. ACTION OF TRANSMITTER

Once released, transmitter diffuses quickly across the synaptic cleft and binds to specific receptor proteins on the postsynaptic membrane. For the neuromuscular junction and other excitatory synapses, the receptor is part of the channel that opens and allows Na^+ and K^+ to cross the membrane. At inhibitory synapses, transmitter interacts with channel proteins that permit either Cl^- or K^+ or both to cross the membrane. And at neuromodulatory synapses, released substances interact with membrane proteins linked to enzymes, thereby initiating biochemical changes in the postsynaptic cell.

3. REMOVAL OF TRANSMITTER

The next step is to rid the synaptic cleft of transmitter. At the neuromuscular junction and other synapses using ACh, an enzyme, acetylcholinesterase (indicated by asterisks in Figure 6.1), rapidly breaks down ACh to its component parts, acetate and choline. The structure of acetylcholine is shown in Figure 6.3, along with schemes for its synthesis and degradation. Neither acetate nor choline can activate the ACh channel; thus, acetylcholinesterase effectively terminates synaptic activity.

Other neuroactive substances are usually cleared from the synaptic cleft by active reuptake of the substance into the presynaptic terminal. This is accomplished by transmembrane proteins called *transporters*. There are specific transporters for different neuroactive substances, and all transporters require Na^+ to function. Indeed, they are driven by the Na^+ gradient across the cell membrane, and Na^+ is co-transported with the neuroactive substance into the terminal. Glial cells may also take up neuroactive substances by such mechanisms and help terminate synaptic activity.

4. RESYNTHESIS AND REPACKAGING OF TRANSMITTER

The last step is to restore the availability of transmitter for the next time it is needed for synaptic transmission. At ACh junctions, choline is actively transported back into the terminal and joined with activated acetate molecules within the cell to re-form acetyl-

6.3 The structure of acetylcholine and how it is synthesized and degraded.

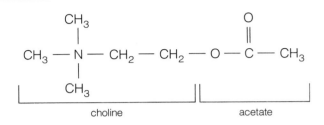

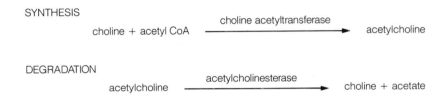

SYNTHESIS

choline + acetyl CoA $\xrightarrow{\text{choline acetyltransferase}}$ acetylcholine

DEGRADATION

acetylcholine $\xrightarrow{\text{acetylcholinesterase}}$ choline + acetate

choline (see Figure 6.3). At other synapses, intact transmitter molecules are brought back into the terminal by the re-uptake system. Synaptic vesicles are re-formed by an infolding of the terminal membrane away from the active release zone and a pinching off of that membrane to form vesicles, as shown in Figure 6.1. Transmitter is concentrated in the synaptic vesicles and synaptic transmission is complete.

Timing

As measured at the frog neuromuscular junction at room temperature, there is a delay of about 0.5 milliseconds or 500 microseconds from depolarization of the presynaptic terminal to a response in the postsynaptic cell. One might suppose that the diffusion of transmitter across the synaptic cleft accounts for most of the synaptic delay. This is not so. It requires only about 50 microseconds for substances to diffuse the 20 to 30 nanometers across the cleft. The bulk of the delay, 300 microseconds, is required for the release of transmitter, and about 150 microseconds are needed to activate and open the channels once transmitter has reached the postsynaptic membrane. In cold-blooded vertebrates operating at temperatures lower than room temperature, synaptic delay can be longer, up to 1–5 milliseconds; in warm-blooded birds and mammals, synaptic delay is about 0.3 milliseconds.

Drugs That Block Synapses

Synapses are very vulnerable to blockade. All sorts of synaptic inhibitors affect virtually every aspect of synaptic transmission. It is believed that the effects of most drugs on the brain result from their ability to alter transmission. For example, lysergic acid, or LSD, a potent hallucinogenic drug, blocks the effects of serotonin, a neuroactive substance released at certain brain synapses; other examples will be discussed later.

A good deal of what we know about synaptic transmission has come from the use of blocking agents or antagonists in investigations of synaptic mechanisms. At the neuromuscular junction, for example, each of the steps in synaptic transmission described earlier can be blocked by one or more agent.

1. RELEASE OF TRANSMITTER

Because Ca^{2+} ions are required for the release of transmitter from presynaptic terminals, reducing Ca^{2+} in the extracellular space—or interfering with Ca^{2+} access into the terminal by raising the extracellular concentration of other divalent cations, such as Mg^{2+} or Co^{2+}—will diminish or stop transmitter release. The release of neuroactive substances from nerve terminals can be prevented by toxins also. Botulinum toxin, for example, will have this effect, but we do not yet understand how it works. This potent toxin, synthesized by an anaerobic bacterium, *Clostridium botulinum,* is found in soil and therefore on fruits and vegetables. If the bacterium is not destroyed by proper heating when food is canned, it will flourish in the anaerobic atmosphere of a can or jar and produce toxin fatal to those who eat the food. Botulinum poisoning was much more of a problem in the early days of canning, but even today occasional incidents occur.

2. ACTION OF TRANSMITTER

Synaptic blockers may bind specifically to protein channels in the postsynaptic neuron and prevent ACh from activating the channel. Two types of inhibitors or antagonists are distinguished. Non-reversible blockers bind so tightly to the channel that they essentially destroy it; reversible inhibitors can be gradually washed away—they will detach from the channel. An example of a non-reversible blocker is α-bungarotoxin, a deadly ingredient of the venom from cobra snakes. Curare, a paralyzing agent, is an example of a reversible blocker. Derived from climbing vines found in South America, the poison was employed by Indians to coat the tips of arrows and spears.

3. REMOVAL OF TRANSMITTER

Disrupting the activity of acetylcholinesterase allows ACh to build up in the synaptic cleft and results in failure of synaptic transmission. Again, two kinds of inhibitors are known: those that interact nonreversibly with the enzyme and those that will separate from the enzyme. Organic phosphates, the major constituent of insecticides and deadly nerve gases, are nonreversible acetylcholinesterase inhibitors; eserine, a derivative of the African calabar bean, is a reversible inhibitor. Eserine was used in some parts of Africa as a truth serum. The lore is that a person who was unjustly accused of some wrongdoing would be without fear and would rapidly drink down the mixture. Since eserine also acts as an emetic, the individual would vomit and survive. But the guilty individual, according to the legend, would drink the potion more slowly, thereby absorbing more of the poison, and usually be done in.

4. RESYNTHESIS OF TRANSMITTER

A chemical called hemicholinium blocks the reuptake of choline and prevents the resynthesis of ACh in the terminal. The ACh is gradually depleted from the terminal and transmission fails. Various agents are known that block the reuptake of other neuroactive substances, resulting in disruption of synaptic transmission.

Myasthenia Gravis: A Neuromuscular Disease

In addition to providing information about the mechanisms underlying synaptic transmission, drugs that block the neuromuscular junction have also been of great value in the study of a disease involving the junction. Myasthenia gravis is characterized in its early stages by muscle weakness and fatigue. Eventually patients may be bedridden and possibly die. The diagnosis, treatment, and analysis of myasthenia gravis requires understanding of the neu-

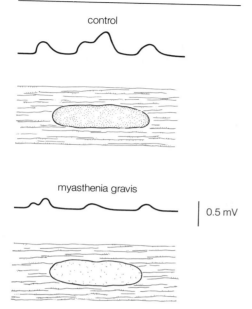

6.4 The min EPSPs recorded from a normal subject are greater than those recorded from a patient suffering from myasthenia gravis, and more α-bungarotoxin binds to the muscle's postsynaptic membrane in the normal subject. The evidence indicates a reduced number of ACh channels in persons with the disease.

romuscular junction's physiology and of the drugs that block transmission at the junction.

Many years ago it was discovered that tiny doses of curare given to a patient with myasthenia gravis caused muscle weakness and debilitation far exceeding that seen in a normal individual. This was some of the first evidence that the disease affected the neuromuscular junction. The excessive sensitivity of a patient with myasthenia gravis to curare or curare-like drugs has been used to diagnose the disease. Also, eserine-like drugs that block acetylcholinesterase aid patients with the disease. Indeed, such drugs remain its primary therapy.

In the mid-1960s, Swedish workers recorded with intracellular microelectrodes min EPSPs from tiny pieces of muscle excised from patients with myasthenia gravis. Finding that the min EPSP amplitude was reduced as compared with min EPSPs recorded from normal muscle cells, they concluded that either a reduced amount of neurotransmitter was available in each synaptic vesicle or there were fewer ACh channels in the postsynaptic membrane. In either case, it was then possible to suggest why inhibition of acetylcholinesterase with eserine-like drugs would help overcome the disease. That is, prolonging the lifetime of ACh in the synaptic cleft causes additional depolarization of the postsynaptic membrane and counters the lowered min EPSP amplitude.

In the early 1970s, Douglas Fambrough and Daniel Drachman at Johns Hopkins University showed that patients with myasthenia gravis have fewer postsynaptic ACh receptors. They discovered this by exposing muscle fibers to radioactively labeled α-bungarotoxin. Muscle fibers from myasthenic patients showed much less binding of α-bungarotoxin to the postsynaptic membrane than was found in control muscle fibers. This is shown schematically in Figure 6.4. It was also demonstrated that myasthenia gravis can be mimicked in an animal by injecting small amounts of α-bungarotoxin, which effectively destroys many of the postsynaptic ACh receptors.

Why are there fewer ACh receptors in patients with myasthenia gravis? Increasing evidence indicates that it is an autoimmune disease: for reasons that are not clear, individuals with myasthenia gravis form antibodies against the ACh receptors. The antibodies, like α-bungarotoxin, bind to the receptors and destroy them. Normally ACh receptors are being continually synthesized by muscle cells; indeed, a muscle cell makes a complete complement of ACh receptors in about a week. In myasthenia gravis, the antibodies inactivate ACh receptors as fast as or faster than they can be synthesized. Because muscle cells are synthesizing ACh receptors continuously, patients experience rapid recovery when antibody levels are lowered. Decreasing antibody levels in the blood can be accomplished by dialysis, but the effect is only temporary; within a few days antibody levels increase again. Unfortunately, we have at present no way to lower antibody levels permanently, but this could be a promising avenue for a cure for the disease.

Neurotransmitters and Neuromodulators

Two types of neuroactive substances are released at synaptic sites: neurotransmitters and neuromodulators. Virtually all of our attention has focused so far on the neurotransmitters, which interact directly with channels in the postsynaptic membrane. When a neurotransmitter activates channels, it changes the conductance of the postsynaptic membrane and this change usually results in the generation of fast EPSPs or IPSPs. The latency for these potential changes is typically on the order of 0.5–1 millisecond, and their duration is about 10–100 milliseconds.

Neuromodulators, though, act in a different way. They interact with specific receptor proteins on the postsynaptic membrane linked to intracellular enzymes. Activating these receptors does not directly change membrane voltage or membrane resistance; rather, the action of neuromodulators is mediated through biochemical changes in the postsynaptic neuron. The physiological changes mediated by neuromodulators typically have long latencies, on the order of seconds, and the changes can last for minutes, hours, or even longer. Slow, long-term changes in the brain are believed to be mediated by neuromodulators.

Some neuroactive substances act exclusively as neurotransmitters, others as neuromodulators, but many can do both. They may act directly on membrane channels, but they also can act on receptors linked to enzymes. Acetylcholine is an example in that it acts on ACh channels at the neuromuscular junction and many other synapses, but it also acts on ACh receptors linked to an enzyme in the membrane. These two actions of ACh can be distinguished pharmacologically; certain drugs will block one action of ACh but not the other, and vice-versa. Curare blocks the activation of the ACh channel but not the activation of the enzyme-linked receptor, whereas atropine blocks the enzyme-linked receptor but not the ACh channel. These two actions of ACh are termed *nicotinic* and *muscarinic,* respectively, for two pharmacologic agents, nicotine and muscarine, that activate specifically one or the other of these ACh actions.

Two or sometimes more neuroactive substances can exist in single terminals, and frequently one of the neuroactive substances is a transmitter and the other is modulator. Thus, both actions can be initiated by the same synapse.

Neuromodulation

Figure 6.5 presents as an example a neuromodulatory system that involves the enzyme adenylate cyclase and an intracellular or second-messenger molecule, cyclic AMP. A neuromodulator released from a presynaptic terminal interacts with a receptor on the postsynaptic membrane that is linked to an intermediate, a so-called G-protein, which in turn is linked to adenylate cyclase. (G-proteins bind GTP, hence their name.) Adenylate cyclase catalyzes the synthesis of cyclic AMP from ATP. Cyclic AMP usually exerts its action in a cell by activating another set of enzymes, kinases, that add a phosphate group to cellular constituents. This process, called *phosphorylation,* is a common cellular mechanism for activating or inactivating biochemical reactions or for modifying proteins.

The cascade of reactions initiated by the neuromodulator is shown in more detail in Figure 6.6. The neuromodulator's interaction with its receptor leads to the binding of a GTP molecule to the G-protein. The activated G-protein in turn interacts with adenylate cyclase, thereby activating it and leading to the formation of cyclic AMP, or cAMP, from ATP. The association of the activated G-protein with adenylate cyclase also results in the breakdown of the bound GTP to GDP. The newly formed cyclic AMP activates a kinase, which leads to phosphorylation of specific cell constituents.

Cyclic AMP and the cAMP-dependent kinases can have effects at several levels in the cell—from the nucleus, to the cytoplasm, to the membrane (see Figure 6.6). In the nucleus, gene transcription can be altered; in the cytoplasm, protein synthesis or enzyme activity can be modified; and at the membrane, ion conductances

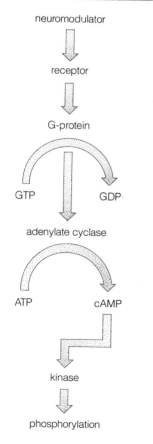

6.5 The cyclic AMP cascade.

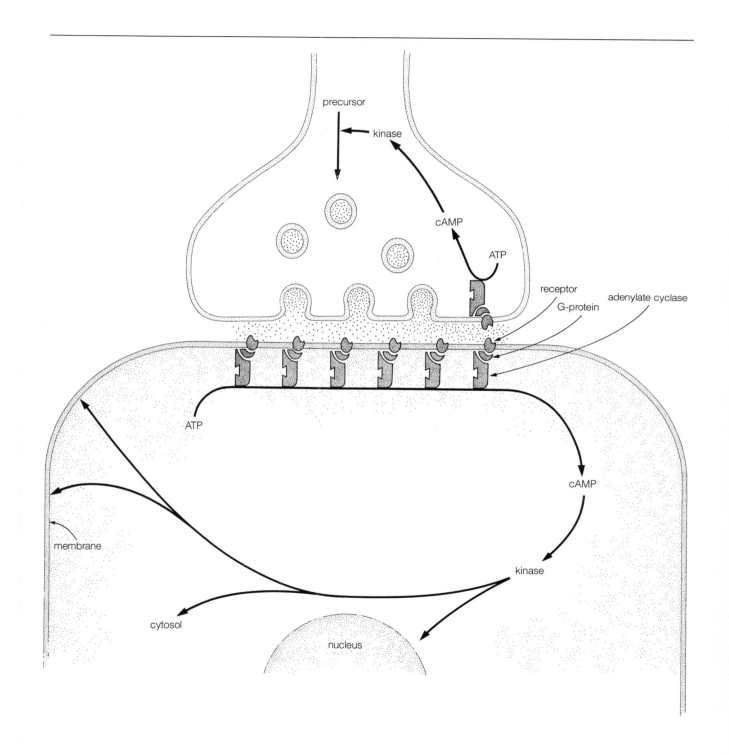

6.6 Actions of a neuromodulator. The neuromodulator released at the synapse interacts with postsynaptic receptors. Receptor stimulation activates a G-protein and an intracellular enzyme—in this case adenylate cyclase, which converts ATP to cyclic AMP. Cyclic AMP activates other enzymes, called kinases, which add or remove phosphate groups from molecules and thereby activate or inactivate them. Cyclic AMP–dependent kinases may exert their effects in the nucleus, in the cytoplasm, or at the cell membrane, thus potentially altering a variety of cellular processes. Neuromodulators can also interact with autoreceptors on the presynaptic terminal membrane and activate enzyme cascades in that cell as well. In this example, transmitter synthesis is regulated by the cyclic AMP cascade.

can be modulated. Second messengers, via kinase phosphorylation, can modify all three types of membrane conductances that we have discussed: voltage-gated conductances, especially those mediated by voltage-sensitive K^+ and Ca^{2+} channels; transmitter-gated conductances; and gap-junctional conductances.

How does phosphorylation affect a channel? In the cases examined so far phosphorylation modifies (1) the frequency of opening of a channel to a given stimulus, (2) the duration of open channel time, (3) the ionic specificity of the channel, or (4) the number of active channels in a piece of membrane. The significance of neuromodulatory effects on neurons will be explored in more detail later, when we consider neural circuits and integrative neuroscience.

Figure 6.6 shows also that neuromodulators can have presynaptic as well as postsynaptic effects. Synaptic terminals can have their own receptors, called *autoreceptors*, that respond to the substance released by that terminal. In the example drawn, the autoreceptors are linked to a cyclic AMP cascade. Activation of a kinase by cyclic AMP can modify enzymes involved in the synthesis of neuroactive substances in the terminal. Thus, the release of a neuromodulator from a synaptic terminal can regulate by a feedback mechanism the synthesis of that or other neuroactive substances.

Cyclic AMP is not the only second messenger known to be generated in neurons by neuromodulators. It is, however, the best known and the most extensively studied second messenger. Figure 6.7 shows another, more complicated system, one involving the activation of a membrane enzyme, phospholipase C. This enzyme, like adenylate cyclase, is activated by a G-protein, which in turn is activated by the interaction of a neuromodulator with a specific membrane receptor. Phospholipase C acts on a specific phospholipid in the cell membrane bilayer, called phosphatidylinositol. The enzyme splits off a part of the phospholipid, called inositol triphosphate, and leaves exposed the rest of the lipid, diacylglycerol. Both inositol triphosphate and diacylglycerol act as second messengers. Diacylglycerol directly activates a specific kinase, protein kinase C, but inositol triphosphate (IP_3) acts in a more complicated fashion. It releases Ca^{2+} from intracellular stores into the cytoplasm, and the released Ca^{2+} binds to a protein called calmodulin. Activated calmodulin interacts with specific kinases that phospho-

6.7 The inositol triphosphate and diacyl-glycerol cascades.

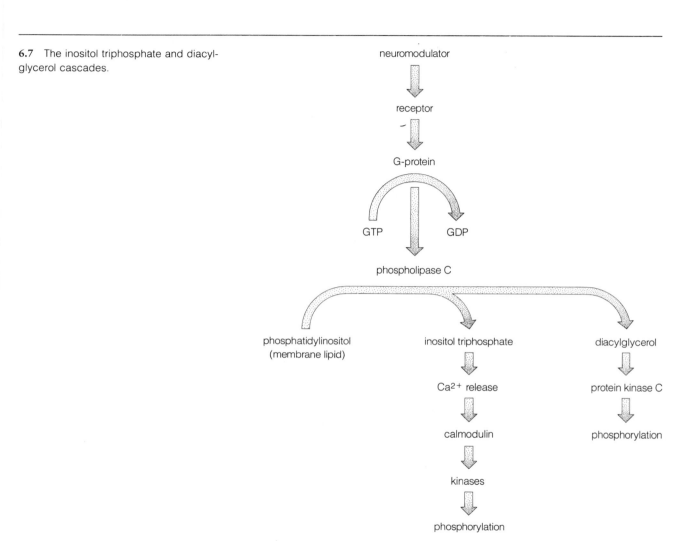

rylate cellular constituents. At the present time, little is known about the actions mediated by the kinases activated by diacylglycerol, IP_3, or calmodulin. Note that in the IP_3 cascade, intracellular Ca^{2+} is released and then acts as a second messenger. In other situations, extracellular Ca^{2+} may enter a neuron via voltage-sensitive channels and also act as a second messenger. An example of extracellular Ca^{2+} serving as a second messenger will be discussed in Chapter 11.

Neuroscientists presume that many second-messenger systems reside in neurons, and it is also becoming clear that many variations exist. For example, transmitter-activated G-proteins may themselves interact with certain membrane channels that are nearby and alter their permeability. Also, a second messenger can directly open membrane channels, as we will see with vertebrate photoreceptors in the next chapter. The only absolute finding so far is that all receptor proteins activated by neuromodulatory substances are linked to a G-protein.

Identifying Transmitters and Modulators

At the present time, over 40 substances have been proposed as neurotransmitters, neuromodulators, or both. Proving that a substance is the transmitter or modulator released by a neuron is not easy. Pharmacologists ask that four criteria be met before accepting a substance as an identified transmitter or modulator.

1. The substance must be present in a nerve terminal and be synthesized by the neuron.

2. The substance must be released from the neuron upon depolarization or other appropriate stimulation of the cell.

3. The substance under test must mimic exactly the effects of the naturally released substance on the postsynaptic membrane. If it is a transmitter candidate, for example, the evoked potential must have the same reversal potential as the naturally elicited synaptic potential.

4. The action of the substance on the postsynaptic neuron should be blocked by appropriate synaptic inhibitors; that is, by antagonists that block the natural activity of the synapse in question.

It is generally believed that neurons release the same substances from all of their terminals. This is known as Dale's law, after Henry Dale, an English pharmacologist who first proposed this idea and who was one of the pioneers of neuropharmacology. As we have noted, two or even more substances can be present in individual nerve terminals and released from synaptic sites. It is supposed that if neuroactive substances coexist in one terminal of a neuron, they reside in all terminals of that neuron; moreover, it is believed that all the coexisting neuroactive substances are released from the terminal. However, there is evidence that some terminals of a neuron release proportionally more of one neuroactive substance than do other terminals of the same cell, indicating that the output of all terminals of a particular neuron may not be identical.

Classes of Neuroactive Substances

The four recognized classes of neuroactive substances differ in terms of their chemistry and their actions. Two classes act mainly as neurotransmitters and the other two mainly as neuromodulators—although some substances can, as pointed out earlier, act as either. Table 6.1 lists the four classes, grouped into the category they best fit, and examples of substances in each class.

ACETYLCHOLINE

Acetylcholine is generally classified as a separate kind of neuroactive substance, though technically it is a monoamine. ACh is the only naturally occurring substance in its class, but synthetic substances are know that mimic ACh activity. Found in the central and peripheral nervous systems, it was the first neuroactive substance to be characterized and has been the most studied and hence is the best known.

Acetylcholine is usually an excitatory neurotransmitter, but it also can exert inhibitory effects. Indeed, its discovery by Otto Loewi in the early 1920s resulted from the observation that a substance released from the vagus nerve slows the heartbeat of an animal (this is explained in Chapter 13). The important point here is that whether a neuroactive substance is excitatory or inhibitory, or acts as a neurotransmitter or as a neuromodulator, depends on the receptor protein in the postsynaptic membrane to which it binds. If the receptor is part of a channel, the substance acts as a neurotransmitter. The channel's ion selectivity then determines whether the effect is excitatory or inhibitory. The substance acts as a neuromodulator if it interacts with a receptor linked to a G-protein. But what action is mediated by this interaction will depend on the enzymes or channels affected by the G-protein, the second

Table 6.1 The four classes of neuroactive substances, with examples of each class

Neurotransmitters		Neuromodulators	
1. Acetylcholine	2. Amino acids	3. Monoamines	4. Peptides

Acetylcholine	L-glutamate
	L-aspartate
	γ-Aminobutyric acid (GABA)
	Glycine
		A. Catecholamines
		  Dopamine
		  Norepineph-
		  rine
		  Epinephrine
		
		B. Indoleamines
		  Serotonin
			A. Hypothalamic
			  Thyrotropin-releas-
			  ing hormone
			  (TRH)
			  Somatostatin
			  Luteinizing hor-
			  mone–releasing
			  hormone (LHRH)
			
			B. Pituitary
			  Vasopressin
			  Adrenocortico-
			  trophic hormone
			  (ACTH)
			
			C. Digestive system
			  Cholecystokinin
			  (CCK)
			  Vasoactive intestin-
			  alntestinal peptide
			  (VIP)
			  Substance P
			
			D. Other
			  Enkephalins

messengers (if any) that are produced, and the kinases (if any) activated by the second messengers.

It is also possible for one transmitter to have an excitatory effect on one neuron and an inhibitory effect on another neuron. An example is the second-order bipolar cell in the retina. One bipolar cell type is depolarized by the transmitter released from the photoreceptors, whereas the other type is hyperpolarized. Some invertebrate neurons have both excitatory and inhibitory potentials induced by the same neuroactive substance released from one

6.8 Schematic drawings of how a neuron can have both excitatory and inhibitory effects on separate postsynaptic neurons *(top)* or on the same postsynaptic neuron *(bottom)*. The transmitter released by the neuron binds to channels whose ion selectivity gives rise to an EPSP *(open circles)* or to an IPSP *(solid circles)*.

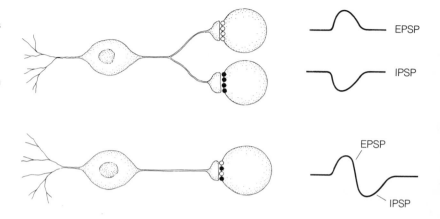

terminal onto the cell. The two potentials do not cancel one another because one follows the other. These effects are illustrated schematically in Figure 6.8.

AMINO ACIDS

Amino acids act mainly as neurotransmitters and they are believed to be the principal excitatory and inhibitory transmitters in the brain. There are two well-established excitatory amino acids, L-glutamate and L-aspartate, and two well-known inhibitory amino acids, γ-aminobutyric acid (GABA) and glycine. These four amino acids are closely related structurally, as shown in Figure 6.9. Other amino acids have excitatory or inhibitory actions, such as L-cysteine and histamine, but not much is yet known about how they function in synaptic transmission.

Powerful and specific inhibitors or antagonists of the inhibitory amino acid transmitters are known. Bicuculline and picrotoxin specifically block GABA channels, whereas strychnine blocks glycine channels. Only recently have specific antagonists for the channels activated by the excitatory amino acids been found, and this has hindered analysis of the excitatory amino acids.

MONOAMINES

Two classes of monoamines serve as neuroactive substances: the catecholamines and the indoleamines. Three catecholamines

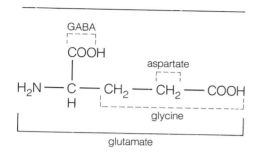

6.9 The molecule as shown in its entirety is glutamate. With one carboxyl group (–COO⁻) removed, the molecule becomes GABA; with one carbon group removed (–CH₂–), the molecule is aspartate; with two carbon groups and the carboxyl group removed, the molecule is glycine.

(dopamine, norepinephrine and epinephrine) and one indoleamine (serotonin) are released at synaptic sites in the vertebrate brain. These agents appear to have something to do with the brain's affective and arousal states. Substances that affect the levels or actions of the catecholamines and indoleamines often alter mood and other mental states, as the following examples show:

1. LSD, a hallucinogenic drug, is a powerful antagonist of serotonin. It appears to interact with and block serotonin receptors.

2. Reserpine is a powerful tranquilizer, an anxiety reliever. It depletes the brain of monoamines and probably exerts its effects in this way.

3. Amphetamines are brain stimulants; they lead to hyperactivity and an inability to sleep. The amphetamines act by interfering with the re-uptake of catecholamines, particularly dopamine, and by releasing catecholamines from nerve terminals.

The range of activities that monoamines may induce in the brain is not completely known. For example, two quite different diseases—Parkinson's disease and schizophrenia—that affect brain function relate to dopamine metabolism and function. Individuals suffering from Parkinson's disease develop a severe tremor of the limbs, along with rigidity of muscle tone. They also have difficulty in initiating movement, and their movements are characteristically slow. It was shown in the late 1950s that the dopamine content of the brain was low in individuals suffering from Parkinson's disease. About 80 percent of brain dopamine is in the region of the basal ganglia, structures involved in the initiation of movement. Subsequently it was found that one enzyme in the pathway of dopamine synthesis, tyrosine hydroxylase, was deficient in people with Parkinson's disease, and this finding appeared to account for the low levels of dopamine.

Tyrosine hydroxylase converts the amino acid tyrosine to L-dopa, the immediate precursor of dopamine (see Figure 6.10). Dopamine can be converted to norepinephrine, and norepinephrine to epinephrine, by additional one-step reactions. Thus, the levels of all three catecholamines are depressed in Parkinson's disease, but it is the decline in dopamine that appears most im-

6.10 The synthesis of the catecholamines—dopamine, norepinephrine, and epinephrine—from tyrosine. The structure of tyrosine is shown below, the group additions or subtractions that take place during the synthesis are indicated by number. For example, step 1, which converts tyrosine to L-dopa, involves the addition of a hydroxyl group (+OH) to the ring of tyrosine, and so forth.

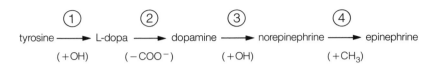

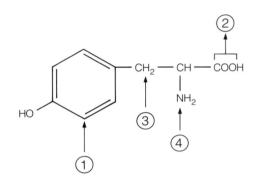

portant. Once the enzymatic deficit was recognized, a therapy was proposed and tried. The administration of L-dopa, the intermediate made by the deficient enzyme, substantially relieves symptoms in many patients and is used to help many suffering from Parkinson's disease.

For schizophrenia a specific biochemical or physiological deficit has not been identified, but dopamine-receptor blockers help many schizophrenics. These drugs are the standard therapy for schizophrenia, the most commonly prescribed being haloperidol. Other evidence that dopamine function is impaired in schizophrenia comes from clinical observations that high doses of amphetamines, which enhance dopamine activity, can induce schizophrenia-like symptoms in normal individuals.

Great care must be taken when administering drugs that affect brain function. Schizophrenia responds to drugs that block dopamine action; thus, the presumption is that a surplus of dopamine is released at some synapses in the brain. Yet Parkinson's disease appears to be due to a deficit of dopamine in one area of the brain, which raises the question whether patients treated with dopamine antagonists develop Parkinson-like symptoms. They do, and this severely limits the benefits of dopamine antagonists for some schizophrenic patients. Conversely, Parkinson's disease patients can exhibit schizophrenia-like symptoms when they are treated with L-dopa.

NEUROPEPTIDES

In the early 1930s, investigators discovered in the brain a peptide that contracts smooth muscle in the digestive system. It was suggested that this peptide, called substance P, might be released at synaptic sites in the brain. Two decades later researchers found that neurons in one part of the brain, the hypothalamus, release small peptides into the bloodstream that affect the release of hormones from the pituitary gland.

The significance of these findings was not appreciated until the 1970s, when, as a result of the development of immunohistochemical techniques, it was observed that many hypothalamic, pituitary, and even gastrointestinal peptide hormones are present in many parts of the brain, where they are localized to specific types of nerve cells. Usually few peptide-containing cells of any one type are located in a specific region of the brain, but peptide-containing neurons usually extend their processes widely. Figure 1.9B shows, for example, a neuron in the retina that contains substance P. Although the area of the micrograph contains dozens of cells, only one stained for substance P, and its processes extended beyond the width of the micrograph.

Neuropeptides are the largest class of neuroactive substances; at least 25 different peptides are now believed to be released at synapses in the brain. They vary in size, containing from 3 to 40 amino acids (see Figure 6.11), and they can be divided into four groups. These groupings and representative examples are listed in Table 6.2.

Little is known about the function of most neuropeptides in the brain, but their range of activities is likely to be considerable. Particularly fascinating examples in this regard are the enkephalins, which behave like natural opiates within the brain.

Opiates, such as morphine, profoundly affect the brain and powerfully relieve pain. They do so at low concentrations, which suggests that they act at specific postsynaptic sites. In the 1970s Candace Pert and Solomon Snyder at Johns Hopkins University showed that brain tissue has specific receptors for opiates; that is, opiates bind to specific receptor proteins. This finding immediately

6.11 A few of the neuropeptides found in the brain and their amino acid structures.

Table 6.2 Peptides found in the brain

Hypothalamic peptides

Thyrotropin-releasing hormone (TRH)	A small, 3-amino-acid peptide that causes the pituitary gland to release thyrotropin
Somatostatin	A 14-amino-acid peptide that inhibits the release of thyrotropin and growth hormone from the pituitary gland
Luteinizing hormone–releasing hormone (LHRH)	A 10-amino-acid peptide that promotes the release of luteinizing hormone from the pituitary

Pituitary peptides

Vasopressin	A 9-amino-acid peptide that causes reabsorption of water by the kidney and blood vessel constriction
Corticotropin (ACTH)	A 39-amino-acid peptide that causes release of steroid hormones from the adrenal glands

Digestive system peptides

Cholecystokinin (CCK)	An 8-amino-acid peptide that initiates the release of bile from the gall bladder
Vasoactive intestinal polypeptide (VIP)	A 28-amino-acid peptide that causes constriction of blood vessels in the intestine
Substance P	An 11-amino-acid peptide that causes the contraction of smooth muscle in the digestive tract

Other peptides

Enkephalins	Two 5-amino-acid peptides, one containing leucine (leu-enkephalin) and the other containing methionine (met-enkephalin)

suggested that natural opiate-like substances exist in the brain and interact with these receptors. They were searched for and soon discovered by two neurochemists working in Scotland, John Hughes and Hans Kosterlitz, and were named the *enkephalins*.

That the brain makes these substances and can release them explains puzzling aspects of pain and how it can be relieved in unconventional ways. For example, during acupuncture treatments enkephalins are released in the brain and relieve pain. Evidence for this comes from experiments in which naloxone, an opiate receptor antagonist, blocks the anesthetic effects of acupuncture. Another example is the placebo effect; pain is often relieved by placebo pills. Again, experiments have shown that the placebo effect can be blocked by naloxone.

Summary

We have distinguished four steps in synaptic transmission. First, transmitter is released when the synaptic terminal depolarizes, which opens voltage-gated Ca^{2+} channels in the membrane. The Ca^{2+} enters the terminal and facilitates the binding of the synaptic vesicles to the presynaptic membrane. The bound vesicles become confluent with the membrane and release their contents into the synaptic cleft. Second, transmitter diffuses to the postsynaptic membrane, where it binds to receptor proteins. Binding initiates channel opening or enzyme activation. Third, synaptic transmission is terminated by the enzymatic breakdown of transmitter in the synaptic cleft or by the reuptake of neuroactive substances into the presynaptic terminal. Fourth, newly synthesized transmitter, or neuroactive substances taken up by the terminal, are repackaged in synaptic vesicles when the terminal membrane folds in and pinches off from the membrane.

Transmission at the neuromuscular junction can be blocked at any of the four steps. Examples are: (1) decreasing Ca^{2+} concentrations, increasing other divalent cation levels, and botulinum toxin all inhibit release of transmitter; (2) α-bungarotoxin and curare bind to acetylcholine (ACh) channels in the postsynaptic membrane so that the channels cannot open; (3) organic phosphates and eserine interfere with acetylcholinesterase and prevent the breakdown of ACh; (4) hemicholinium and other substances block the reuptake of transmitter precursors or the transmitter molecules themselves, thereby causing the depletion of neuroactive substances from presynaptic terminals.

Myasthenia gravis is an autoimmune disease of the neuromuscular junction. Antibodies bind to and inactivate the ACh receptors and thereby reduce postsynaptic potentials. Eserine-like drugs, by prolonging the lifetime of ACh in the synaptic cleft, are an effective therapy for the disease.

Two types of neuroactive substances, neurotransmitters and neuromodulators, are released at synapses in the brain. Neurotransmitters act directly on channels and induce a conductance change across the postsynaptic membrane; fast EPSPs and IPSPs result. Neuromodulators interact with receptor proteins linked to enzyme systems. Second-messenger molecules are usually produced and kinases activated. Biochemical changes, usually phosphorylations, are induced in the neurons that alter membrane channels, enzymatic activity, and perhaps gene transcription. Substances serving as second messengers include cyclic AMP, inositol triphosphate, diacylglycerol, and Ca^{2+} ions.

Four classes of neuroactive substances are recognized; acetylcholine and the amino acids usually act as neurotransmitters; the monoamines and the neuropeptides are most often neuromodulatory in their effects. Whether a substance acts as transmitter or modulator depends on the nature of the receptor protein to which it binds. Some substances, such as acetylcholine, act as both.

Two amino acids, L-glutamate and L-aspartate, serve as excitatory neurotransmitters; GABA and glycine, also amino acids, are inhibitory transmitters. Four monoamines—dopamine, norepinephrine, epinephrine, and serotonin—are well established as neuromodulatory substances. They particularly influence the affective and arousal states of the brain. Two brain diseases have been linked to dopamine: Parkinson's disease results from a deficiency of dopamine in one part of the brain; schizophrenia may be caused by excessive release of dopamine from certain synapses elsewhere in the brain. Some 20–30 neuropeptides have been identified in neurons in the brain, most of which act as neuromodulators. Many of these neuropeptides also serve as hormones released from the hypothalamus, pituitary gland, or within the digestive tract. Other neuropeptides, like the enkephalins, which are endogenous opiates, may be unique to the brain.

Further Reading

BOOKS

Cooper, J. R., F. E. Bloom, and R. H. Roth. 1991. *The Biochemical Basis of Neuropharmacology,* 6th ed. New York: Oxford University Press. (A readable overview of neuropharmacology and the chemistry of substances released at synapses.)

Hall, Z. W., J. G. Hildebrand, and E. A. Kravitz. 1974. *Chemistry of Synaptic Transmission.* Newton, MA: Chiron Press. (A collection of the classic papers on neurotransmission.)

Katz, B. 1969. *The Release of Neural Transmitter Substances.* Liverpool: Liverpool University Press. (Katz's Sherrington lectures.)

Nestler, E. J., and P. Greengard. 1984. *Protein Phosphorylation in the Nervous System.* New York: John Wiley and Sons.

Snyder, S. H. 1986. *Drugs and the Brain.* New York: Scientific American Books.

ARTICLES

Berridge, M. J. 1985. The molecular basis of communication within the cell. *Scientific American* 253(4):142–152. (Part of a special issue on biological molecules.)

Drachman, D. B. 1981. The biology of myasthenia gravis. *Annual Review of Neuroscience* 4:195–225.

Heuser, J. E., T. S. Reese, M. J. Dennis, Y. Jan, and L. Evans. 1979. Synaptic vesicle exocytosis captured by quick freezing and correlated with quantal transmitter release. *Journal of Cell Biology* 81:275–300.

Iversen, L. L. 1979. The chemistry of the brain. *Scientific American* 241(3):70–81.

Iversen, L. L. 1983. Neuropeptides—what next? *Trends in Neuroscience* 6:293–294. (Introduction to a special issue devoted to neuropeptides and other neuroactive substances.)

Katz, B., and R. Miledi. 1967. The timing of calcium action during neuromuscular transmission. *Journal of Physiology* 189:535–544.

Nathanson, J. A., and P. Greengard. 1977. "Second messengers" in the brain. *Scientific American* 237(2):108–119.

Receptor Potentials

AN ORGANISM is in contact with the outside world through its sensory receptors. Without them, it would be isolated from its environment and fellow organisms. Most animals have a variety of sensory receptors, each responding to a specific type of stimulus—light, sound, touch, chemical compound, heat, electricity—and all of them transducing the stimulus into a sustained, graded receptor potential. Not all animals have all kinds of receptors. Many fish have electroreceptors that detect surrounding electrical fields, and snakes have infrared detectors that sense warm objects nearby. We do not possess either of these receptors.

All neurons have characteristics of sensory receptors in that they possess specialized regions of membrane that respond selectively to specific chemicals—the neurotransmitters and the neuromodulators. The effects of neurotransmitters on neurons, producing conductance and potential changes across the membrane, are similar to the effects of stimuli on touch and auditory receptors. Photo- and olfactory receptors respond to stimuli as neurons respond to neuromodulators; enzyme cascades are activated and second-messenger levels are altered.

In this chapter, I examine in detail two types of receptors: mechanoreceptors and photoreceptors. In mechanoreceptors, the deformation of specialized membrane or membrane channels results directly in the generation of receptor potentials. The absorption of light by photoreceptors, in contrast, leads to changes in second-messenger levels and the second messengers generate the receptor potentials.

Mechanoreceptors

The Pacinian Corpuscle

Mechanoreceptors include touch, pressure, and auditory receptors. Our model is the Pacinian corpuscle, a large pressure receptor present in skin, muscle, joints, and tendons. The Pacinian corpuscle has been studied extensively because of its large size, and so we know more about it than any other touch or pressure receptor. Other touch and pressure receptors are thought to work much like the Pacinian corpuscle. Under the microscope the corpuscle looks like a slice of onion (Figure 7.1), with many concentric layers of flattened nonneural (epithelial) cells surrounding a bare nerve ending. The nerve ending becomes myelinated upon emerging from the corpuscle. This corpuscle and a length of nerve fiber can easily be excised and kept functioning for some time; hence, they are convenient for study.

Figure 7.1 also shows what a Pacinian corpuscle receptor cell looks like. It is a typical sensory cell, being monopolar in nature—that is, only one process comes from the cell body. This process divides into two myelinated processes, one to the sensory endings and the other to the spinal cord. Action potentials generated in the sensory endings are propagated directly through the myelinated processes to terminals in the spinal cord. The cell bodies of sensory cells innervating the limbs and trunk lie just adjacent to the spinal cord in structures called dorsal ganglia, which are discussed in Chapter 12.

With a preparation consisting of the corpuscle and a short length of nerve hooked with a wire, it is easy to stimulate neural activity with a fine needle or stylus (see Figure 7.2). Compression of the capsule by a tiny amount, just a fraction of a micrometer (0.2–0.5 μm), induces a small receptor potential. With more compression, provided by a stronger stimulus, a larger receptor potential is generated. A depolarization of 10 mV or so generates an action potential superimposed on the receptor potential.

Action potentials produced by the Pacinian corpuscle are blocked by tetrodotoxin (TTX), but the receptor potential is not. Thus, receptor potentials, like synaptic potentials, are not generated by voltage-sensitive Na^+ channels. By bathing the preparation in Ringer's solution containing TTX, one can study the receptor potential in isolation, without the intrusion of action potentials. Intensity-response relations for the receptor potential can then be worked out; an example is shown in Figure 7.2C. Many types of receptors have similar S-shaped intensity-response relations.

How is this receptor potential generated? Compression of the myelinated fiber coming from the capsule yields no response unless

7.1 A typical pressure receptor. The neuron is monopolar; a short process from the cell body divides into two, one myelinated process extending to the sensory endings (Pacinian corpuscles) and the other to the cell terminals. A Pacinian corpuscle (enlargement on right) consists of a bare nerve ending surrounded by a capsule of flattened nonneural cells.

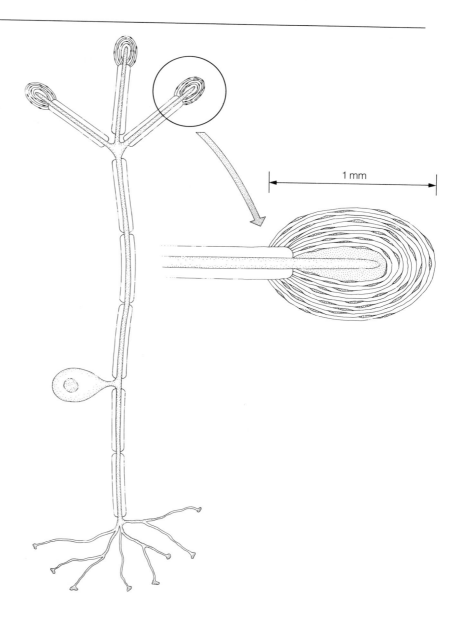

1 mm

7.2 When a Pacinian capsule is depressed with a stylus (**A**), depolarizing receptor potentials are evoked in the nerve that, if large enough, generate an action potential (**B**). An S-shaped relation typically exists, as shown in (**C**), between receptor voltage *(V)* and stimulus strength.

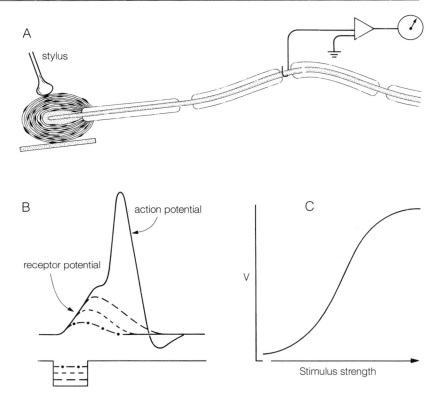

a large distortion of the membrane (10–15 μm) is induced. Then a change in membrane potential is recorded that probably represents damage to the nerve. (This is analogous, probably, to the excitation of the ulnar nerve when you hit your elbow on a hard object; that is, when you hit your crazy or funny bone). Removing the onion-like capsule does not affect the response. Indeed, when a preparation is stripped of the capsule, small deformations all over the bare nerve ending generate small receptor potentials. These small receptor potentials summate and produce a larger receptor potential. If the first node of the myelinated part of the fiber is mechanically blocked (by applying pressure to the node), no action potentials are generated but receptor potentials are still recorded.

The experimental results suggest the following model. Con-

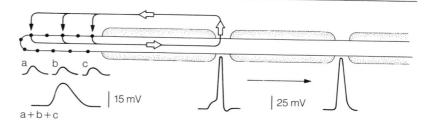

7.3 When the Pacinian capsule is removed, leaving a bare nerve ending, deformations of the membrane *(arrows)* result in small depolarizing potentials *(a, b,* and *c)* that can summate. Current *(open arrows)* entering the fiber through the mechanosensitive channels *(filled circles)* results in depolarization of the membrane at the first node of Ranvier and the generation of action potentials, which then travel down the fiber.

tained within the membrane of the bare nerve ending are specialized channels whose conductance is altered when they or the surrounding membrane are deformed. At rest, few if any of the channels are open, but when the membrane is stretched, the channels change configuration and allow ions to cross the membrane. Because of the conductance change, net positive charge accumulates inside the cell, depolarizing the membrane at the first node in the fiber's myelinated part. Here voltage-sensitive channels are located, and, with sufficient depolarization, action potentials are generated that propagate down the fiber. Figure 7.3 is a current-flow diagram of the Pacinian corpuscle based on this model. Current enters the bare nerve ending where the membrane is deformed and flows internally down the fiber. This depolarizes the first node in the fiber's myelinated part and generates action potentials that travel down the cell.

What ions generate the receptor potential? The nerve ending is too small to penetrate with a micropipette, and so the experimenter cannot easily measure a reversal potential for the response. But we can answer this question by altering ion concentrations in the extracellular Ringer's solution. Changes of Na^+ and K^+ affect the receptor response; hence the channels, when open, are permeable to Na^+ and K^+, as is the case for channels that give rise to EPSPs.

Hair Cells

Hair cells are also mechanoreceptors. They are the sensory receptors in the inner ear—for both the auditory and vestibular systems—and in the lateral line organ. The vestibular system mediates the sense of balance and is related physically and functionally to the auditory system. The lateral line organ, found in aquatic animals, provides information about the flow of water over the animal; in some animals it gauges water pressure, temperature, and even salinity. Hair cells, like the Pacinian corpuscle, have channels that open when surrounding membrane or the channels themselves are stretched or deformed. These channels are found at the tips of hair-like projections that extend from the top surface of the cell, and they appear to be connected directly by fine filaments that run between adjacent "hairs." When sound impinges on the fluid-filled inner ear, as shown schematically in Figure 7.4, the hair cells move relative to the tectorial membrane to which the hairs are attached. The hairs bend, increasing the tension on the filaments and causing the channels to open, which allows positive ions (cations) to cross the membrane. The fluid surrounding the hairs in the inner ear is high in K^+; indeed, K^+ is higher outside the hairs than inside. Thus, K^+ flows into the hair cell and depolarizes it when the channels are opened.

Action potentials are not generated by hair cells. Rather, the graded receptor potential leads to the release of transmitter from the cell's synapses; it is the second-order cells that first generate action potentials in the auditory and vestibular systems. A similar situation is found in many invertebrate visual systems. In the vertebrate visual system, action potentials are first generated by third-order cells. That is, in the vertebrate retina excitation and synapse activation is via graded potentials in the receptor *and* the second-order cells.

Adaptation

The Pacinian corpuscle illustrates dramatically an important property of all receptors, namely adaptation. With a sustained stimulus, the receptor potential declines. It may do so rapidly or slowly, completely or partially. Sensory adaptation represents a distortion of the real world to an organism, in the sense that receptors do not provide a faithful representation of the stimuli impinging on the organism. Even so, sensory receptor adaptation can have significant advantages: it is unlikely that we could wear clothes without being severely distracted by the continuous response of our touch receptors if they did not rapidly adapt.

Three types of adapting receptors are distinguished: fast, medium, and slow. Fast-adapting or phasic receptors respond only to a change in stimulus level. Pacinian corpuscles are fast-adapting receptors, as are other touch receptors and olfactory receptors.

7.4 (**A**) Schematic drawing of a hair cell in the inner ear *(top)*, and intracellular responses recorded from a hair cell *(bottom)*. When sound impinges on the ear, the cells move relative to the tectorial membrane in the fluid-filled inner ear, bending the hairs on the cell. The movement stretches the filaments connecting the hairs and opens channels, allowing K^+ to flow into the cell and causing cell depolarization. The path of current flow is indicated by the thin lines and arrows. When a pure tone is presented to the ear, a sinusoidal potential is recorded in the cell that closely matches the tone's waveform. (**B**) Scanning electron micrograph of the bundle of hairs on a hair cell *(left)*, and a higher-magnification micrograph showing the fine filaments that extend from the tip of one hair to the side of the next hair *(right)*.

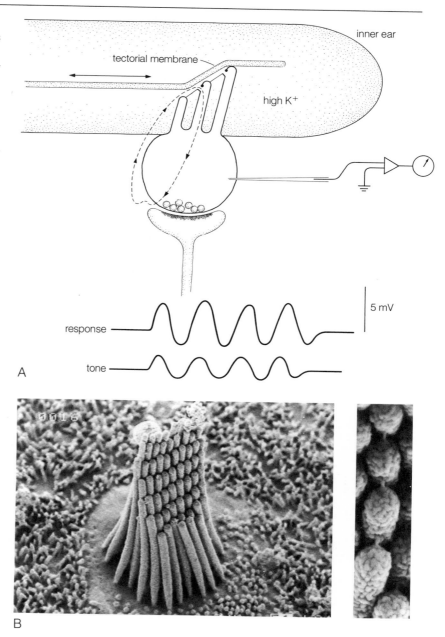

7.5 Sensory receptor adaptation. With fast-adapting receptors, a response at stimulus offset (the off-response) is often observed *(dashed line)*.

fast-adapting   medium-adapting   slow-adapting

Figure 7.5A shows the response of a Pacinian corpuscle to a prolonged stimulus. Upon application of pressure to the receptor, it rapidly depolarizes. But within 10 milliseconds or so the voltage decays to zero even though the pressure is maintained. The Pacinian corpuscle displays another feature of many phasic receptors in that it responds at the offset of the stimulus as well as at the onset of it.

Medium-adapting receptors decline much more gradually in potential during the presentation of a long stimulus. If the stimulus is sufficiently prolonged, the response can decay almost completely, as illustrated in Figure 7.5B. Taste and hearing receptors are of this type. Photoreceptors and deep pressure receptors are slow-adapting or tonic receptors (Figure 7.5C). Photoreceptors respond to a prolonged stimulus in two stages, an initial transient potential that decays to a smaller plateau potential, which is maintained for as long as the stimulus continues. These receptors respond for as long as the stimulus is applied, but, as with other receptors, they give a maximal response at stimulus onset and a smaller response thereafter. All receptors, even tonic ones, respond maximally to changes of stimulus level.

What mechanisms underlie adaptation? They vary according to the receptor. The mechanism may be intrinsic to the receptor cell, as in photoreceptors. Or the mechanism may involve accessory structures related to the receptors, of which the Pacinian corpuscle is an example. If the capsule surrounding the bare nerve ending of the Pacinian corpuscle is removed, the response to a sustained stimulus dramatically alters (Figure 7.6A). The receptor potential no longer adapts rapidly, and the off-response is not observed.

What is going on? How does the capsule, which is made of nonneural epithelial cells, contribute to the receptor's adaptation? Figure 7.6B suggests likely mechanisms. When the capsule is com-

7.6 Removal of the capsule changes the Pacinian corpuscle from a fast-adapting to a medium-adapting receptor. (A) Records from a preparation with and without the capsule. (B) A possible explanation: with continued depression of the capsule, the underlying tissue adjusts in shape to allow reformation of the nerve cell membrane.

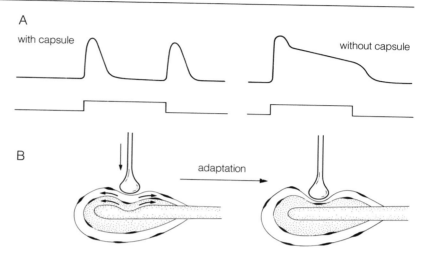

pressed, all of the cells underlying the impinging stimulus are compressed and membrane deformation extends to the nerve ending. With time, though, cytoplasm in the compressed capsule cells moves laterally, or fluid between the capsule cells is squeezed out, or some capsule cells slip sideways, or all three. Whatever is the case, the capsule thickness under the region where pressure is being applied becomes thinner and the nerve cell membrane returns to its original shape, closing the membrane channels.

When pressure is relieved, the capsule elastically rebounds. The rebound deforms the nerve ending membrane and triggers the off-response. On the other hand, when the capsule is removed, the membrane nerve ending remains deformed for as long as the stimulus continues, the receptor potential is maintained for the duration of the stimulus, and no off-response is seen. Why the stripped nerve ending preparation still adapts to some extent is not clear; it may reflect a mechanism intrinsic to the receptor.

Invertebrate Photoreceptors

A typical invertebrate photoreceptor cell is illustrated in Figure 7.7. A characteristic of all photoreceptor cells is a specialized region, often called the outer segment, which consists of elaborated

7.7 An invertebrate photoreceptor cell has a specialized outer segment region, consisting of elaborated membrane, which contains the light-sensitive visual pigment molecules *(small dots)*.

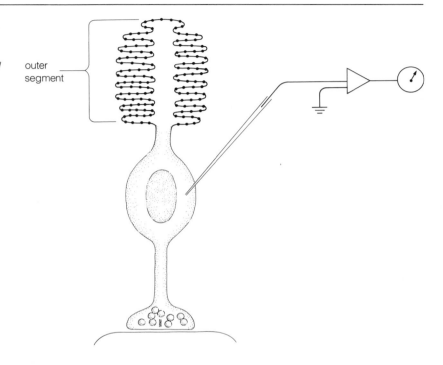

membrane that is highly infolded. Imbedded in this membrane are light-sensitive molecules, the visual pigments. The concentration of visual pigment molecules in the outer segment membrane makes for an efficient photodetector. A photon of light passing through these membranes is extremely likely to be absorbed by one of the light-sensitive molecules and initiate visual excitation.

Many invertebrate photoreceptors can be penetrated by micropipettes to record intracellular photoreceptor potentials. The resting potential of invertebrate photoreceptors is typically −60 mV, and illuminating the cell's outer segment region generates depolarizing potentials. Figure 7.8 shows some typical records. With dim illumination, the response is small and quite square in form. With brighter illumination, the photoresponse is larger and it shows some adaptation. With bright, saturating light stimuli, responses may be 70–75 mV in amplitude, overshooting zero membrane voltage. Adaptation of the response from the initial peak to

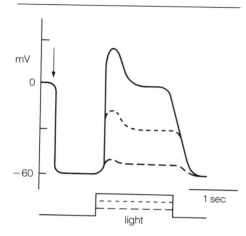

7.8 Intracellular records from a photoreceptor cell of the horseshoe crab. The cell has a resting potential of about −60 mV. Light depolarizes the cell, and the responses typically show adaptation; the initial peak response decays to a steady plateau potential. The arrow indicates electrode penetration.

a steady plateau is also most prominent in responses elicited with bright lights.

What underlies the light responses of photoreceptors? When a photoreceptor cell is probed with a small spot of light, receptor potentials are generated only when the light is applied to the cell's outer segment region. Furthermore, if the cell is voltage-clamped, we find that current flows into the cell when the outer segment is illuminated. These experiments indicate that, upon illumination, channels open in the outer segment membrane and allow positive ions to enter the cell.

The reversal potential for many invertebrate photoreceptor responses is about +15 mV. This suggests that the membrane channels allow both Na^+ and K^+ to cross the membrane. Yet, measurements of internal Na^+ concentration in some invertebrate photoreceptors reveal high levels of Na^+ ions. Thus, the equilibrium potential for Na^+ in many invertebrate photoreceptors is close to +15 mV, and so it may be that the channels in certain invertebrate photoreceptors are permeable mainly to Na^+.

Quantal Responses

From calculations we know that 10^5–10^6 Na^+ ions enter the cell per absorbed quantum of light, but can we see evidence of these quantal events in intracellular recordings from photoreceptors? In thoroughly dark-adapted preparations that are highly sensitive to light, small (1–3 mV) depolarizing responses are recorded from many photoreceptors when they are illuminated with dim light. Estimates of the number of photons in the dim light flashes match well the number of responses observed, meaning that these are quantal responses. The responses are frequently referred to as quantal bumps (see Figure 7.9). In many ways, the quantal bumps resemble min EPSPs recorded from the neuromuscular junction.

When brighter flashes are presented to the photoreceptor, more quantal bumps are generated. They summate and create a sustained depolarization of the cell. With yet brighter stimuli, so many quantal bumps are produced that individual events cannot be distinguished; the response has the waveform of a typical receptor potential—made up of summed quantal bumps (Figure 7.9).

But why does the response adapt in moderate and bright lights? In other words, why is the response to the onset of the light larger

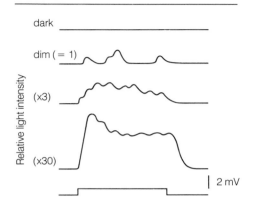

7.9 Quantal "bumps" recorded from a dark-adapted photoreceptor cell of the horseshoe crab eye. The number of quantal responses generated depends on stimulus intensity, and the receptor potential appears to be made up of summed quantal responses.

than the maintained response if the light intensity is kept constant and the number of quanta absorbed over time remains the same? Clues come from comparing quantal bumps recorded from dark-adapted photoreceptors with photoreceptors recently exposed to light (see Figure 7.10A). The quantal responses recorded from light-adapted photoreceptors are considerably smaller than those recorded from thoroughly dark-adapted receptors. Following exposure of a photoreceptor to light, fewer Na^+ ions are admitted into the cell per quantum caught. Why is this?

The channels in the outer segment membrane admit mainly Na^+ into the cell when they open. Some Ca^{2+} also enters the cell, and the Ca^{2+} appears to regulate the efficiency of the Na^+ channels. How Ca^{2+} exerts this effect is not known, but adaptation of the response can be circumvented if a rise in intracellular Ca^{2+} is prevented. This is done by injecting a Ca^{2+} buffer into the cell; records from an experiment are shown in Figure 7.10B. Following the injection, the trace of the photoresponse to a bright stimulus is square—it does not sag from peak to plateau.

As already noted, most invertebrate photoreceptor cells do not generate action potentials. Thus, it is the receptor potential that activates the photoreceptor synapse—with some variations. In squid, a long axon extends from the photoreceptor cell into the brain, and this axon generates action potentials in response to the receptor potential. In the horseshoe crab, the photoreceptor cells are coupled via electrical (gap) junctions to a second-order cell that fires action potentials (see Figure 3.2). The receptor potential passes directly into the second-order cell via the gap junctions and generates action potentials in the second-order cell that travel along its axon and transmit the visual signal to the brain. In most cases, however, the receptor potential itself leads directly to the release of transmitter from the photoreceptor.

Vertebrate Photoreceptors

Figure 7.11 gives examples of typical vertebrate photoreceptors. All vertebrates, with few exceptions, have two kinds of photoreceptors, rods and cones, distinguished by the shape of their outer segments. Rods, which are more sensitive to light than are cones, mediate dim light or night vision. Cones are the receptors for day vision, and many animals have several types of cones that differ

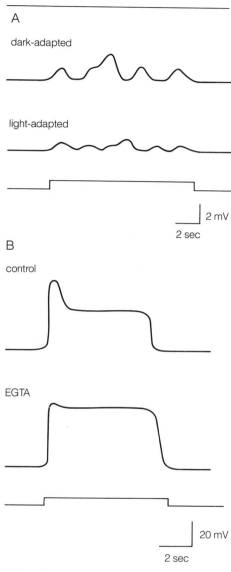

A

dark-adapted

light-adapted

2 mV

2 sec

B

control

EGTA

20 mV

2 sec

7.10 (A) Light adaptation decreases the amplitude of the quantal responses recorded from the horseshoe crab eye. (B) Injection of a Ca^{2+} buffer (EGTA) prevents adaptation—no sag in receptor potential from peak to plateau is observed. These results implicate Ca^{2+} in the adaptation process.

in their responses to colored light. Thus, cones also mediate color vision.

Rods and cones are elongated cells consisting of an outer segment, an inner segment, a nuclear region, and a synaptic terminal. The outer segment consists of numerous infoldings from the surrounding outer segment membrane. In rods, the infolded membrane pinches off from the surrounding membrane and forms flattened membranous disks piled one on top of another, enclosed in the structure. Only a few disks at the base of rod remain in continuity with the surrounding membrane. In cones, though, the individual disk membranes tend to remain connected to the outer segment membrane along the structure's entire length. Visual pigment molecules are imbedded in the disk membranes and are concentrated there to a high degree. Because the visual pigments absorb light, they have color—which often can be seen in a dark-adapted retina brought into the light.

As we noted, many disks in rods are physically separate from the outer segment membrane. But light changes the potential across the surrounding outer segment membrane. The anatomy thus indicates that a signal is transmitted from the disk membrane, where the light is caught by a visual pigment molecule, to the outer segment membrane, where the receptor potential is generated. As we shall see, this signal is carried by a second-messenger molecule.

The outer segment connects via a thin (cilial) stalk to the inner segment, where most of the cell's metabolic machinery resides. In the inner segment are numerous mitochondria that generate the ATP needed to run membrane ion pumps. (Why vertebrate photoreceptors require so many mitochondria will be explained later.) The cell's nucleus, below the inner segment, is usually connected to the synaptic terminal by a short process.

Vertebrate photoreceptors are more difficult to penetrate with micropipettes than invertebrate photoreceptors, primarily because they are usually thinner. The first convincing intracellular recordings from vertebrate photoreceptors were obtained in the mid-1960s by Tsuneo Tomita in Japan; his recordings were surprising. Tomita discovered that the photoreceptors' resting potential is less negative than the resting potential of most neurons, typically about -30 mV to -40 mV rather than -60 mV to -70 mV. Furthermore, when the cell was illuminated, it hyperpolarized and the

7.11 (**A**) In both rod and cone photorecep-tors, the light-sensitive pigments are found in the outer segments of the cells. (**B**) A portion of a cone outer segment of a lizard. The lizard cones are long, and over a short portion of their length the cone shape is not obvious. Note the disks on the left side that are in continuity with one another *(arrow)*.

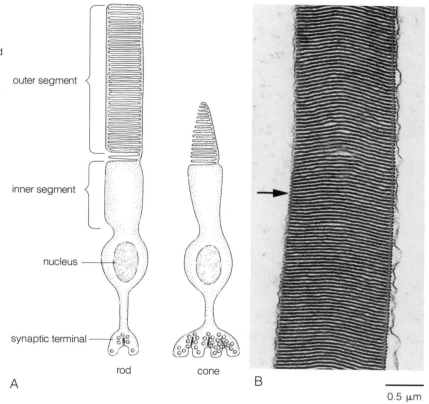

outer segment

inner segment

nucleus

synaptic terminal

rod   cone

A   B

0.5 μm

membrane potential became more negative (see Figure 7.12). With bright lights, the cell hyperpolarized initially by as much as 30 mV, bringing the cell to a voltage of −60 mV, or close to the resting potential of most neurons. Like invertebrate photoreceptor responses, the initial voltage was not maintained; it recovered partially to a maintained plateau. Hence, vertebrate photorecep-tors also adapt.

Why does the vertebrate photoreceptor hyperpolarize upon il-lumination, and what does this response mean? Neuron hyper-polarization generally indicates inhibition and the opening of chan-nels that allow Cl^-, K^+, or both to cross the cell's membrane. Is the vertebrate photoreceptor response similar to an IPSP? This

7.12 Typical intracellularly recorded responses from a vertebrate photoreceptor. The resting potential of these cells is about −30 mV, and the cells hyperpolarize to steps of light in a graded fashion.

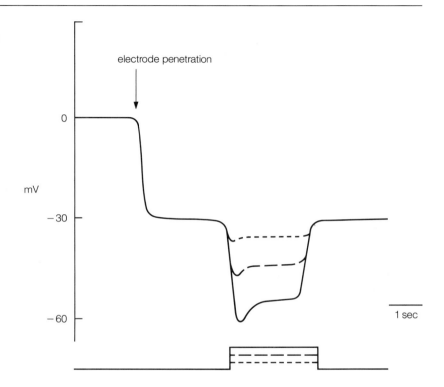

possibility was explored by determining the cell's resistance before, during, and after the presentation of a light stimulus. Current pulses were injected into the cell and voltages across the membrane in response to the current were measured. The amplitude of the voltage across the membrane is, of course, a reflection of the membrane's resistance.

The results are presented in Figure 7.13. Rather than falling during the light stimulus, membrane resistance rose, as evidenced by an increase in the voltages elicited in response to the current pulses. This finding indicates that channels close in the vertebrate photoreceptor membrane when it is illuminated. This observation, coupled with the fact that the vertebrate photoreceptor's resting potential is low in the dark (−30 mV), lead to the conclusion that photoreceptor channels are open in the dark and that illumination closes them. In essence, it is as though darkness is the stimulus for

7.13 Measurements of membrane resistance of a photoreceptor cell before, during, and after a light stimulus. The membrane resistance increases in the light, as shown by the increased voltage deflection of the membrane in response to the constant current pulses (*I*). These results indicate that light closes channels in the membrane.

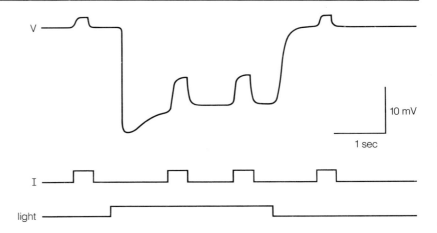

the vertebrate photoreceptor cell: current flows into the cell in the dark (the response to the stimulus) and light shuts down the current.

Studies of the "dark current" of vertebrate photoreceptors and how it is affected by light were first carried out in the late 1960s by William Hagins at the National Institutes of Health in Bethesda, Maryland. He recorded the dark current with extracellular electrodes placed at either end of a group of rod photoreceptors taken from a rat. He measured a potential difference between the electrodes that reflects the dark current. The inward dark current is restricted to the outer segment. By contrast, current flows out of the cell, through leak channels, along the inner segment, the nuclear region, and the terminal membrane—a loop of current flowing around the cell in the dark, as shown in Figure 7.14A. Because positive ions are flowing into the outer segment, there is a small excess of negative ions around this region of the cell. The flow of ions from the other end of the cell means that there is excess positive charge there. Electrodes positioned on either end of the cell detect this difference in charge, as Figure 7.14B shows. A potential difference of about 300 μV is recorded in the dark between the two electrodes. Light decreases the dark current in a graded fashion, and with bright light the dark current is completely turned off. Hence when the outer segment is brightly illuminated all of the channels giving rise to the dark current are shut down.

7.14 (A) The extracellular measurement of the potential difference around a vertebrate photoreceptor cell in the dark and (B) a typical record of the differences in voltage between the two ends of the cell. A voltage of about 300 μV can be measured between the two ends of the cell. Light (λ) decreases this voltage in a graded fashion. Cyanide (CN⁻) poisons the pumps that exude Na⁺ from the cell, resulting in the gradual loss of the dark voltage. The dark voltage can be reestablished by washing the Na⁺ from the photoreceptors with Na⁺-free Ringer's solution.

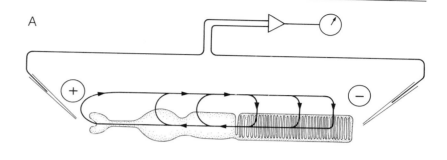

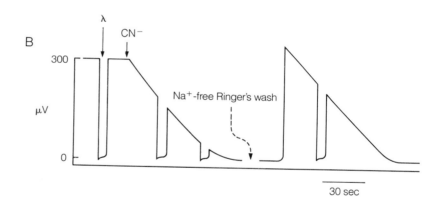

With this preparation, Hagins showed that the dark current was carried mainly by Na^+ ions and that each quantum caught by a visual pigment molecule prevented about 10^5 Na^+ ions from entering the cell. He also discovered that if the pumps that rid the cell of inside Na^+ were blocked by a metabolic poison such as cyanide, which prevents the production of ATP by mitochondria, the cell filled with Na^+ in just a few minutes. As the Na^+ built up inside, the dark current decreased because the concentration gradient for Na^+ between outside and in was lowered. Thus, the driving force on Na^+ to enter the cell was abolished.

As shown in Figure 7.14, it is possible to reestablish the sodium gradient and dark current in a cyanide-poisoned preparation by washing the cells in Na^+-free Ringer's solution and then reintroducing normal Ringer's to the outside of the cells. The Na^+ washes out of the cells as long as they are kept in the dark and the Na^+ channels remain open. The Na^+ runs down its concentration gradient, in this case from inside to out. Once the Na^+ is washed away and normal Ringer's is present on the outside, Na^+ once again moves into the cell and reestablishes the dark current. If the pumps remain blocked, the dark current decreases as Na^+ accumulates inside, but light responses can be recorded for a few minutes.

These experiments emphasize the importance of the metabolic pumps in maintaining photoreceptor function, and they also suggest why so many mitochondria are needed in the inner segment. That is, because the Na^+ channels are continuously open in the dark, Na^+ must be continually pumped out of the cell, which places a significant metabolic demand on the cell. To maintain the pumps, ATP is required; hence the numerous mitochondria in the inner segment.

Regulation of the Na^+ Channels

What regulates the Na^+ permeability of the outer segment? This was a vexing question that took nearly twenty years to answer. The definitive experiment was carried out by a group of Russian scientists led by E. Fesenko in 1985. They showed that cyclic GMP, a close analogue of cyclic AMP, directly opens Na^+ channels when applied to a piece of outer segment membrane removed from a photoreceptor by a patch-clamp pipette (see Figure 1.16). No phosphorylation is involved; the second messenger itself binds to the channel and opens it. Examples of current traces reflecting single-channel openings induced by cyclic GMP are shown in Figure 7.15.

When light strikes a vertebrate photoreceptor, Na^+ channels in the outer segment membrane close. This means that cyclic GMP levels must fall in the outer segment in the light. A cascade of reactions that leads to decreases in cyclic GMP in the light has been worked out (Figure 7.16). Upon the absorption of a quantum of light, the visual pigment becomes activated. In the next chapter, I examine the chemistry of the visual pigment molecules and how they are excited by light. For now, it is sufficient to note that photoactive visual pigment interacts with a G-protein, called transducin. A GTP molecule binds to transducin and activates it. Activated transducin now interacts with another enzyme, called phos-

7.15   Current traces reflecting the opening (downward deflections) of channels, by cyclic GMP, in the outer segment membrane.

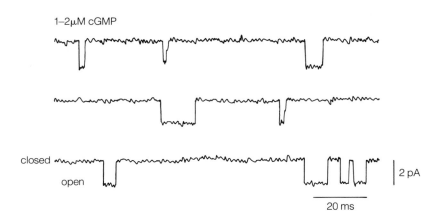

phodiesterase (PDE), which breaks down cyclic GMP to an inactive form. PDE, once activated by transducin, rapidly reduces cyclic GMP concentration in the vicinity of the absorbed light quantum, and nearby Na^+ channels in the outer segment membrane close. With the channels closed, Na^+ can no longer enter the cell and the cell hyperpolarizes.

Why this elaborate biochemical cascade? An important reason is amplification. One activated visual pigment molecule activates about 500 transducins and each transducin about 100 phosphodiesterase molecules. A single phosphodiesterase molecule can break down 2,000 cyclic GMP molecules. In other words, one visual pigment molecule can break down up to 10^6–10^8 cGMP molecules. Thus, the absorption of a single quantum leads to the closure of many Na^+ channels and a significant reduction in Na^+ current across the cell's membrane.

Invertebrate photoreceptors also respond to single quanta. Is there a cascade of biochemical steps in these photoreceptors between the absorption of light by visual pigment and the opening

7.16 The cyclic GMP cascade in the photo-receptor outer segment. Light-activated rhodopsin *(Rh*)* activates transducin *(T)* a G-protein, which in turn activates the enzyme phosphodiesterase *(PDE)* Phosphodiesterase breaks down cyclic GMP to an inactive product *(GMP)*; in the absence of cGMP, which opens channels in the outer segment membrane, the channels close and the cell hyperpolarizes. Also, Ca^{2+} levels in the cell decrease, allowing guanylate cyclase *(GC)*, which is normally inhibited by the Ca^{2+}, to synthesize cyclic GMP from GTP. With more cGMP available, more channels open; this partially counters the effects of light. Ca^{2+} thus plays a role in photoreceptor adaptation.

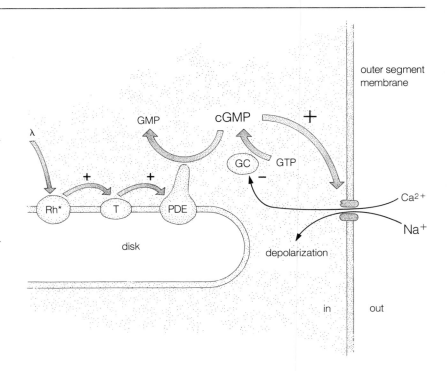

of membrane channels? From indirect evidence we think this is the case, but the second messengers have not yet been identified unequivocally. Some workers believe the cascade involves inositol triphosphate, others cyclic GMP. In olfactory receptors, cyclic AMP is produced when the cells are stimulated, and the cyclic AMP directly opens channels in the membrane, resulting in depolarization and excitation of the cells.

Adaptation

Vertebrate photoreceptors adapt; that is, the photoresponse decays from an initial peak to a sustained plateau level in response to a step of light. As with invertebrate photoreceptors, Ca^{2+} appears instrumental (see Figure 7.16). In the dark, the channels in the outer segment membrane also admit Ca^{2+}. Indeed, these channels are more permeable to Ca^{2+} than to Na^+. But since there are many more Na^+ ions in the extracellular fluid than Ca^{2+} ions, many more Na^+ ions move through the open channels than do Ca^{2+} ions; this is why channels are commonly referred to as Na^+ channels. Na^+ is the major charge carrier passing through the channels and is most responsible for depolarizing the photoreceptor in the dark. However, Ca^{2+} appears to have an important function in regulating certain enzymes in the cyclic GMP cascade.

Cyclic GMP is synthesized by an enzyme, guanylate cyclase, that converts GTP to cyclic GMP. Ca^{2+} inhibits this enzyme. In the dark, when Ca^{2+} is entering the cell, the synthesis of cyclic GMP is depressed. In the light, the influx of Ca^{2+} ceases as the outer segment membrane channels close and intracellular Ca^{2+} levels fall. Guanylate cyclase is then activated and cyclic GMP levels increase in the outer segment, countering the effects of light. With an increase in cyclic GMP levels, channels reopen and the cell partially depolarizes. By preventing Ca^{2+} levels from falling in the outer segment during illumination (by blocking Ca^{2+} pumps), experimenters have demonstrated that the photoreceptor response does not decay in steady light, thereby providing evidence that Ca^{2+} is important in photoreceptor adaptation.

Summary

A variety of sensory receptors transduce specific stimuli into graded receptor potentials. Some stimuli induce direct conductance changes in the receptor cell membrane, similar to the action of neurotransmitters on postsynaptic membranes. Stimuli to other receptor cells activate enzyme cascades and the production of second messengers, like the action of neuromodulators on neurons.

The Pacinian corpuscle is a mechanoreceptor whose endings respond to pressure. Minute deformations of the capsule surrounding the nerve ending lead to the generation of graded, depolarizing receptor potentials. With sufficient depolarization, action potentials are generated at the first node in adjacent myelinated membrane and the sensory message is transmitted along the cell. Specialized channels in the membrane of the nerve ending open upon membrane deformation and allow Na^+ and K^+ to cross the membrane. This receptor potential is thus similar to an EPSP.

Receptor potentials are not maintained during prolonged stimuli. Rather, the potential typically declines or adapts. The Pacinian corpuscle is a fast-adapting receptor; it gives a transient response at the onset and at the cessation of a long stimulus. Other receptors are medium- or slow-adapting. Photoreceptors are examples of the latter. They continue to respond for as long as the stimulus is maintained, although the initial response to a stimulus is larger than the maintained response. Adaptation can involve accessory structures related to the receptor or be intrinsic to the receptor. The Pacinian corpuscle is an example of the former; stripping away the capsule surrounding the receptor nerve ending transforms the receptor response into one that more slowly adapts, and it abolishes the off-response. Photoreceptor responses adapt because of intrinsic receptor cell mechanisms.

Invertebrate photoreceptors respond to light with graded depolarizing potentials of up to 75 mV. Absorption of light by visual pigment molecules found in specialized membrane leads to the influx of Na^+ into the receptor cell. In weakly illuminated dark-adapted receptors, small depolarizing responses to single quanta are detected. In brighter light the quantal responses, or bumps, summate, giving rise to the receptor potential. The quantal bumps of dark-adapted receptors are larger than those recorded from receptors recently exposed to light, which explains the decline (adaptation) of the receptor potential in continuous light. The Ca^{2+} ions that enter the cell in the light through the Na^+ channels appear to regulate the efficiency of the Na^+ channels. When intracellular Ca^{2+} levels are increased, fewer Na^+ ions enter the cell per quantum of light caught.

Vertebrate photoreceptors are depolarized in the dark and they hyperpolarize in response to illumination. Channels in the cell's membrane remain open in the dark, admitting Na^+ and causing cell depolarization. Light shuts down the channels, preventing Na^+ from entering the cell, and the cell hyperpolarizes. In other words, current flows into the vertebrate photoreceptor in the dark and this dark current is turned off by light. The light-sensitive channels in the cell's outer membrane are directly opened by cyclic GMP; no phosphorylation takes place. A cascade of biochemical reactions between activated visual pigment molecules and cyclic GMP lead to the lowering of cyclic GMP levels in the light and the closing of the light-sensitive channels. Light-activated visual pigment interacts with a G-protein (called transducin) that activates phosphodiesterase, an enzyme that breaks down cyclic GMP.

Ca^{2+} appears to be necessary for adaptation in vertebrate photoreceptors. In the light, Ca^{2+} levels in the cell fall, allowing increased synthesis of cyclic GMP. This effect counters the action of light on the cell and permits the membrane potential to recover partially.

Further Reading

BOOKS

Fein, A., and Szuts, E. Z. 1982. *Photoreceptors: Their Role in Vision.* Cambridge: Cambridge University Press.

ARTICLES

Fesensko, E. F., S. S. Kolesnikov, and A. L. Lyubarsky. 1985. Induction by cyclic GMP of cationic conductance in plasma membrane of retinal rod outer segments. *Nature* 313:310–313.

Hagins, W. A., R. D. Penn, and S. Yoshikami. 1970. Dark current and photocurrent in retinal rods. *Biophysical Journal* 10:380–412.

Hudspeth, A. J. 1985. The cellular basis of hearing: The biophysics of hair cells. *Science* 230:745–752.

———— 1989. How the ear's works work. *Nature* 341:397–404

Loewenstein, W. R. 1960. Biological transducers. *Scientific American* 203(2):98–111.

Loewenstein, W. R., and M. Mendelson. 1965. Components of receptor adaptation in a Pacinian corpuscle. *Journal of Physiology* 177:377–397.

Miller, W. H., F. Ratliff, and H. K. Hartline. 1961. How cells receive stimuli. *Scientific American* 205(3):222–238.

Nakamura, T., and G. H. Gold. 1987. A cyclic nucleotide-gated conductance in olfactory receptor cilia. *Nature* 325:442–444.

Schnapf, J. L., and D. A. Baylor. 1987. How photoreceptors respond to light. *Scientific American* 256:40–47.

Stryer, L. 1986. Cyclic GMP cascade of vision. *Annual Review of Neuroscience* 9:87–119.

von Bekesy, G. 1957. The ear. *Scientific American* 197(2):66–78.

Membrane Channels and Receptor Proteins

ALL INFORMATION passes from one neuron to another via membrane channels or receptors. Needless to say, neuroscientists are intensely interested in learning more about these membrane proteins, how they function, and how stimuli modify them. Because they are highly insoluble, they have been difficult to study: they reside mainly in the hydrophobic domain of the cell membrane and cannot be easily extracted and dissolved in aqueous solutions. Solubilizing agents may be used to extract membrane proteins, but they often alter or denature the molecules. It is now possible to isolate the genes for many of the proteins, deduce their amino acid structure, synthesize and alter them, and incorporate them into cells and even artificial membranes. Over the next few years we will undoubtedly see spectacular progress in our knowledge of membrane proteins.

Already substantial progress has been made in determining the general structure of many receptors and channels and how they fit into the membrane. We know much less about how they are altered or excited by stimuli. One class of receptor proteins, the light-sensitive visual pigments, has been known for over 100 years. We understand some details of how this molecule absorbs light and how photon capture leads to changes in the molecule. The story of how we arrived at this understanding is an interesting one, and so the visual pigments will be our model of a membrane receptor. To begin, though, I will discuss membrane channels. I will focus first on the acetylcholine channel, for it has been extensively studied, and then describe one type of voltage-sensitive K^+ channel.

The Acetylcholine-Gated Channel

As noted in Chapter 3, there are at least four distinct classes of membrane channels—voltage-sensitive, ligand-sensitive, mechanosensitive, and gap-junctional—in addition to the nongated (and not well characterized) leak channels. Several types of channels are in each of these classes, but each member of a class has a strong homology to other members of the class. In other words, each class type is a member of one gene family and shares common amino acid sequences and other structural similarities. So, for example, voltage-sensitive Na^+ and Ca^{2+} channels are made up of subunits that are approximately the same size and are significantly homologous with regard to amino acid composition and sequence. The same holds true for ligand-sensitive channels, whether they are activated by acetylcholine, glycine, or GABA; all three are made up of subunits of similar size and have similar amino acid sequences. In comparison, the subunits of the voltage-gated channels are distinctly different from the subunits of the ligand-gated channels. For example, the size of the major subunit of the voltage-gated channels is about four times the size of the subunits of the ligand-gated channels.

We know most about the structure of the channel that is sensitive to acetylcholine—the nicotinic ACh channel. A large membrane protein with a molecular weight of 270,000 daltons, it consists of 4 polypeptide subunits, each coded by a different gene. One of the subunits, termed α, binds ACh. Each ACh channel has 2 α subunits and 1 each of the β, γ, and δ subunits ranging in size from 50,000 to 60,000 daltons and traversing the membrane 4 times. Membrane-spanning regions of proteins are typically coiled in a helical structure; hence the wavy depiction of these regions in Figure 8.1A.

Three of the membrane-spanning regions are hydrophobic while the fourth is hydrophobic on one side and hydrophilic on the other. This suggests that one of the membrane-spanning regions from each subunit contributes to a water-filled pore through which ions can pass. The subunits are arranged circularly around the central pore in the sequence α, γ, δ, α, β (Figure 8.1B).

The ACh channel's shape has been visualized by electron microscopic and X-ray diffraction studies of membranes containing the purified channel. Surprisingly, the molecule extends well out of the membrane on the extracellular side; the channel has a length of about 11 nm, approximately 4 nm of which juts outside the cell. The molecule also flares out on the membrane's extracellular side and has a maximal width there of about 8.5 nm. A central pit of

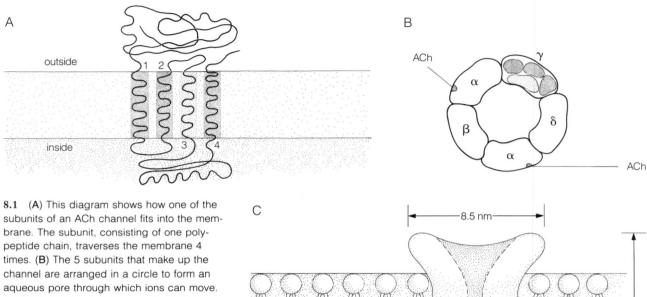

8.1 (A) This diagram shows how one of the subunits of an ACh channel fits into the membrane. The subunit, consisting of one polypeptide chain, traverses the membrane 4 times. (B) The 5 subunits that make up the channel are arranged in a circle to form an aqueous pore through which ions can move. Acetylcholine binds to the α subunits. How the membrane-spanning regions are arranged in a subunit is indicated in the γ subunit. (C) Overall shape and structure of the acetylcholine channel. It is 11 nm long and up to 8.5 nm wide on the extracellular side. The central pore is constricted at the level of the two phospholipid head groups of the membrane's bilayer.

2 nm diameter in the molecule's extracellular portion leads to the pore that traverses the membrane. The pore appears to be constricted sharply in diameter twice, at the level of the phospholipid head groups in the membrane bilayer. Hence, the channel's ionic specificity may be determined at these regions, the narrowest points in the channel. The ACh channel's structure is illustrated in Figure 8.1C.

With techniques now available, the structure of the subunits can be altered down to the level of single amino acids—that is, one amino acid can be substituted for another. It is hoped that these experimental approaches will help us uncover the relations between structure and function. For example, which amino acids are essential for ACh binding, and which for ion permeation? Eventually, we want to know what happens to the α subunit when ACh binds to it, and how binding leads to the opening of the channel so that Na^+ and K^+ can pass through the aqueous pore.

Shaker Flies and Potassium Channels

Less is known about the overall structure of the other classes of membrane channels. Voltage-gated Na^+ channels have a molecular weight of about 340,000 daltons, and they consist of one principal subunit of molecular weight 200,000–260,000 daltons and two or three smaller subunits of between 30,000 and 40,000 daltons. Gap-junctional channels are made up of 12 identical subunits of about 28,000 daltons; the total molecular weight of the gap-junctional complex thus being about 340,000 daltons.

A genetic mutant of the fruitfly, *Drosophila melanogaster*, has been used for an extensive analysis of one kind of potassium channel, and this analysis has yielded important insights concerning structural-functional relationships of voltage-gated channels. The mutant yielding this information is called the *Shaker* fly because it violently shakes its legs, wings, and abdomen when exposed to an anesthetic. (To examine flies in the laboratory for possible mutations, investigators routinely anesthetize them; hence the discovery of the *Shaker* mutation.) Neurophysiological examination of *Shaker* flies showed that some of their neurons and muscle cells generate abnormal action potentials; they are greatly prolonged compared with action potentials produced in normal, wild-type flies.

The site of the mutation was localized to one of the 4 chromosomes of *Drosophila*, and after much work the *Shaker* gene was isolated by Lily and Yuh Jan and their colleagues at the University of California, San Francisco. The *Shaker* gene codes for a voltage-dependent K^+ channel, a somewhat different channel from the one found in the squid giant axon. This channel (termed the A-type K^+ channel) activates relatively rapidly and it also inactivates (see Chapter 4). Examples of the important roles that inactivating K^+ channels play in neural function are given in Chapters 9 and 10.

A particularly interesting feature of the *Shaker* gene is that it can give rise to somewhat different proteins. This occurs because the *Shaker* gene, like other genes in higher organisms, contains both coding regions (exons) and noncoding regions (introns) (see Chapter 1). During the production of messenger RNA (mRNA) leading to protein synthesis, the coding regions are spliced together whereas the noncoding regions are eliminated. At the *Shaker* gene locus, the splicing can vary and therefore produce somewhat different mRNAs. As a result, the proteins synthesized have the same central region but different ends. All of the proteins have a molecular weight of about 70,000 daltons, but when channels containing the different proteins are expressed—as when messenger RNA is injected into oocytes, for example—the electrical properties of the channels differ somewhat. All give rise to transient K^+ currents, but the rates of activation and inactivation are different, or the time it takes to recover from inactivation varies.

The construction of a functional channel requires 4 of the *Shaker* proteins, but whether the normal channel complex is made up of 4 identical proteins or a mix of the *Shaker* proteins is not known. There is a close structural homology between the *Shaker* proteins and some parts of the Na^+ and Ca^{2+} channels. These parts of the Na^+ and Ca^{2+} channels are believed to contain the voltage-sensing regions of the channels and to form the channel pores. Thus, it has been postulated that the A-type potassium channel may represent the ancestral voltage-sensitive ion channel; all voltage-sensitive channels may be derived from this simpler, smaller protein. Since the principal subunit of the Na^+ channel can form a functional channel by itself and is about 4 times larger than a *Shaker* protein, investigators suppose that the principal subunit of the Na^+ channel is made up of 4 *Shaker*-like proteins.

The *Shaker* proteins have about 650 amino acids and consist of 6 coiled and hydrophobic, membrane-spanning regions, termed S1–S6 (Figure 8.2A). One of these regions, the S4 region, contains positively charged amino acids mixed with hydrophobic amino acids. It has been postulated that this region of the molecule serves as a voltage-sensor. When voltage across the membrane changes, this region of the protein may move, resulting in channel opening. (Presumably, gating currents result from the movement of this region of the protein; see Chapter 4.) Recent evidence, based on the binding of channel-blocking agents to specific regions of the

8.2 **(A)** A diagram of a *Shaker* protein, its functional parts, and how it fits into the membrane. The transmembrane regions (S1–S6) are coiled in a tight helical fashion. **(B)** Single-channel recordings from membranes of oocytes injected with messenger RNA. The records on the left are from an oocyte injected with normal *Shaker* mRNA; the one in the center from an oocyte injected with mRNA coding for a *Shaker* protein that lacks 41 amino acids on its amino end. On the right is a recording from a membrane of an oocyte injected with mutant mRNA but to which was added a synthetic peptide identical in sequence to the ball part of the protein. **(C)** A model of the *Shaker* channel. Upon depolarization of the membrane, the activation gate opens, allowing K⁺ to pass through the membrane. Shortly thereafter, the channel is inactivated by the ball obstructing the channel's pore.

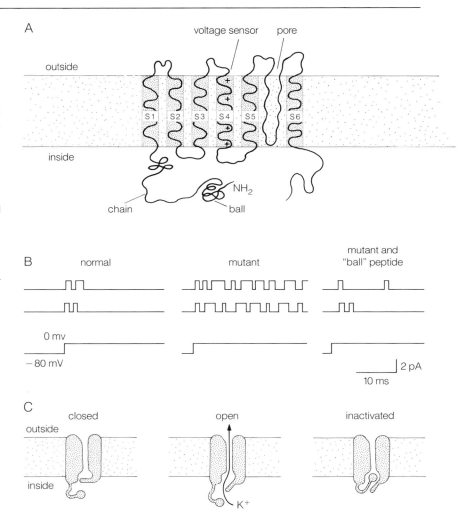

Shaker proteins, suggests that the pore of the channel is formed by a region of noncoiled protein between S5 and S6 that dips into the membrane, as shown in Figure 8.2A.

One of the most interesting results obtained from the analysis of the *Shaker* proteins concerns inactivation of the channel. As is the case for the Na^+ channel, inactivation of the *Shaker* channel can be eliminated by treating the channel with an enzyme that partially digests protein. This enzyme must be applied to the cytoplasmic side of the channel, suggesting that a part of the protein on the inside of the membrane is critical for inactivation. By preparing mutants of the *Shaker* gene that add or subtract amino acids from the amino (NH_2) end of the protein, investigators have significantly modified or entirely eliminated inactivation. For example, deletion of the first 20 amino acids from the amino end of the molecule prevents inactivation. These amino acids are thought to form a ball-like structure that physically blocks the channel's pore. The ball is connected to the rest of the protein by about 20 additional amino acids, a chain. Deletion of amino acids in the chain region speeds up inactivation, presumably by shortening the chain; whereas addition of amino acids slows inactivation, by lengthening the chain.

Figure 8.2B shows the effects of deleting amino acids from the amino end of the *Shaker* protein on single-channel records obtained from membranes of oocytes injected with messenger RNA. When oocytes were injected with normal mRNA, the recorded channels rapidly inactivated; they opened only once or twice before inactivation. When oocytes were injected with mutant mRNA coding for a protein that lacked the first 41 amino acids on the amino end of the molecule, the channels continuously opened and closed; they did not inactivate. Evidence that the deleted part of the protein is responsible for inactivation was provided by adding to the inside of the membrane a 20-amino-acid synthetic peptide whose amino acid sequence was identical to the inactivation ball peptide; this restored inactivation.

These results suggest that there is a binding site in the channel's pore for the inactivating ball. They also suggest that the affinity of the binding site for the pore increases once the channel is activated. That is, as shown in Figure 8.2B, inactivation occurs generally only after activation. A simplified model of the *Shaker* channel is shown in Figure 8.2C. Three states of the channel exist—closed, open, and inactivated. In the closed state, the activation gate is closed and the inactivation gate is open. In the open state both the activation and inactivation gates are open. In the inactivated state, the activation gate may be open or closed, but the inactivation gate is closed; the ball obstructs the channel's pore.

Visual Pigments and Receptor Proteins

Just as there is homology among the members of a class of membrane channels, strong homology exists among the different membrane receptor proteins that interact with G-proteins. Thus muscarinic ACh receptors, norepinephrine (β-adrenergic) receptors, and visual pigment molecules all are closely related and from the same gene family. The visual pigments will be our model for G-protein–linked receptor proteins.

As pointed out in Chapter 7, visual pigment molecules are highly concentrated in the outer segment region of the photoreceptor. The most extensively studied photoreceptors are rods, and rhodopsin, the rod visual pigment, is packed in the disk membranes of the rod outer segment at a density of 50,000 molecules per μm^2. In contrast, there are about 10,000 ACh channels per μm^2 at the neuromuscular junction and about 12,000 voltage-gated Na^+ channels per μm^2 at nodes of Ranvier in rabbit axons. A single outer segment of a rod photoreceptor contains between 10^8–10^9 molecules of rhodopsin, and 80–90 percent of the protein in outer segment membranes is rhodopsin.

Rhodopsin is a transmembrane protein, approximately 8 nm long and 3 nm wide. It consists of one polypeptide chain of 348 amino acids and has a molecular weight of about 35,000 daltons. Figure 8.3A shows its general shape and how it fits into the membrane of the disks inside the rod cells. The protein extends from the membrane further on the cytoplasmic side, where it presumably interacts with G-proteins (transducin) in the rod. The peptide chain crosses the membrane seven times (Figure 8.3B), as do all G-protein–related receptor proteins. Figure 8.3C shows how the transmembrane segments are clustered together.

The genes for visual pigments have been isolated; as expected, all show a strong homology. Later I discuss color vision in humans and the homology between the rod pigment rhodopsin and the three cone visual pigments of the human retina. A strong homology also holds between the visual pigments of different species and even between vertebrate and invertebrate visual pigments. For

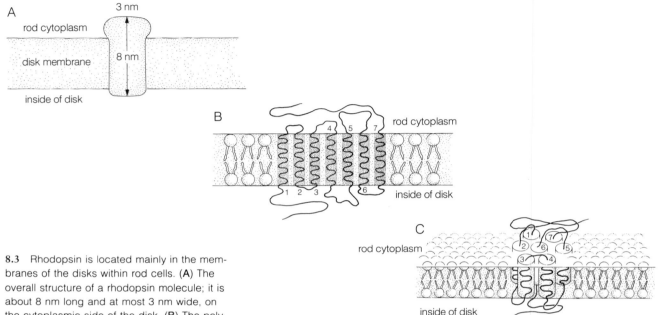

8.3 Rhodopsin is located mainly in the membranes of the disks within rod cells. (A) The overall structure of a rhodopsin molecule; it is about 8 nm long and at most 3 nm wide, on the cytoplasmic side of the disk. (B) The polypeptide chain of rhodopsin traverses the membrane seven times, but the transmembrane segments are not linearly arrayed as in this diagram. (C) Rather, the segments are clustered together in a compact group.

example, cow rhodopsin and the principal visual pigment found in the fruit fly, drosophila, are about 50 percent homologous.

Discovery of the Visual Pigments

Visual pigment molecules absorb light and hence have color. This led to their discovery over a century ago. Franz Boll, a German physiologist, noted that retinas containing many rods, such as a frog retina, had a reddish-purple color in the dark-adapted state but that the color gradually faded in the light. Boll proposed that the pigment giving rise to the color in the dark-adapted retina was the light-sensitive pigment initiating vision.

F. Wilhelm Kühne, a German biochemist known for his discovery of myosin, the major protein of muscle, seized upon and quickly extended Boll's observations (too quickly in the view of some scientists; the work generated a bitter debate). Kühne and his students studied the color changes taking place in rod-dominated retinas when they were exposed to light, and from these observations they proposed a visual cycle, the basic outlines of which are still valid today.

The pigment in the dark-adapted red-dominated retina was called visual purple (now *rhodopsin*). When exposed to light, visual purple bleaches rapidly to a yellow intermediate, termed visual yellow. With time, the retina's yellow color disappears and it becomes entirely colorless—the final bleached stage, called visual white. Kühne and his students also determined that light was required for visual purple to go to visual yellow; this is a photochemical reaction. But they observed that retinas in the visual yellow state could go to visual white either in the light or in the dark; it appears to be an ordinary chemical reaction. They further found that some visual purple regenerates if a fresh retina is promptly put back into the dark after it has reached the visual yellow state, but regeneration usually will not take place after the visual white stage of bleaching has been reached. Kühne's visual cycle, presented in the 1870s, is shown in Figure 8.4.

Vision and Vitamin A

Early in the twentieth century vitamin A was implicated in the visual process. Indeed, the ancient Egyptians knew that a dietary factor was needed to maintain vision, but the nature of this factor was unknown until the discovery of the vitamins. By the 1920s, vitamin A deficiency and night blindness (loss of visual sensitivity in dim light) had been linked and vitamin A found in the retina.

What does vitamin A have to do with vision, and how does it connect with the visual pigments and Kühne's visual cycle? George Wald, an American, carried out the critical experiment in the early 1930s while a postdoctoral student in Germany. Wald tested dark-adapted retinas, and retinas in the visual yellow and visual white states, for the presence of vitamin A and compounds similar to vitamin A. He began using mild organic solvents that extract free vitamin A and found abundant vitamin A in retinas in the visual white state of bleaching. In visual yellow retinas he discovered no vitamin A but, instead, a new substance he called retinene (now called *retinal*). Retinal has many properties similar to vitamin A but it is distinct from vitamin A; for example, it is yellow in color whereas vitamin A is colorless. When Wald treated dark-adapted retinas with mild organic solvents, he detected neither vitamin A nor retinal, but if he used stronger solvents—ones that denature

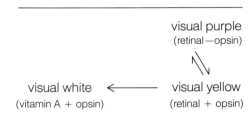

visual purple
(retinal—opsin)

visual white   **visual yellow**
(vitamin A + opsin)   (retinal + opsin)

8.4 Kühne's visual cycle. Rod visual pigment molecules consist of retinal (a form of vitamin A) bound to opsin (a protein) and appear reddish-purple in the dark-adapted state. Visual purple bleaches in the light to visual yellow; during the bleaching process, retinal separates from opsin in a photochemical reaction. Then the retinal is converted to vitamin A by a simple chemical reaction and the pigment reaches the visual white stage.

proteins and extract lipids bound to proteins—he found retinal in his extracts.

Having made these observations, Wald proposed in the mid-1930s that visual pigment molecules such as rhodopsin consist of two components, retinal and a protein called *opsin*, which are bound together in the dark-adapted state. Light releases retinal from the protein opsin and bleaches visual purple to visual yellow, Wald concluded, and retinal is converted chemically to vitamin A. When the retinal is gone, all color is lost in the retina and the visual white stage of bleaching is reached, as indicated in Figure 8.4. For this seminal discovery and subsequent work on the chemistry of the visual pigments, Wald was awarded the Nobel Prize in 1967.

Unknown at the time of Wald's experiments was the relation of vitamin A to retinal, but in the 1940s R. A. Morton in Liverpool showed that retinal is a slightly oxidized form of vitamin A (see Figure 8.5). In the late 1930s the structure of vitamin A had been worked out; it consists of a ring at one end of the molecule, a long side-chain, and a terminal hydroxyl group. Morton reasoned that the terminal hydroxyl group was most vulnerable to oxidation and would likely be converted to an aldehyde under mild oxidizing conditions. He hypothesized, therefore, that retinal was vitamin A aldehyde—which turned out to be correct.

Shortly after, Ruth Hubbard, a graduate student of Wald's at Harvard, identified the enzymes that convert vitamin A to retinal, making it possible to assemble the components in a test tube and attempt to synthesize visual pigment in the laboratory. Vitamin A extracted from fish liver, the appropriate enzymes to convert vitamin A to retinal, and opsin from a frog retina were incubated together in the dark. Some light-sensitive pigment that had many properties of visual pigments was formed, but only a small amount, and its color was not exactly that of native rhodopsin. Nevertheless, the experiment showed that visual pigment–like molecules would form spontaneously in the dark in the presence of retinal and opsin.

These experiments led indirectly to the next important advance in our understanding of the visual pigment molecules. A Swiss organic chemist, Otto Isler of the Hoffman-LaRoche Company, had achieved a synthesis of vitamin A that was commercially feasible in the late 1940s. Earlier workers had synthesized vitamin

8.5 The structure of vitamin A, all-*trans* retinal, and 11-*cis* retinal. At every corner of the molecules is a carbon atom with associated hydrogen atoms.

A, but the syntheses were difficult and costly. Isler knew of the work of Wald, Hubbard, and their co-workers on the synthesis of visual pigment in the test tube, and he sent them some of his synthetic vitamin A for testing. Much to everyone's surprise, Isler's synthetic vitamin A did not work; no pigment was formed!

Why didn't Isler's vitamin A work? A clue lay in the structure of vitamin A. The long side-chain of vitamin A consists of carbon atoms linked by alternating single and double bonds. When two carbon atoms in a molecule are linked by a double bond, the molecule can exist in two isomeric forms. Even though the atoms are identical in the two isomers, the molecules have distinct shapes and are called *cis-trans* isomers. Because several double bonds link the carbon atoms in the side-chain of vitamin A, several *cis-trans* isomers are possible. *Trans*-isomers are more balanced and stabler molecules; hence Isler had made the all-*trans* configuration. This means that the molecule's side-chain is straight, as in Figure 8.5. When a double bond is in the *cis* configuration, the molecule bends and is less stable.

These considerations immediately suggested that synthesis of visual pigment requires a certain *cis*-isomer of retinal. But which one? Since four double bonds are in the vitamin A side-chain, many isomers are theoretically possible—not all of them likely, however. Some *cis*-isomers bend the molecule in such a way that one part of the molecule interferes with another part and makes a twist. Called stearic hindrance, this twist makes the molecule even more unstable. The obvious *cis*-isomers were prepared and tested, but none of them worked. A fifth one, thought unlikely to exist because it is bent and slightly twisted (it has some stearic hindrance), has turned out to be the precursor for all visual pigments known: the 11-*cis* isomer, pictured in Figure 8.5.

With 11-*cis* retinal, it is possible to synthesize rhodopsin molecules identical to the native pigment, as proven in an experiment carried out by Paul Brown, a co-worker of Wald at Harvard (see Figure 8.6). Opsin, obtained from cow retinas, was incubated in the dark with 11-*cis* retinal and absorption spectra readings were taken periodically to monitor the synthesis of the visual pigment. Rhodopsin absorbs light maximally at 500 nm wavelength, whereas retinal absorbs maximally at 380 nm. After just 5 minutes of incubation, the spectrum exhibited little absorption at 500 nm and strong absorption at 380 nm; that is, little synthesis of rho-

8.6   Absorption spectra—records of the light absorption properties of molecules—determined during the synthesis of rhodopsin from 11-*cis* retinal and opsin *(left)* and the bleaching of rhodopsin by light *(right)*. Opsin was incubated in the dark with 11-*cis* retinal and as rhodopsin (which absorbs radiation maximally at 500 nm wavelength) formed, absorbance at 380 nm, the peak absorption level for retinal, fell. The newly synthesized rhodopsin was then exposed to light, bleaching it to all-*trans* retinal and opsin. The absorbance at 500 nm decreased and that at 380 nm increased.

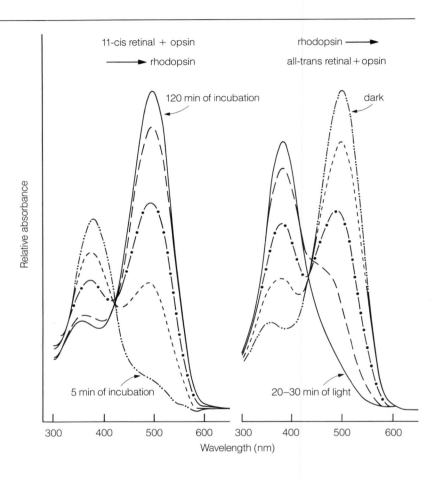

dopsin had occurred as yet. With time, as rhodopsin formed, absorption at 500 nm increased while absorption at 380 nm decreased. After 120 minutes of incubation, the 380 nm peak was essentially gone and absorption at 500 nm was maximal; in other words, synthesis was complete.

The right side of Figure 8.6 shows the absorption spectra when the newly synthesized rhodopsin was bleached by light. As bleaching progressed, the absorption peak at 500 nm decreased while that at 380 nm increased. After 20–30 minutes of bleaching, the 500 nm peak was gone, replaced by strong absorption at 380 nm. Hence, retinal had been released from the protein.

8.7 The rhodopsin cycle. Rhodopsin bleaches in the light to opsin and all-*trans* retinal. All-*trans* retinal is converted to all-*trans* vitamin A, which is then isomerized via an enzyme (isomerase) to 11-*cis* vitamin A. The 11-*cis* form is converted to 11-*cis* retinal, which spontaneously combines with opsin to form rhodopsin.

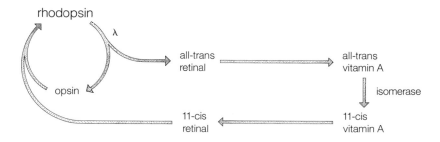

A most important feature of bleaching and of the visual cycle is also illustrated in Figure 8.6. The peak at 380 nm after bleaching is significantly higher than it was before bleaching. This is because the retinal released upon bleaching is in the all-*trans* form, whereas 11-*cis* retinal is required for the synthesis of rhodopsin. In other words, light induces an isomerization of retinal, and it is this change in the shape of retinal that initiates bleaching and releases retinal from opsin. For a summary of the visual cycle in biochemical terms, elucidated by the mid-1950s, see Figure 8.7.

The Action of Light and Visual Pigment Intermediates

Once the biochemical makeup of the visual pigments was understood, the focus of research turned to the action of light on these molecules. This research soon indicated that all light does to initiate vision is to isomerize retinal. This causes changes in the protein, opsin, which eventually activates transducin, the visual pigment–linked G-protein.

What is the nature of these changes in opsin and how are they studied? By cooling extracts of visual pigment to various temperatures and by studying carefully the absorption changes following the presentation of a flash of light to the extract, Wald, Hubbard, and other investigators identified a series of intermediate steps that occur between the absorption of a photon of light by rhodopsin and the release of retinal from opsin. These intermediates suggest that conformational or shape changes take place in the opsin molecule following the isomerization of retinal. For example, following bleaching of rhodopsin, there is less coiling of the opsin peptide chain, two sulfhydryl groups are exposed that are not accessible to reagents in rhodopsin, and a proton is taken up by

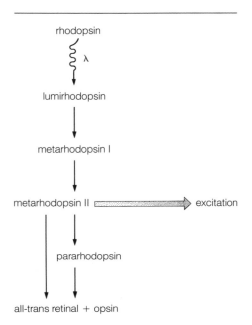

8.8 Intermediates in the bleaching of rhodopsin. Light converts rhodopsin to lumirhodopsin, which decays rapidly to metarhodopsin I and metarhodopsin II. Metarhodopsin II is photoactive rhodopsin; it activates transducin (a G-protein). From metarhodopsin II, a fourth intermediate may be formed (pararhodopsin) or metarhodopsin II may break down directly to retinal and opsin.

the molecule. In extracts, intermediates can be stabilized at one temperature or another and studied at leisure. At least six intermediates have been identified in extracts, and three of these intermediates have been observed in the intact functioning retina.

Figure 8.8 presents a partial sequence of events proceeding from the absorption of a quantum of light by rhodopsin to the release of retinal from opsin. Within a fraction of a microsecond after the isomerization of retinal, an intermediate, lumirhodopsin, is formed. Lumirhodopsin is converted to metarhodopsin I within just a few microseconds, and this intermediate decays to metarhodopsin II over the next few hundred microseconds. Metarhodopsin II is much longer lived, with a half-life of up to a few minutes at room temperature. Thereafter the sequence of events is less clear; at least some metarhodopsin II is converted to yet another intermediate, pararhodopsin, but some can break down directly to retinal and opsin. Eventually all the retinal is released from the protein.

The key intermediate in the excitation process is metarhodopsin II; that is, metarhodopsin II is the molecule that activates transducin and triggers the excitation of the visual cell. Metarhodopsin II can interact with many transducin molecules, perhaps as many as 500. Thus, it contributes substantially to the amplification of the visual signal. Metarhodopsin II can be inactivated by phosphorylation; after a short time, a phosphate group is added to metarhodopsin II and prevents it from further interaction with transducin.

The picture derived from these and other studies is that the activation of a receptor protein like rhodopsin, or a channel like the ACh channel, induces conformational or shape changes in the molecule. These shape changes enable the molecule to interact with other molecules (G-proteins), or they allow ions to pass through the molecule. The shape changes have a time dependence—they are not instantaneous nor do they last permanently. Furthermore, the receptor's activity or the channel's conductance can be modified by phosphorylation.

Color Vision

In most retinas rods are much more common than cones. The human retina, for example, has about 120 million rods but only 6 million cones. Furthermore, rods in many species have larger outer segments, so we know much more about the rod pigment rhodopsin than about the cone pigments. It is also the case that not all vertebrate animals have good color vision. Teleost (bony) fishes, birds, and reptiles have good color vision, but other vertebrates, including most mammals, appear to have poor color vision. Among the mammals, most primates, such as ourselves, have good color vision. Our retina has three types of cone cells, each with a different visual pigment: one pigment absorbs maximally in the blue region of the spectrum (~430 nm); one in the green (~530 nm); and one in the yellow (~565 nm) (see Figure 8.9). This latter pigment, although absorbing maximally in the yellow region of the spectrum, mediates red-sensitive vision and is commonly referred to as the red-sensitive cone visual pigment.

The biochemistry of the cone visual pigments is virtually identical to that of the rod pigment rhodopsin. Cone pigments consist of 11-*cis* retinal bound to cone opsins. Upon bleaching, they go through a similar series of intermediates and release all-*trans* retinal. Furthermore, they activate transducin (G-protein) molecules associated with the disk membranes of the cone outer segments. The differences between rhodopsin and the cone pigments lie in the protein (opsin) part of the molecule.

The genes encoding for the opsins of rhodopsin and the red-, green-, and blue-sensitive cone pigments in humans have been identified and isolated by Jeremy Nathans and David Hogness at Stanford University. Their experiments proved that each opsin is

8.9 The absorption spectra of the four visual pigments in the primate. The cone pigments *(solid lines)* absorb maximally at about 565 nm, 530 nm, and 430 nm. The rod pigment (rhodopsin, *dashed line*) absorbs maximally at 500 nm.

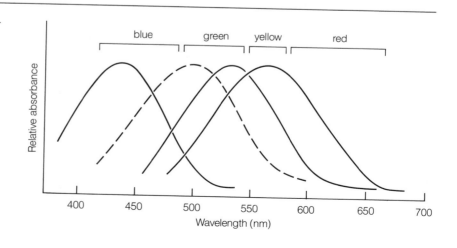

specified by a different gene but that significant homology exists among the opsins. The deduced amino acid sequence of the cone opsins has approximately 40 percent identity with human rod opsin. The red- and green-sensitive pigments, whose genes are on the same chromosome (the sex-related X chromosome), are very similar (95 percent) to one another. The gene for the blue-sensitive pigment is on another chromosome, and it has only about 43 percent identity with the red- and green-sensitive pigments. One guesses that separate red- and green-sensitive pigments developed relatively late in mammalian evolution, whereas the blue-sensitive pigment separated from the red-green pigments much earlier. It is of interest that some mammals, such as the ground squirrel and the cat, do not appear to have red-sensitive cones but only blue- and green-sensitive ones, so the red-sensitive cone is likely to have been the latest to have evolved in mammals.

Color blindness in humans is inherited and has long been thought to be caused by an absence or a modification of one or another of the cone visual pigments. Red-blind individuals are missing or have a modified red pigment; green-blind individuals are lacking or have an altered green pigment; and blue-blind individuals are missing or have a different blue pigment. Because the genes for the red- and green-pigments are on the X chromosome, color blindness is also sex-linked; red- and green-blind individuals are much more likely to be male than female. (Males have just one X chromosome whereas females have two.)

Studies of the color vision pigment genes in red- and green-blind individuals have now shown that color blindness is caused by a loss or alteration of one or another of the genes. Red-blind individuals have an altered gene for the red-sensitive pigment, whereas green-blind individuals have an altered green-sensitive pigment gene or are lacking the gene altogether. When the gene for the red- or green-pigment is altered, usually no red- or green-sensitive cone cells develop in the retina. Sometimes, red- or green-sensitive cones form, but the pigment within the cell, and therefore color vision, is not normal.

Summary

Channel and receptor molecules play a key role in neuronal function. With recent technical advances, our understanding of these membrane proteins is rapidly increasing. At least five distinct classes of membrane channels exist, including leak, voltage-sensitive, ligand-sensitive, mechanosensitive, and gap-junctional channels, and one class of G-protein–related receptor proteins. Several types of channels or receptors fall in each of these classes, but each class appears to be derived from one gene family, and all members of the class show significant homology.

The membrane channel sensitive to acetylcholine is perhaps the best-known channel. It is a large membrane protein (270,000 daltons) consisting of 4 polypeptide subunits of molecular weight between 50,000 and 60,000 daltons that span the membrane. The polypeptide subunits are arranged circularly around a central aqueous pore that forms the channel. On the extracellular side the structure flares out and forms a central pit that leads into a membrane-spanning aqueous pore.

The voltage-sensitive Na^+ channel is even larger—it has a molecular weight of about 340,000 daltons. It consists of one large subunit of 200,000–260,000 daltons and 2 or 3 smaller subunits. Gap-junctional channels are made up of 12 small, identical subunits of molecular weight 28,000 daltons.

Analysis of a genetic mutant of the fruitfly has provided important information on the structural-functional relationships of voltage-gated channels. The mutant, the *Shaker* fly, has defective K^+ channels that cause the fly to shake when anesthetized. The channels, termed A-type K^+ channels, inactivate as Na^+ channels do. The *Shaker* gene codes for proteins of about 70,000 daltons, 4 of which are required to form a channel. By removing amino acids from the amino end of the *Shaker* protein, channel inactivation can be eliminated. The first 20 amino acids at the amino end of the protein form a ball that plugs the channel's pore and inactivates the channel; the next 20 amino acids form a chain connecting the ball with the rest of the protein. Mutants that alter the chain or ball alter or entirely prevent channel inactivation.

The visual pigments are models for G-protein–linked membrane receptors. They have been known for over a century and have been well studied because they are densely packed in the outer segment membranes of the photoreceptor cell and are colored. Rhodopsin, the rod visual pigment, is a transmembrane protein consisting of one polypeptide chain of molecular weight about

35,000 daltons. All visual pigments from all types of photoreceptors and all species show a strong homology with one another.

Rhodopsin and other visual pigments are a complex of retinal (vitamin A aldehyde) bound to a protein called opsin. Light releases retinal from the protein and bleaches the molecule from a reddish-purple color to yellow. When retinal is converted to vitamin A, all color is lost. A particular shape (*cis*-isomer) of retinal is required to synthesize rhodopsin, but when rhodopsin is bleached, another isomer (the all-*trans* form) is released. Thus, the action of light on rhodopsin is to isomerize (alter the shape of) retinal, which initiates conformational or shape changes in the protein. The pigment molecule goes through a series of intermediates before retinal is released. One of these intermediates, metarhodopsin II, interacts with transducin, the G-protein in the photoreceptor cell outer segment, and initiates visual excitation.

Cone pigments have a biochemistry similar to that of rhodopsin. They consist of 11-*cis* retinal bound to cone opsin. Cone and rod pigments differ, therefore, on the basis of their opsins. The genes encoding for rod and cone opsins in the human have been isolated; all show a strong homology, particularly the red- and green-sensitive pigments, which have 95 percent identity. The blue-sensitive pigment and rhodopsin have about 40 percent identity to each other and to the red-green pigments.

In persons with red- or green-blindness, one or another of the cone visual pigments is missing or modified because the genes encoding for the opsins are absent or are altered. Thus, red-blind individuals are missing or have altered red-sensitive pigments, whereas green-blind individuals are missing or have altered green-sensitive pigments.

Further Reading

BOOKS

Molecular Neurobiology. 1983. Cold Spring Harbor Symposium on Quantitative Biology, Vol. 48. Cold Spring Harbor, NY: Cold Spring Harbor Laboratory. (A collection of research articles emphasizing the molecular biology of channels.)

ARTICLES

Catterall, W. A. 1988. Structure and function of voltage-sensitive ion channels. *Science* 242:50–61.

Jan, L. Y., and Y. N. Jan. 1989. Voltage-sensitive ion channels. *Cell* 56:13–25.

MacNichol, E. F., Jr. 1964. Three-pigment color vision. *Scientific American* 211(6):48–57.

Miller, C. 1989. Genetic manipulation of ion channels: A new approach to structure and mechanisms. *Neuron* 2:1195–1205.

Miller, C. 1991. 1990: Annus Mirabilis of potassium channels. *Science* 252:1092–1096.

Nathans, J. 1989. The genes for color vision. *Scientific American* 260(2):42–49.

Numa, S., M. Noda, H. Takahashi, T. Tanabe, M. Toyasato, Y. Furatani, and S. Kikyotani. 1983. Molecular structure of the nicotinic acetylcholine receptor. *Cold Spring Harbor Symposium on Quantitative Biology* 48:57–69.

Wald, G. 1955. The photoreceptor process in vision. *American Journal of Ophthalmology* 40:18–41. Appears in slightly modified form in *Handbook of Physiology*, Sec. 1, *Neurophysiology*, Vol. 1, ed. J. Field, 671–691. Washington, D.C.: American Physiological Society, 1959. (A superb review of Wald's classic studies.)

N E T W O R K S

Integrative Neuroscience

A great deal has been learned about the processing of information by individual nerve cells, but we are only beginning to understand the neurobiological basis of behavior. This is the realm of integrative neuroscience—the study of aggregates, or networks, of neurons and how they produce the range of behaviors we associate with the higher brain functions, from perception and movement initiation to memory and creative thought. Increasing emphasis has been given to integrative mechanisms in recent years. Both invertebrate and vertebrate nervous systems are being analyzed in complementary studies. A particularly intriguing problem is how a brain develops and establishes its complex circuitry.

NINE

Invertebrate Nervous Systems

SO FAR we have considered the function of individual nerve cells: how a neuron receives stimuli at synapses or at specialized sensory receptor sites; how neurons carry information with sustained, graded potentials or with action potentials; and how neurons transmit information at chemical or electrical synapses. Now we are ready to look at how aggregates of nerve cells produce meaningful patterns of neural activity and, ultimately, regulate the behavior of an animal.

Much of present-day neurobiological research concentrates on the relation of neural activity and behavior, but so far neuroscientists are just scratching the surface of this enormous subject. Undoubtedly these studies will occupy neuroscientists for many decades to come, and there are those who hold that we will never understand the behavior of humans in strictly neurobiological terms. Nevertheless, all agree that we have much to learn, and inklings of what can be learned are beginning to appear in the literature.

One big problem is that most animal brains are extraordinarily complex and consist of neurons that differ anatomically from place to place and presumably interact differently in different parts of the brain. How do we approach the problem? How can we break down the complexity so we can begin to understand how brains work? One obvious approach is to study simpler nervous systems or simpler parts of complex brains, and both approaches have been taken.

For simpler nervous systems, workers have turned to invertebrates, whose brains consist of fewer nerve cells than are found in vertebrates. Furthermore, invertebrate neurons are usually larger and more accessible for analysis than those of vertebrates. Invertebrates do not have the behavioral repertoire of vertebrates, but they do exhibit many neural and behavioral phenomena similar to those of higher organisms. In the vertebrate brain, investigators have turned to well-defined and relatively simple systems, such as the visual system. Even for the visual system, though, the analysis is still incomplete; beyond the retina in the eye, our detailed knowledge of the underlying neuronal and synaptic mechanisms is scanty.

In this section of the book, I review some of this work in integrative neuroscience, starting with invertebrate nervous systems. I evaluate the advantages and disadvantages of invertebrates for neurobiological research and present examples of how invertebrate systems can provide insights into behavioral and perceptual phenomena, including how nervous systems enable organisms to learn and perhaps remember things. Later, I provide an overview of the vertebrate brain and describe in some detail what is known about the visual system and how this system has elucidated vertebrate neural mechanisms. Many other examples could have been chosen from what is already a vast literature on integrative neuroscience. The ones selected are those that seem to me particularly seminal.

Nervous System Design

There are many kinds of invertebrates and their nervous systems vary widely in design and extent. At one end of the spectrum is the sponge, an organism with no nervous system; at the other end is the octopus, whose nervous system approaches that of vertebrates in its complexity. The invertebrates most often used in integrative neurobiological studies are arthropods (insects, crustacea, and related creatures), annelids (worms, leeches, and the like) and molluscs (snails, clams, and their relatives). In these invertebrates, the nervous system is generally distributed throughout the animal; in contrast, in vertebrates virtually all of the nervous system concerned with complex neural behavior is collected together in one place—the head. Stated differently, these invertebrates display less encephalization. Their nervous systems typically consist of paired ganglia that extend much of the animal's length along its ventral side. Individual ganglia control relatively complex behaviors, including feeding, swimming, escape, and so on. From individual ganglia, nerve bundles, called *roots*, extend to the animal's periphery—to the muscles and skin. *Connectives* link individual ganglia, and the entire complex is referred to as

9.1 In a number of invertebrates, paired ganglia extend the length of the animal to form a ventral nerve cord. Connective axons run between ganglion pairs and root axons extend to the periphery of the animal. In the head region of the animal, ganglia frequently fuse to form a brain.

ganglion

head ganglia   connectives   roots

the *ventral nerve cord* (see Figure 9.1). Fusion of ganglia in the animal's head region to form a rudimentary brain is typical in many of these animals, and species with more complex nervous systems generally have larger brains; that is, they have more and larger fused ganglia in the head region of the animal.

The number of neurons in invertebrates is usually limited, and furthermore in some instances it appears to be constant among individual animals of the same species. Yet the range of neuronal numbers in different species is enormous. The nematode *Caenorhabditis elegans* (a roundworm) has 302 neurons; every normal *C. elegans* has precisely this number of neurons. Genetic mutants may have different numbers, and they have been used to study the function of certain neurons. The behavior of *C. elegans* is primitive, though, so this organism has not furthered our understanding of complex neural behavior.

The octopus, on the other hand, has the largest brain of all invertebrates. It is roughly the same size as a fish brain (one of the smallest vertebrate brains—see Figure 11.4). The octopus exhibits many complex behaviors, including short- and long-term memory. The number of neurons in the octopus brain is on the order of 10^8, a far cry from the 10^{11}–10^{12} neurons in the human brain. Nevertheless, neurobiological analysis of the octopus brain has not progressed very far because of its complexity (the sheer number of cells) and technical difficulties (for example, the neurons are difficult to record).

Some animals offer a compromise with respect to neuronal number and behavioral repertoire. The nervous system of a marine snail or a leech has on the order of 10^4 neurons, and snails exhibit interesting behaviors that can be linked to neuronal function.

Furthermore, they have a manageable number of neurons (400–1,500) in each ganglion. This makes it possible to map individual ganglia in some detail and thereby identify the function of many neurons and explore connections between cells.

Many other invertebrates have been used profitably for integrative neurobiological studies, including crayfish, lobsters, cockroaches, locusts, grasshoppers, and horseshoe crabs. Although these organisms have more neurons than marine snails, and although their nervous systems are more complex, they may have large neurons that are accessible for study or other features that elucidate certain aspects of neuronal function. An example is the lateral eye of the horseshoe crab, which provided the first recordings from individual axons coming from an eye, the first intracellular recordings from photoreceptors, and a model for the effects of lateral inhibition in a neural circuit.

Cellular Organization

The organization of invertebrate and vertebrate neurons is distinctly different. Whereas most neurons found in the brains of vertebrates are typically bipolar or multipolar, with one or several dendrites extending from one side of the cell body and an axon from the other side, most invertebrate neurons are monopolar. That is, a single process extends from the cell body and dendrites and axons derive from this single process. Figure 9.2 compares schematically the structure of typical vertebrate and invertebrate neurons. Axons arise from within the dendritic tree of the invertebrate neuron, and one neuron can have several axons. Each axon has its own spike-initiating zone (equivalent to the axon hillock in the vertebrate neuron), and it is thought that parts of the dendritic tree, with associated axon, act independently of other parts of the cell. Hence, individual neurons in invertebrates can be considerably more complex than are typical vertebrate neurons.

It seems to be the case that invertebrates often develop more complex neural circuitry by making individual neurons more complex, whereas vertebrates make more complex neural circuitry by adding more neurons to the circuit. Given the relative success of vertebrate brains as compared with invertebrate brains, it is reasonable to conclude that the former approach may be limiting. It may be that an individual neuron can be made only so complex,

9.2 A typical vertebrate bipolar neuron *(left)* and an invertebrate monopolar neuron *(right)*. The structure of invertebrate neurons may be quite complex: each may have several axons and spike-initiating zones.

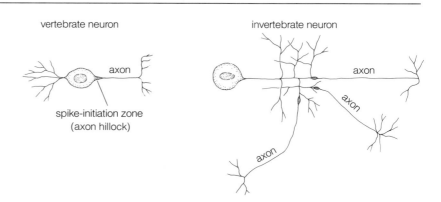

whereas adding more neurons to a brain does not overextend any one neuron. The large size of the cell body of many invertebrate neurons is probably due to the large number of processes that the cell must support. Although its size makes intracellular recording relatively easy, the position of the cell body off to one side—not directly in the mainstream of information flow from the synaptic input on the dendrites to the axon hillock, as is the case for vertebrate neurons—means that the signals recorded from the invertebrate cell body may be small and difficult to interpret.

Another reason why individual invertebrate neurons are large, and why invertebrate nervous systems have not developed as much as have vertebrate brains, is that invertebrate axons are not myelinated. Although the axons of invertebrate neurons are in close association with glial cells or glial processes, no myelin sheath surrounds the axons (except in a few crabs). To maximize conduction velocity, invertebrate neurons make larger axons, the extreme case of which is the squid giant axon. Even so, a squid giant axon does not conduct action potentials as rapidly as a myelinated axon of 20 μm diameter in the vertebrate, and 2,000 of these axons will fit into the space occupied by a single giant squid axon of 1 mm diameter! As pointed out in Chapter 4, the human brain without myelin would need to be ten times larger to function as it does, and we would need to eat ten times as much to maintain it.

Ganglion Organization

The paired ganglia of a snail or lobster have a characteristic organization, as shown in cross section in Figure 9.3. The cell bodies are arranged around the ganglion's periphery, whereas the ganglion's center consists solely of processes. The synaptic contacts, therefore, are made mostly in the center of the ganglion, a neuropil region that is quite intricate. It is difficult to sort out by anatomical analysis the synaptic connections made between the neurons in this region, but they can be elucidated by recordings taken from two or more cells simultaneously. As we shall see, spectacular insights to neural circuitry in invertebrates can be revealed in this way.

The cell bodies in a ganglion are not all identical in appearance; they vary in size and color. When ganglia of different individuals of the same species are studied, we recognize specific neurons located in virtually the same position in every ganglion. Furthermore, when the electrical responses of these cells are recorded, we find that they have the same physiological properties. If they are sensory neurons in one ganglion, they are sensory neurons in every ganglion. In addition, the pattern of action potential discharge to a stimulus can be characteristic of a cell, and in every ganglion the same pattern is observed for that cell. Thus, we can map in an individual ganglion a cell's position, its type (motor neuron, sensory neuron, or interneuron), and its physiological properties. Moreover, this map holds for ganglia in other animals—a tremendous advantage for the experimenter. Examples of identified neurons in the leech are shown in Figure 9.4.

As might be expected, the cells in one of a pair of ganglia are mirror images of the cells in the other; they innervate comparable areas on the animal's two sides. When identified cells are intracellularly stained, the cells are remarkably similar to one another in shape and distribution of processes. Certain cells usually have a unique shape, one that is readily recognized in all animals but distinctly different from the other neurons in the ganglion. The form of individual invertebrate neurons is quite invariant among individuals.

Command Neurons and the Hierarchy of Cells

When an individual neuron is stimulated in an invertebrate ganglion, it can simply activate a single muscle or gland or it can

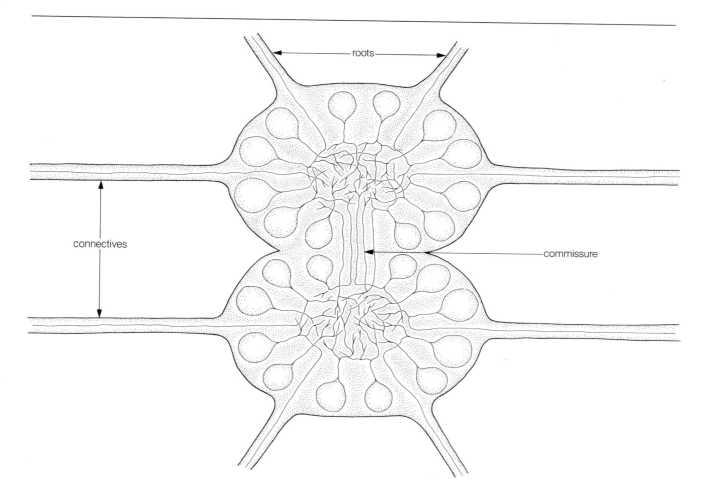

9.3 General organization of a pair of invertebrate ganglia. Neuronal cell bodies are arranged around the periphery of a ganglion and their processes extend into its center, where synaptic interactions occur. Axons arising in the central neuropil region of a ganglion extend to other ganglion pairs via connectives or to the periphery of the animal via roots. The two members of a ganglion pair are mirror images of one another and are connected by a commissure.

initiate a complex behavior. Hence, neurons in invertebrates have a hierarchy with respect to the effects they initiate. Neurons that initiate a complex movement or behavior in an invertebrate are often called command neurons. In one invertebrate, activation of a command neuron initiates an escape mechanism that lasts for 30 seconds or so and involves a circuit consisting of more than a dozen neurons. The activity initiated is prepatterned in that once the sequence begins, it usually continues for some time. Command neurons may be part of the circuit whose output results in a

9.4   A ganglion pair from a leech. The location of three types of sensory cells are indicated: *T*, touch cells; *P*, pressure cells; and *N*, cells that respond to noxious (strong) mechanical stimuli. Note that the cells sit in mirror-image positions on the two sides of the ganglion pair. Furthermore, each type of cell gives a distinctive response. *T* cells respond with a burst of fast impulses and they rapidly adapt. The impulses generated by *P* and *N* cells are larger and longer-lasting. Typically, the latency for firing of the *P* cell is shorter than the *N* cell's latency.

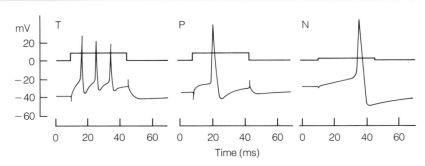

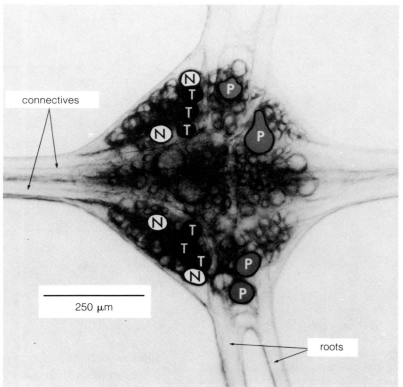

behavior, and command neurons may receive feedback from the circuit. An example will be discussed in the next chapter.

Activation of a command neuron often requires a strong stimulus and a precise one. Different command neurons can cause similar but somewhat different behaviors, and for any one behavior several command neurons may be involved. Thus, activation of a single command neuron may initiate a behavior, but activation of that neuron may not always be necessary for the behavior. For a behavior of any complexity, a system of command neurons usually participates, and command neurons appear to have extensive interactions among themselves.

Mechanisms Underlying Behavior

Neuronal Properties: Inking in Aplysia

The specialized physiological properties that individual neurons can have underlie certain simple behaviors. An example is the inking response in the marine snail, Aplysia (see Figure 11.1). If a strong and prolonged tactile stimulus is applied to the animal, it releases a cloud of ink after a few seconds that blinds a predator. The ink is toxic even to itself, so the Aplysia does not release ink unless it is sufficiently disturbed for a few seconds.

The circuitry underlying the inking response is simple. Sensory neurons from the skin either directly or through interneurons innervate three motor neurons that activate the ink gland. It is possible to record the activity of both the sensory and the motor neurons participating in the behavior. When a strong stimulus is applied to the animal, impulse activity is immediately recorded from sensory neuron axons; the activity continues for as long as the stimulus is applied. Impulse activity recorded from motor neurons involved in the inking response begins only about 3 seconds after the stimulus is applied to the animal, but thereafter it is accelerating and vigorous. The delay of about 3 seconds between the initiation of activity in the sensory neurons and in the inking motor neurons corresponds to the delay in the release of ink following the presentation of a strong stimulus to the animal (see Figure 9.5).

Why the long delay in the initiation of impulses in the motor neuron if it receives direct synaptic input from the sensory neurons? Intracellular recordings from the motor neuron show that

9.5 **(A)** Circuit diagram for the inking response in *Aplysia*. **(B)** Records of a sensory neuron's response to a long (5 sec) stimulus applied to the sensory endings of the neuron *(left)* and the discharge of a motor neuron innervating the ink gland *(right)*. Note that the motor neuron does not begin to discharge impulses for about 3 seconds after the stimulus is applied to the sensory neuron.

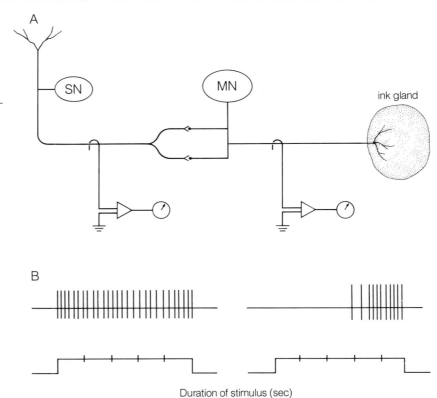

Duration of stimulus (sec)

following the injection of depolarizing current into the cell, the membrane potential of the motor neuron depolarizes quite slowly. Indeed, spike-firing threshold is reached only about 3 seconds after the depolarizing current injection begins. As discussed in Chapter 3, when depolarizing current enters a neuron, it takes a little time for the membrane to charge up and to depolarize fully, but only a few milliseconds, not 3 seconds or so! This discrepancy of a thousandfold suggests that there is something special here about the motor neuron membrane.

Voltage-clamp analysis of currents flowing across the motor neuron's membrane in response to the transmitter released by the sensory neurons and to depolarization of the membrane (by steps of current injected into the cell) provided the answer (see Figure

9.6 **(A)** Intracellular recording *(Vm)* from an ink gland motor neuron following injection of current *(I)* into the neuron. Note that the neuron depolarized very slowly and reached spike-firing threshold only after about 3 seconds. **(B)** Voltage-clamp records of currents flowing across the motor neuron membrane in response to synaptic input *(Is)* or depolarization of the membrane *(Ik)*. The sum of these current flows *(IT)* results in the slow depolarization of the motor neuron in response to a stimulus.

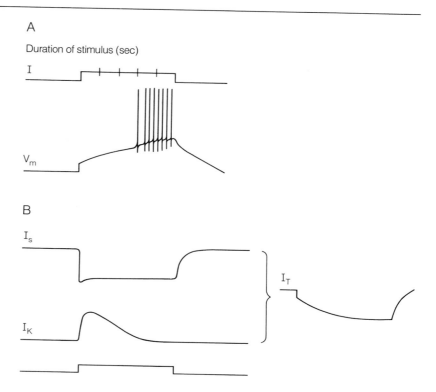

A

Duration of stimulus (sec)

I

V_m

B

I_s

I_K

I_T

9.6). When transmitter was released from the sensory neurons, inward current flowed across the motor neuron membrane in a sustained fashion, an expected result. However, depolarization of the membrane (in the presence of TTX and TEA to block voltage-gated Na^+ and K^+ channels) was accompanied by an early outward current across the cell membrane that gradually subsided over the course of several seconds. The ion carrying this current was K^+, indicating that the motor neuron's membrane has a special voltage-sensitive K^+ channel (not sensitive to TEA) that slowly inactivates. When transmitter released by the sensory neurons causes positive current to flow into the motor neuron, thus depolarizing the membrane, this K^+ channel opens and lets K^+ ions leave the cell. This outward current counters the effects of the inward flow; thus, the cell initially depolarizes only minimally and action potential threshold is not reached. Only as the K^+ channels inactivate does the motor neuron depolarize sufficiently to fire action potentials.

In this simple case, the specific behavior can be linked to the presence of a special channel in the motor neuron membrane. Inactivating K^+ channels (discussed in Chapter 8) are not limited to motor neurons; they are in other cells and can be blocked by an antagonist, 4-aminopyridine, that does not affect the voltage-gated K^+ channels activated during action potential generation. They are also present in spontaneously active (beating) neurons found in many invertebrate ganglia, and in these neurons they control the rate of action potential generation. These cells have a high resting Na^+ conductance that drives the membrane toward spike-firing threshold. As the neuron depolarizes, however, the fast K^+ channels open, opposing the depolarizing Na^+ current and delaying impulse generation. As the fast K^+ channels inactivate, the Na^+ current predominates and generates an action potential. The membrane's hyperpolarization during action potential recovery reactivates the K^+ channels and the process begins over.

Neural Circuitry: Lateral Inhibition in the Limulus *Eye*
Studies of the lateral eye of the horseshoe crab, *Limulus*, provide an elegant example of how simple neural circuitry can explain complex perceptual phenomena. This animal is in the same subphylum as the spiders and is not technically a crab. Like many other arthropods, *Limulus* has two lateral compound eyes (see Figure 9.7A and 9.7B), each consisting of about 1,200 separate photoreceptive units called *ommatidia*. (*Limulus* also has two simple eyes and even a collection of photoreceptor cells on its ventral surface.) Each ommatidium contains about 15 photoreceptor cells (called *retinular cells*), which surround the distal process of a second-order cell called the *eccentric cell*. The photoreceptors are all electrically coupled to the eccentric cell's distal process, so the graded receptor potentials generated when light is absorbed by the photoreceptors pass directly into the eccentric cell, where they generate action potentials that carry the visual signal to the rest of the animal's brain. A diagram of an ommatidium of *Limulus* and records from retinular and eccentric cells are presented in Figure 9.7C.

One might wonder why *Limulus* was chosen for vision studies. What led investigators to this animal? *Limulus* was "discovered"

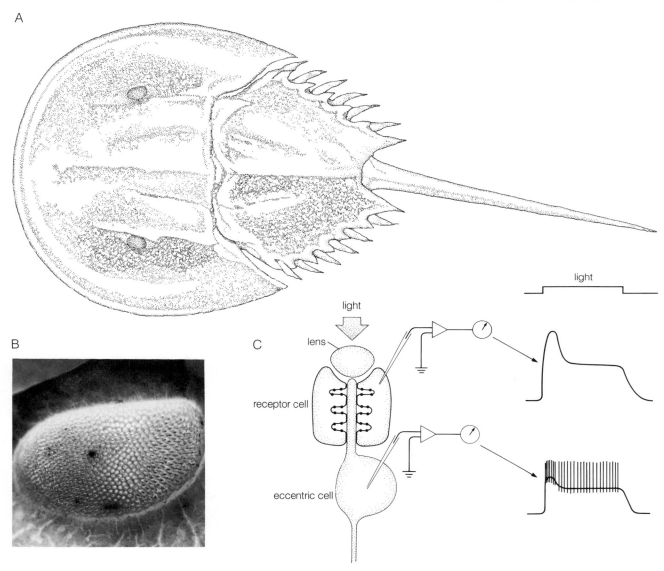

9.7 **(A)** The horseshoe crab, *Limulus poly-phemus*. **(B)** The compound eye of *Limulus*. Each ommatidium has its own lens and receives input from a different part of the visual field. **(C)** When an ommatidium is stimulated by light, the receptor cell does not generate action potentials; rather, generation of action potentials first occurs at the axon hillock region of the second-order eccentric cell.

by H. K. Hartline, who as a young investigator in the 1920s surveyed the eyes of several invertebrates in search of a visual system from which he could record the activity of single axons coming from the eye. He noted that the horseshoe crab's eyes responded with a vigorous field potential when an electrode was placed on the eye's surface and a second electrode was placed elsewhere on the animal. (This potential is called the electroretinogram.) But he also found that the relatively long optic nerve of

Limulus was quite easy to dissect. Unlike in most nerve bundles, connective tissue did not tightly bind together the individual axons.

In subsequent work, which led to a Nobel Prize in 1967, Hartline dissected the axons of the eccentric cells and recorded their action potentials. These were the first single-cell recordings made from a visual system, and Hartline demonstrated that the responses recorded from the individual axons are remarkably parallel to several features of human visual performance. For example, the frequency of impulses recorded from the axons was roughly proportional to the logarithm of the stimulating light intensity. A logarithmic relation is characteristic of a human's perception of intensity when presented with different light intensities. Hartline also found that with short flashes (less than 1 second in duration), the number of action potentials generated in the eccentric cell axon was strictly proportional to the number of light quanta in the flash (Figure 9.8). The same holds true for a human observer; for flashes of short duration, reciprocal changes in intensity and duration cannot be distinguished. Thus, a short, intense flash of light looks identical to a longer, dimmer light flash so long as both contain the same number of quanta. The visual system's response under these conditions depends strictly on the number of visual pigment molecules activated, a phenomenon known as Bloch's law. Hartline's findings that some of the simpler features of human vision are reflected in the activity recorded from axons coming from single ommatidia of the *Limulus* eye suggested that these visual phenomena originate in the photoreceptor cells themselves. More recent work on the activity of single vertebrate photoreceptor cells has amply confirmed this idea.

When Hartline began his work, and for many years thereafter, he thought that the ommatidia were completely independent units with no cross-talk between them. One day he noticed, however, that when recording the response of a single ommatidium to a spot of light focused on it, additional light falling on the eye caused a decrease in the recorded response. He, along with Floyd Ratliff, proved that stimulation of surrounding ommatidia was responsible for the decreased activity of the recorded ommatidium, and they called this *lateral inhibition*. Neuroscientists have come to realize that lateral inhibition is present in all visual systems and most other sensory systems as well.

Hartline, Ratliff, and their colleagues at the Rockefeller Institute

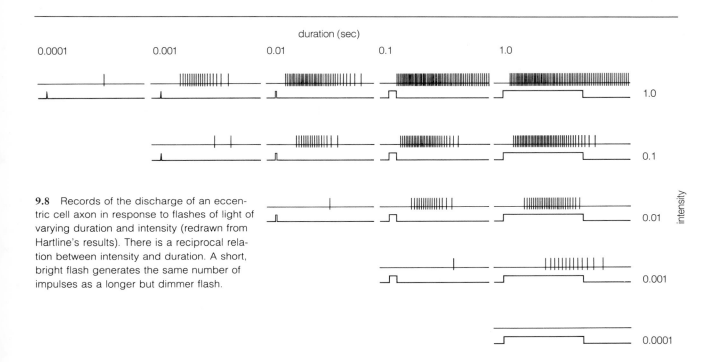

9.8 Records of the discharge of an eccentric cell axon in response to flashes of light of varying duration and intensity (redrawn from Hartline's results). There is a reciprocal relation between intensity and duration. A short, bright flash generates the same number of impulses as a longer but dimmer flash.

in New York began an extensive analysis of lateral inhibition in the *Limulus* lateral eye and showed its significance for the performance of a visual system. The inhibitory effects are mediated by a plexus of neural processes located just proximal to the ommatidia. This plexus, consisting of fine branches that arise from the eccentric cell axons, extends laterally to form inhibitory synapses on adjacent eccentric cell axons. The inhibition is reciprocal; that is, an axon that inhibits a neighboring axon is in turn inhibited by it.

The effects of lateral inhibition on the firing frequency of an eccentric cell are presented in Figure 9.9. In the idealized experiment illustrated, the action potentials generated in the eccentric cell axon marked *X* are recorded with wire electrodes hooked around the axon. When a light stimulus is confined to the recorded ommatidium, the axon's firing frequency is proportional to the intensity of light falling on the ommatidium. The firing frequency shown in the figure indicates stimulation with a moderate light

9.9 **(A)** In the *Limulus* eye, a plexus of axons connects the eccentric cell axons arising from the ommatidia. Records from an axon extending from ommatidium X show that the axon generates the most spikes when it is illuminated alone; spikes are less frequent when ommatidium Y, X's neighbor, is also illuminated, and when ommatidia X, Y, and Z are illuminated. **(B)** Firing frequency of a single eccentric cell in response to a sharp bright-dim edge moving in stepwise fashion across the ommatidial array of a *Limulus* eye. Retinal distance is measured in numbers of ommatidia between the eccentric cell being recorded and the position of the edge. Because of lateral inhibition, the differences in firing frequency of the cell were accentuated when the stimulus was adjacent to the cell being measured; differences in firing frequency were less extreme when the stimulus was some distance away from the cell (that is, approximately 10 ommatidia away).

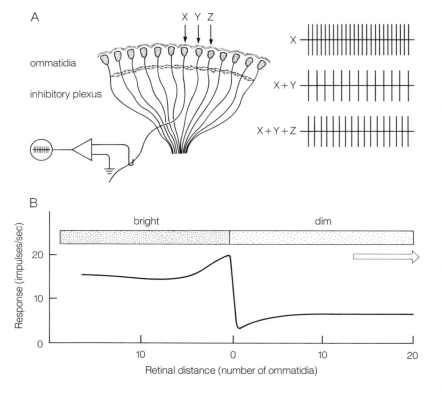

intensity. Lateral inhibition between axons is demonstrated by illuminating ommatidium *Y* and ommatidium *X* simultaneously. Under these conditions, the steady-state firing frequency of the eccentric cell axon of ommatidium *X* is significantly reduced because of lateral inhibition. The strength of the inhibition between eccentric cells depends on the level of activity in the interacting cells and on the distance between them: the stronger the illumination, the stronger the inhibition exerted; and near neighbors affect one another more strongly than do distant ones. One consequence of the inhibitory interaction among eccentric cells is a phenomenon called *disinhibition*, also shown in Figure 9.9. When ommatidium *Z* is illuminated at the same time as ommatidia *X* and *Y*, the firing frequency of the eccentric cell of ommatidium *X* is now greater than that observed when only X and Y are illuminated. This comes about because the eccentric cell of Z inhibits the eccentric cell of Y, which relieves partially the inhibitory effect of Y on X.

Hartline and Ratliff demonstrated that the simple lateral and reciprocal inhibition in the *Limulus* eye shapes the neural signals passing down the eccentric cell axons in such a way as to enhance contrast at an edge or border. As illustrated in Figure 9.9B, the discharge rates of eccentric cells are higher along the brighter side of a dim-bright border and lower along the darker side. This happens because an eccentric cell in bright light that is adjacent to a dark border is inhibited weakly by its neighbors in the dimmer light. Conversely, eccentric cells along the dimmer side are inhibited strongly by neighbors over the border in brighter light. Because of this interplay between excitatory and inhibitory influences, the differences in firing frequency for eccentric cells adjacent to a dim-bright border are significantly greater than they are for eccentric cells away from the border. Thus, border or edge enhancement can be fashioned by a simple lateral and reciprocal inhibitory network among neurons.

Enhancement of borders by the human visual system was recognized by Ernst Mach, an Austrian physicist and psychologist in the late nineteenth century; consequently, it is often called the *Mach band phenomenon*. Figure 9.10 depicts a series of intensity steps that illustrates the phenomenon. Although each step is even in intensity from one edge to the next, it appears as if the steps are lighter along the margins bordering the darker steps and darker along the margins bordering the lighter steps. We now believe that the Mach band phenomenon in the human visual system can be explained by a similar lateral and reciprocal inhibitory interaction occurring within the retina.

Inking in *Aplysia* shows how a simple behavior can be linked to the special neuronal properties of an invertebrate neuron, whereas the *Limulus* experiments illustrate how a perceptual phenomenon can be explained by a simple neuronal circuit. It is important to keep in mind that most neural processing and more complex behaviors are achieved through both kinds of mechanisms. Thus, both neural circuitry and the special properties of individual cells underlie the output of many neurons and, ultimately, behavior.

9.10 Although the reflected intensity of each square is constant from one border to the next, it appears as if each one is lighter along the border with a darker square and darker along the border of a lighter square. The Mach band phenomenon also accentuates the intensity differences between the steps of the series. If the border between two steps is obscured by a pencil or other thin object, the adjacent steps will appear much more similar in intensity; when the same steps meet at a distinct border, the difference in intensity appears greater.

Summary

The nervous systems of many invertebrates are simpler than those of vertebrates and are more amenable for detailed analysis. The nervous system of a number of invertebrates is distributed along the animal in ganglia that control individual behaviors. Invertebrates have fewer neurons than vertebrates, but the range of neuronal numbers varies enormously, from a few hundred in a nematode to about 200 million in the octopus. Even so, the neurons of invertebrates tend to be larger than vertebrate neurons, and more complex. One neuron can have several axons, for example. Invertebrate neurons are usually monopolar; dendrites and axons arise from a single process that extends from the cell body. Invertebrate axons are generally not myelinated; for faster conductance of action potentials, axons are made larger in diameter.

The ganglia of invertebrates typically consist of cell bodies arranged around the periphery, with cell processes in the center forming a complex neuropil where most synaptic junctions are made. Neurons found in one location in a ganglion are identical among different animals with regard to their type, physiological properties, and function. As a result, invertebrate ganglia can be mapped and neurons specified. When stained intracellularly, neurons are often seen to have a unique morphology. The hierarchy of neuronal function in invertebrates is such that some neurons innervate a single muscle or patch of skin; but others, when activated, initiate complex behaviors involving many other neurons. These latter cells are called command neurons.

Some rudimentary behaviors in invertebrates can be related to the specific properties of the neurons. The inking response in the sea snail *Aplysia* is an example. The delayed response of the behavior following sensory stimulation can be attributed to a special voltage-sensitive K^+ channel in the motor neuron membrane that opposes the cell's depolarization by sensory input. As the channel slowly inactivates, the motor neuron depolarizes gradually and then stimulates secretion of ink from the animal.

In the *Limulus* lateral eye, a simple lateral and reciprocal inhibitory network among the axons of the output (eccentric cell) neurons of individual ommatidia provides a model for perceptual phenomena such as Mach bands. Adjacent eccentric cell axons inhibit each other, and eccentric cell axons along the brighter side of a dim-bright border are inhibited weakly by their neighbors in the dimmer light. Thus, they generate more action potentials than do eccentric cell axons away from the border. Conversely, eccentric cell axons along the dimmer side of the border are strongly inhibited by the adjacent axons in brighter light; they fire less vigorously than axons coming from ommatidia that are away from the border. The differences in firing rates of the eccentric axons adjacent to the border are greater than the differences in firing rates of eccentric cell axons away from the border. As a result, perception of the border between light and dark is enhanced.

The *Limulus* studies provide a model of perception based on neuronal circuits. This model is in contrast with the explanation of inking behavior in *Aplysia*—namely, that neural responses to stimuli may be shaped by the special properties of individual neurons. Most information processing and the behaviors that result from it are under the control of mechanisms at both levels—the cell and the circuit. Further examples of this are described in the next chapter.

Further Reading

BOOKS

Kandel, E. R. 1976. *Cellular Basis of Behavior.* San Francisco: W. H. Freeman. (A nice introduction to the use of invertebrates in neurobiological research.)

Ratliff, F. 1965. *Mach Bands: Quantitative Studies on Neural Networks in the Retina.* San Francisco: Holden-Day. (Describes the classic studies of Hartline and Ratliff on lateral inhibition in the horseshoe crab eye.)

ARTICLES

Byrne, J. H. 1980. Analysis of ionic conductance mechanisms in motor cells mediating inking behavior in *Aplysia californica*. *Journal of Neurophysiology* 43:630–650.

Nicholls, J. G., and D. A. Baylor. 1968. Specific modalities and receptive fields of sensory neurons in the CNS of the leech. *Journal of Neurophysiology* 31:740–756.

Ratliff, F. 1972. Contour and contrast. *Scientific American* 226(6):90–101.

Control of Rhythmic Motor Behavior

THE ABILITY to move and carry out purposeful acts distinguishes animals from plants. Movement requires effector systems, such as muscles, and also neural circuitry to control the muscles. The way in which these systems work in tandem is of special interest to neuroscientists.

Much motor activity, such as swimming, walking, or flying, is rhythmic; it requires a stereotyped and repetitive sequence of neural outputs and a correspondingly coordinated sequence of muscle contractions. Invertebrates display a variety of rhythmic motor actions, and thus neuroscientists have spent years studying the neural basis for these movements with the hope that they may serve as useful models for rhythmic motor behaviors.

The neural circuitry that underlies a rhythmic activity is termed a *central pattern generator*. It consists of a group of neurons that generate a sequence of neural outputs. These outputs lead to muscle activity that is coordinated temporally and spatially. A central pattern generator can be activated by a command neuron that responds to sensory input. Figure 10.1 illustrates this sequence of events. This simple progressive scheme describes many rhythmic systems but certainly not all. For instance, there may be feedback between command neurons and central pattern generators; hence the command neurons may be part of the central pattern generator, as may be both sensory and motor neurons. Furthermore, as we shall see, neurons in a central pattern generator may participate in more than one motor circuit.

The Heartbeat in the Leech

An especially simple central pattern generator that has been extensively studied, particularly by Ronald Calabrese and his colleagues at Emory University, controls the heartbeat in the leech. Blood is pumped through the leech's circulatory system by two

10.1   Scheme of the initiation of rhythmic motor activity. Sensory input activates command neurons, which initiate activity in the central pattern generator (CPG). The output of the CPG drives the motor neurons that result in the rhythmic behavior.

lateral heart tubes, one on each side of the animal. Although each tube in isolation will rhythmically contract by itself, *in vivo* the heart tubes contract alternately: first the left tube contracts, then the right tube, then the left, and so forth. How is this alternating rhythmic activity established and controlled?

The heart muscle in each tube is innervated by motor neurons, termed heart-excitatory or HE cells, in the ganglia that make up the ventral nerve cord. The leech has 21 ganglia; HE motor neurons are found in ganglia 3 to 18, on both sides of each ganglion. For this reason they are designated HE(R) or HE(L) cells. HE(L) motor neurons innervate the left heart tube and HE(R) neurons innervate the right heart tube. In addition, seven pairs of heart interneurons are located in the first seven ganglia of the nerve cord, designated HN(1–7) cells. These interneurons control and integrate the output of the HE cells and hence the beating of the heart tubes. They make up the central pattern generator for the heartbeat.

Of the seven pairs of heart interneurons, the first four—the HN(1–4) cells—are responsible for regulating the heartbeat rhythm. They are able to reset and entrain the entire network, whereas the last three pairs of interneurons—the HN(5–7) cells—cannot reset the system. These posterior interneurons mainly coordinate motor neuron activity between ganglia. I will concentrate here on the four anterior interneurons that regulate the timing and rhythmicity of the system.

Figure 10.2 presents the general circuitry connecting the four anterior interneurons with the motor neurons (the HE cells) and each other. All of the motor neurons are directly innervated by HN(3) cells, but motor neurons HE(5–18) receive input from both HN(3) cells and HN(4) cells. HN(1) and HN(2) cells do not directly innervate motor neurons, but they synapse reciprocally with the HN(3) and HN(4) cells. Their principal role is to coordinate the activity of the HN interneurons in ganglia 3 and 4.

10.2 The circuitry that controls the beating of the lateral heart tubes in the leech. Motor neurons (HE cells) found in ganglia 3–18 innervate the heart tubes. The motor neurons in ganglia 3 and 4 are innervated by interneurons found in ganglion 3. The motor neurons in ganglia 5–18 are innervated by interneurons present in both the third and the fourth ganglia. Interneurons in ganglia 1 and 2 synapse reciprocally with the interneurons in ganglia 3 and 4. The interneurons in ganglion 3 interact reciprocally with one another, as do those in ganglion 4, but there is no direct interaction between the interneurons in ganglia 3 and 4. All of the interneuronal synapses in the circuit (represented by solid dots) are inhibitory.

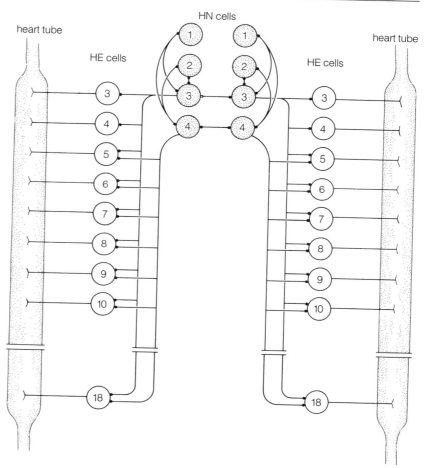

Thus it is the HN(3) and HN(4) interneurons that are most responsible for establishing the heartbeat rhythm.

Figure 10.3 shows intracellular recordings from the two HN interneurons in the fourth ganglion and from one motor neuron in the fifth ganglion, the one that innervates muscles in the right heart tube. The two interneurons fire bursts of action potentials out of phase; when HN(L,4) is active, HN(R,4) is silent, and vice versa. The bursts of activity in each interneuron last for about 6

10.3 Intracellular recordings from heart interneurons on either side of the fourth ganglion, HN(L,4) and HN(R,4), and a motor neuron, HE(R,5), in the fifth ganglion in leech. The two interneurons fire bursts of impulses out of phase with one another. The bursting of the motor neuron is also out of phase with the interneuron driving it, because all the synapses in this system are inhibitory.

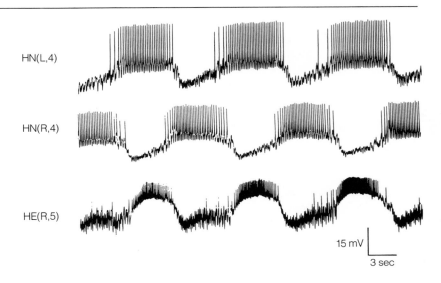

HN(L,4)

HN(R,4)

HE(R,5)

15 mV

3 sec

seconds and then the neuron is silent for about 6 seconds. The HN(R,4) cell synapses directly on the HE(R,5) motor neuron and controls its rhythm. The synapse from HN(R,4) to HE(R,5) is inhibitory, as are all the synapses between the interneurons and motor neurons. As a consequence of this design, the activity in the motor neuron is out of phase with the interneuron's activity; when the interneuron is active, the motor neuron is silent, and vice versa.

Of special interest here is the control of the motor neurons. For example, how does the interneuron entrain the activity of the motor neuron by way of an *inhibitory* synapse? It is easy to see how activity would be controlled if the interneurons excited the motor neurons. In this case, however, the motor neurons were found not to need excitatory input to be activated. When the synapses from the interneurons to the motor neurons are blocked, the motor neurons are tonically active. In other words, the rhythm is established in the motor neuron by strong inhibitory input to the cell from the interneuron. This inhibition hyperpolarizes the motor neuron to below spike-firing threshold, so the motor neuron stops firing. Without this periodic inhibitory input, the motor neurons fire continuously.

The next question one may ask is how the alternating and

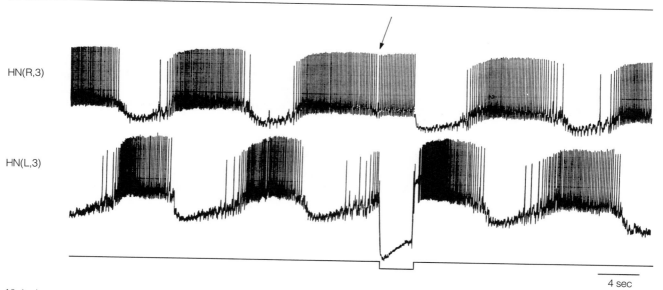

HN(R,3)

HN(L,3)

4 sec

10.4 Intracellular recordings from the heart interneurons on either side of the same ganglion in the leech. Because of the inhibitory reciprocal innervation between the two interneurons, they fire bursts of impulses out of phase. A hyperpolarizing current pulse was injected into one of the interneurons, HN(L,3), just as it began to burst and inhibit HN(R,3) (arrow). This allowed HN(R,3) to continue to fire, which it did until the hyperpolarizing current pulse was terminated. HN(L,3) then became active and the alternate bursting of the two cells resumed.

rhythmic bursting activity in the interneurons is established. When pairs of interneurons are recorded simultaneously (as in Figure 10.3) and one cell is depolarized, hyperpolarization is observed in the other cell. As with the motor neurons, when synaptically isolated the interneurons are spontaneously active, but in the living animal they are inhibited from firing tonically by input from the other interneurons. The pairs of interneurons, thus, reciprocally inhibit one another. When one cell is active, the other is quiescent, and vice versa.

Evidence that inhibition underlies the rhythmic activity of the interneurons is shown in Figure 10.4. In this experiment, a hyperpolarizing current pulse is injected into one interneuron, HN(L,3), just as it is beginning a burst cycle. The start of inhibition of the interneuron on the other side of the ganglion, HN(R,3), can be seen in the record (arrow), but with hyperpolarization of HN(L,3), HN(R,3) quickly resumes firing and continues to fire at a high rate until the hyperpolarizing pulse is terminated in HN(L,3). Thereafter, HN(L,3) rapidly depolarizes and fires a vigorous burst of impulses that strongly inhibit HN(R,3). The alternating bursts of activity in the two interneurons resumes, but the timing is reset.

That is, the bursting pattern is delayed in both neurons by the hyperpolarizing current pulse injected into HN(L,3). This indicates that the bursting pattern is not endogenous to the cells but depends on inhibitory interactions between the two interneurons.

Can the rhythmic activity be explained entirely by the circuitry? The answer is no. Something is needed to terminate the inhibition after a period of time. Otherwise, the circuit will be stuck in one configuration or the other: one interneuron will always be active and the other always inhibited. What limits the duration of the inhibition? A closer look at the traces in Figure 10.4 shows that the current pulse injected into HN(L,3) is constant, but with time the resultant hyperpolarization of the cell decays. This implies that the neuron possesses an intrinsic mechanism that over time opposes hyperpolarization of the membrane and allows the membrane to return to a depolarized state. Given sufficient time, therefore, an interneuron should begin to fire again even if its inhibiting partner is continuously depolarized by injected current, and this has been observed.

But how does this intrinsic mechanism permit interneurons to escape from hyperpolarization and inhibition? Voltage-clamp studies show that as the interneuron membrane hyperpolarizes below about −45 mV, an inward current begins to flow across the membrane. This current has a slow time period, requiring 5–6 seconds to reach its peak. Furthermore, it increases in magnitude with greater membrane hyperpolarization, up to about −80 mV. Hence, a voltage-gated channel in the membrane is activated by hyperpolarization. This channel has a reversal potential of about −20 mV, indicating that it allows both Na^+ and K^+ to cross the membrane. But, when the channel is open, at membrane voltages between −45 mV and −80 mV, much more Na^+ than K^+ will flow across the membrane, thus depolarizing the cell (see discussion of EPSPs and reversal potentials in Chapter 5). Because of the slow activation of the channel, the cell depolarizes only over the course of several seconds. This accounts for the slow escape of the interneuron from inhibition. When the interneuron is sufficiently depolarized it begins to fire again and to inhibit the interneuron on the other side of the ganglion. This sequence of events is illustrated in Figure 10.3. Note that in both interneurons, following the initial hyperpolarization induced in the cell by the burst of spikes in the other cell, the membrane potential gradually depolarizes, even

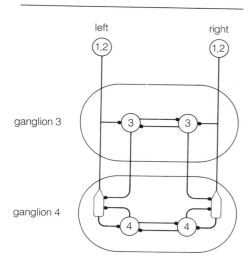

10.5 Circuitry of the heart interneurons in the first four ganglia of the leech. The interneurons in ganglia 1 and 2 coordinate the activity of the interneurons in ganglia 3 and 4. The interneurons in the first two ganglia make identical connections and are thus shown here as a single cell. These interneurons receive input in ganglion 4, but they make synapses onto the interneurons in both ganglia 3 and 4. The right and left interneurons make reciprocal connections with each other in the same ganglion, but direct connections between the interneurons in ganglia 3 and 4 do not exist.

though the firing rate in the inhibiting interneuron remains constant.

As I mentioned earlier, the interneurons in the third and fourth ganglia innervate the motor neurons, and the motor neurons in all but the third and fourth ganglia receive input from both pairs of interneurons. HN(3) and HN(4) cells, however, do not synapse on one another. It is obvious that HN(3) and HN(4) neurons must fire in a coordinated fashion, the control of which is the job of the interneurons found in the first two ganglia. The HN(1) and HN(2) cells form reciprocal inhibitory synapses with the HN(3) and HN(4) cells. They receive input in the fourth ganglion from both HN(3) and HN(4) cells, and they make synapses back onto the HN(3) and HN(4) interneurons in ganglia 3 and 4. This circuitry is shown in Figure 10.5. In these ganglia HN(1) and HN(2) cells make identical connections, and for the purposes of the circuit diagram they are depicted as a single cell. This circuitry ensures coordination between the interneurons in ganglia 3 and 4. When HN(R,3) and HN(R,4) cells are active, HN(L,3) and HN(L,4) neurons are inhibited, as are the HN(R,1) and HN(R,2) cells. In the opposite phase, HN(R,3) and HN(R,4) cells are inhibited along with the HN(L,1) and HN(L,2) cells.

Is there a cell pair that orchestrates the transitions from activity to inactivity? In the animal, the HN(4) cells appear to assume the lead. That is, the system's oscillation rate depends on how fast the HN(4) neurons escape from inhibition. Why the HN(4) interneurons escape from inhibition most rapidly is not entirely clear. An obvious possibility is that the time constant for the inward current flow induced by hyperpolarization is shorter in HN(4) cells than in the other interneurons.

Figure 10.6 shows the sequence of activation and inhibition of the neurons during a bursting cycle. Open circles represent active cells; shaded circles, inhibited cells. Beginning at the top (*A*), the first cell to switch activity is HN(L,4) (arrow). As it becomes active (*B*), it inhibits HN(R,4) and HN(L,1 and 2). HN(L,3) now becomes active, inhibiting HN(R,3). HN(R,1 and 2) are released from inhibition at this point (*C*) and become active. The phase begins to switch again when HN(R,4) overcomes inhibition and begins to fire (*D*). The reverse sequence now starts, and the cells return to state (*A*).

Of the four states of activity presented in Figure 10.6, two states

10.6 The sequence of activation and inhibition of the leech heart interneurons during a bursting cycle. Shown here as starting out in the inhibited phase (indicated by shading), HN(L,4) is the first cell to switch activity (marked by the small arrow in state A). As this interneuron becomes active (in state B), it inhibits HN(R,4) and HN(L,1 and 2). HN(L,3) now becomes active and inhibits HN(R,3); this results in releasing HN(R,1 and 2) from inhibition (C). The phase transition begins to switch when HN(R,4) overcomes inhibition and becomes active (D). This results in the inhibition of HN(L,4) and the circuit reverts to state A. States A and C are relatively long-lived; states B and D are much shorter.

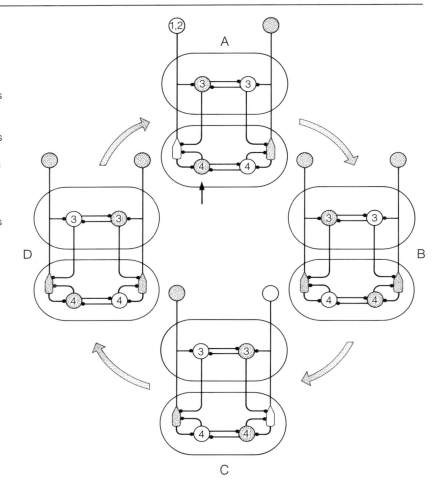

(A and C) are long-lived ($\sim$ 6 seconds) whereas the other states (B and D) are much shorter. It is evident, then, that once the HN(4) cells begin to change activity state, the other cells follow very quickly. This means that the interneuron pairs in ganglia 3 and 4 are closely, but not exactly, in phase with one another. These phase differences may have important effects; for example, they may help to establish proper coordination of motor neuron activity along the animal.

Swimming in *Tritonia*

The central pattern generator that regulates the heartbeat in the leech requires both neural circuitry and special membrane properties in some of its neurons if it is to function properly. The same holds true for the central pattern generator that controls swimming in the sea slug, *Tritonia*. When attacked by a predator, such as a starfish, the slug escapes by making a series of alternating ventral and dorsal flexions; this simple behavior is called escape swimming. Two groups of motor neurons found in one ganglion, dorsal flexion neurons (DFN) and ventral flexion neurons (VFN), drive the muscles that cause this rhythmic behavior. Figure 10.7 shows the relation between the swimming movements and the output of the flexion motor neurons. The top trace shows the output of a position monitor on the animal's tail. Upward deflections indicate dorsal flexions; downward deflections, ventral flexions. Below are intracellular recordings from a DFN and a VFN cell. They fire in alternation, and the dorsal and ventral flexions of the animal are closely linked with the motor neurons' outputs. Firing of the DFN results in dorsal movement, bursting in the VFN results in ventral movement.

The central pattern generator underlying swimming in *Tritonia* has been localized to 12 interneurons, which can be classified into three types of cells. Cells of the same type behave quite similarly,

10.7   The sequence of swimming movements and motor neuron activity in *Tritonia*. The top trace is the output of a position monitor showing tail movements (upward deflection indicates dorsal tail flexion; downward deflection shows ventral flexion). The motor neurons fire alternatively: bursting in the dorsal flexion neuron (DFN) is followed by dorsal flexion; bursting in the ventral flexion neurons (VFN) is followed by ventral flexion.

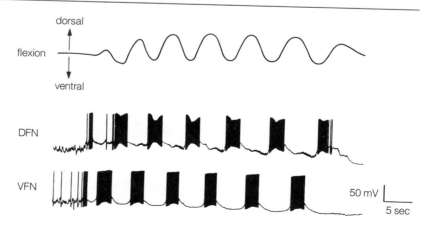

10.8 Intracellular recordings from neurons in the central pattern generator that drives the swimming motor neurons in *Tritonia*. The dorsal swim interneurons (DSI) and ventral swim interneurons (VSI) fire bursts of impulses out of phase with one another. The C2 (cerebral) cells fire just after the DSI cells have begun their bursting activity; C2 firing overlaps the transition between DSI and VSI activity.

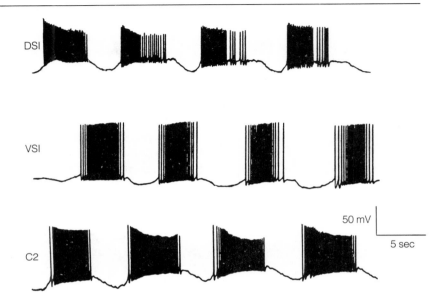

so we consider each type as a single cell. Six dorsal swim interneurons (DSI) provide the primary drive for the DFN motor neurons, whereas four ventral swim interneurons (VSI) provide the primary drive for the VFN cells. Two subtypes of VSI cells are recognized, VSI-A cells and VSI-B cells. In addition, two cells, called the cerebral cells 2 or C2 cells, are an important component of the circuit.

As expected, during swimming the DSI and VSI neurons fire in alternating bursts, as shown in Figure 10.8. The C2 cells, by contrast, begin to fire after the DSI cells become active and their firing continues beyond the transition from DSI to VSI activity.

By recording from pairs of cells involved in the central pattern generator, Peter Getting and his colleagues at the University of Iowa deduced the basic circuitry of the system, shown in Figure 10.9. The DSI cells make inhibitory synapses (indicated by solid circles) on VSI cells and excitatory junctions (open circles) on C2 cells. The VSI cells make inhibitory synapses on both DSI and C2 cells. The C2 cells, which synapse only on the VSI cell, have junctions that are only excitatory or both excitatory and inhibitory on the VSI cells (half-filled circle).

10.9 The basic circuitry of the central pattern generator underlying swimming in *Tritonia*. Open circles indicate excitatory synapses; filled circles indicate inhibitory synapses. The synapse between the C2 and VSI cells is a mixed excitatory-inhibitory junction (*half-filled circle*).

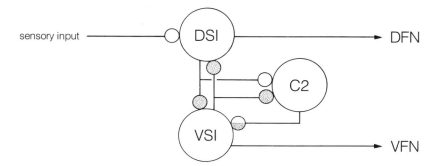

How does the system work? The circuit is turned on by sustained depolarizing input—most often sensory input—to the DSI cells, thereby activating them. The DSI cells then inhibit the VSI cells but excite the C2 cells. The C2 cells, after a delay, excite the VSI cells, and these, when active, inhibit the DSI *and* C2 cells and turn them off. When the C2 cells stop firing, they cease to excite the VSI cells, which stop firing and hence no longer inhibit the DSI cells. The DSI cells are then able to resume firing and the cycle can repeat itself. The circuit shuts down when depolarizing input to the DSI cells is not maintained.

The key to the prolonged oscillatory activity set up by the central pattern generator is the delay between excitation of the C2 cells and activation of the VSI cells. This delay is substantial, lasting 1–2 seconds—far beyond normal synaptic delays of 1–2 milliseconds. Two mechanisms are responsible for the delay; these mechanisms are what differentiate VSI-A and VSI-B cells. The synapse between C2 cells and VSI-A cells is a mixed excitatory-inhibitory junction (see Figure 6.8 and corresponding text). The synapse is initially inhibitory, but after 1 or 2 seconds it becomes excitatory and VSI-A cells are activated. The synapse between the C2 cells and VSI-B cells, by contrast, is always excitatory. But the membrane of the VSI-B cells contains a voltage-sensitive K^+ channel that rapidly opens upon depolarization and inactivates in 1 or 2 seconds. This channel allows K^+ to flow out of the cell and oppose the inward excitatory current induced by the C2 synapse. Only after this channel inactivates will the excitatory input from the C2 cell be sufficient to activate the VSI-B cells. This inactivating K^+

10.10 The two activity states of *Tritonia*'s central pattern generator for swimming. In the early state the DSI cells are active; in the late state the VSI cells are active.

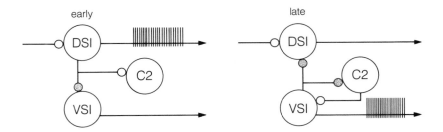

channel is quite similar to the channel in the motor neuron of the *Aplysia* ink gland, discussed in Chapter 9.

The central pattern generator for swimming in *Tritonia* can be thought of as having two states: an early state, when the DSI cells are active; and a late state, when the VSI cells are active. This is shown schematically in Figure 10.10. In the early state, active DSI cells inhibit the VSI cells but excite C2 cells. After a delay the C2 cells activate VSI cells, which inhibit the DSI cells and eventually the C2 cells. Motor neuron activity and dorsal-ventral muscle flexion result from the activity of the DSI and VSI cells.

This system illustrates several general principles. First, as with the control of the leech heart, synaptic connectivity and intrinsic membrane properties are both essential for the central pattern generator to function. Second, the major synaptic interaction underlying the oscillatory behavior of the system is reciprocal inhibition, again like the leech heart central pattern generator. And third, to initiate swimming, the DSI cells must be activated, for their activation leads directly to the oscillatory behavior of the central pattern generator. The DSI cells can thus be considered the command neurons for the generator. Activation of the DSI neurons initiates activity in the circuit, and the circuit continues to oscillate as long as the DSI cells are receiving depolarizing input.

Defensive Withdrawal

When *Tritonia* is attacked by a predator, it assumes a defensive position before it begins to swim away. Not only does it pull in its extremities, but it contracts all over; these actions serve to take the animal out of the predator's reach. Then swimming ensues and the animal escapes. During this defensive withdrawal, the dorsal and ventral flexion muscles simultaneously contract. How does this come about? And how does this behavior relate to swimming?

By recording from central pattern generator neurons in the intact animal, neurobiologists can observe the neural activity that initiates withdrawal and swimming. When an animal is resting and not disturbed, DSI and VSI-A neurons fire slowly, at a rate of 1–3 spikes per second, whereas the C2 and VSI-B cells are silent. Upon tactile stimulation of the animal, the DSI and VSI neurons become active and fire simultaneously, causing dorsal and ventral flexion muscles to contract. The C2 neurons typically fire a single spike at the onset of stimulation, but then they remain silent until the withdrawal reflex is almost over. When C2 cells resume firing, they signal the beginning of swimming.

These observations indicate that C2 neurons play no role in defensive withdrawal. Further investigation of the synaptic interactions during defensive withdrawal showed that DSI cells inhibit one another via an inhibitory interneuron, called the I-neuron. By contrast, during swimming no such inhibition between the DSI neurons is observed; indeed, DSI neurons excite one another during swimming. How might these observations be interpreted? Stimulation of C2 neurons during withdrawal provided a clue, for it altered the interaction between DSI neurons from inhibitory to excitatory. This change indicates that the C2 neurons inhibit the I-neuron, and as a result the I-neuron is not functional during swimming.

The neural events surrounding the initiation and termination of the defensive withdrawal reflex, the initiation and termination of swimming, and the interactions within and between interneuronal groups in *Tritonia* are complex and still not entirely understood; nevertheless some important generalizations can be made. First, individual interneurons can participate in more than one behavior; that is, in more than one circuit. Hence, networks can assume different configurations and mediate different behaviors depending upon the behavioral context. For *Tritonia*'s defensive withdrawal and swim systems, two circuit configurations are proposed. In the defensive withdrawal circuit, the C2 neurons are silent and DSI-DSI interactions are inhibitory. Furthermore, excitatory sensory input to DSI and VSI neurons simultaneously activates the cells. In the swimming circuit, the C2 cells are active—they mediate oscillatory bursting of the DSI and VSI cells. Moreover, the C2 cells inhibit the I-neuron and change the DSI-DSI interactions from inhibitory to excitatory. In addition, swimming is thought to be initiated in the animal primarily by excitatory input to the DSI neuron.

Modulation of Neural Circuits

How can a network be switched from one configuration to another? The activity generated by a neuronal network depends on the input to the system, the synaptic interactions between the neurons in the circuit, and the intrinsic membrane properties of the neurons themselves. An alteration of any of these factors could change patterns of activity and the circuit's basic configuration. At the present time, neurobiologists are keenly interested in neural circuit modulation and have presented evidence for all three mechanisms outlined. I review here some of the data that have been obtained for one motor system, the stomatogastric ganglion in lobsters and crabs. But modulatory effects have been observed in many motor systems, including the leech heart, and *Tritonia* swimming central pattern generators, and also in several sensory systems, including the *Limulus* eye; indeed, neuromodulation is a general feature of neural systems (see Chapter 14).

The Pyloric Central Pattern Generator

In the stomatogastric ganglion of lobsters and crabs there are about 30 neurons that control the digestive system. The muscles innervated by this ganglion mix and grind the food and move it through the stomach. The ganglion produces rhythmic motor discharges in accordance with two central pattern generators: one that controls the muscles in the anterior part of the stomach, the gastric muscles, and another that controls the muscles of the stomach's posterior or pyloric region. The pyloric central pattern generator is particularly amenable for analysis; it has been studied intensively by Allan Selverston of the University of California, San Diego, Frederic Nagy of the University of Bordeaux, Eve Marder of Brandeis University, and their colleagues. It consists of 14 neurons of 6 types, which are listed in Table 10.1 and which are shown in a schematic circuit in Figure 10.11. Of these cell types, five are motor neurons and only one, the AB cell, is an interneuron.

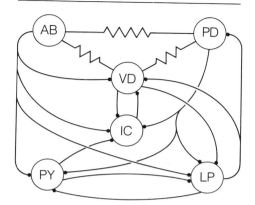

10.11   The circuitry of the central pattern generator that controls the pyloric region of the stomach in the crab. AB is an interneuron, the other cells are motor neurons. All of the chemical synapses that make up the circuit are inhibitory *(filled circles)*; the synapses between the AB, PD, and VD neurons are electrical *(resistor symbols)*.

Table 10.1   The neurons of the pyloric central pattern generator in lobsters and crabs

Neuron	Number	Function	Approximate phase shift
Anterior burster (AB)	1	Interneuron	0
Pyloric dilator (PD)	2	Dilator	0
Lateral pyloric (LP)	1	Constrictor	0.50
Ventricular dilator (VD)	1	Dilator	0.75
Inferior cardiac (IC)	1	?	0.50
Pyloric (PY)	8	Constrictor	0.65

Note: A repeating pattern that is in phase with another has a phase shift of 0 or 1 relative to it; if it is out of phase with the other its phase shift is 0.5. In this case a phase shift of 0.5 indicates that the neuron is active while the other is silent; for a phase shift of 0.75, the activity is shifted by three-quarters of a cycle and there is usually some overlap of activity.

Two types of synapses are found in the circuit: electrical junctions between the AB, PD, and VD neurons, and inhibitory chemical synapses between the rest of the neurons, including reciprocal synapses between many pairs. Not all of the synapses have equal strengths. For example, the electrical coupling between the AB and PD neurons is very strong, whereas the electrical coupling between the AB and VD neurons and the PD and VD neurons is usually weak.

Some features of the circuit are as follow: The AB neuron is spontaneously active, firing bursts of impulses rhythmically. The strong electrical coupling of the AB neuron to the PD neurons means that the AB and PD cells depolarize synchronously and fire in phase. The AB and PD neurons inhibit the LP and PY constrictor neurons, which consequently fire out of phase with the AB and PD neurons. Because reciprocal inhibitory synapses are located between the LP and PY neurons, these cells are also out of phase with one another. Table 10.1 presents the approximate phase relations of the neurons in the circuit.

Although only 30 neurons are in the stomatogastric ganglion, between 100 and 250 axons are in the stomatogastric nerve that link the ganglion with the rest of the animal. These axons contain

a variety of neuroactive substances, including a number of monoamines and peptides thought to act as neuromodulators, such as
dopamine, octopamine (a dopamine-like catecholamine), serotonin, acetylcholine, histamine, a proctolin-like peptide, and a
cardioexcitatory-like peptide called FMRFamide. Many of these
substances influence the pyloric rhythm, each in a different way.
In what follows I describe the various effects the substances have
on the spiny lobster's pyloric system.

- *Monoamines*   Immediately after dopamine is applied to a stomatogastric ganglion, the frequency of the pyloric rhythm decreases because the dopamine prompts the PD neurons to hyperpolarize. Somewhat later (5–10 min), the rhythm increases,
  as does the intensity of the pyloric bursting. This comes about
  by late excitatory effects of dopamine on the LP and PY neurons.
      Octopamine also has time-dependent effects on the circuit.
  Initially it simply increases the frequency of the pyloric
  rhythm, but after 5 minutes or so it disrupts the normal pyloric rhythm by making the LP neuron fire in long bursts. The
  subsequent inhibition of the PD neuron then alters the whole
  system's firing pattern.
      Serotonin also affects the LP neuron. In the crab, it has the
  surprising effect of changing the alternating PD-LP pattern of
  activity to one in which long LP bursts are interspersed by
  several PD bursts of short duration. Serotonin also silences the
  PY neuron. These dramatic effects are shown in Figure 10.12.
  Histamine, on the other hand, completely inhibits rhythmic
  pyloric activity, and only the LP neuron continues to fire, but
  it does so in a sustained fashion after histamine is applied to
  the ganglion.

- *Acetylcholine*   Muscarinic cholinergic agonists typically increase the frequency of pyloric neuron bursting. Acetylcholine
  affects many of the pyloric neurons by changing their membrane characteristics. That is, following the addition of acetylcholine to a ganglion, many of the neurons generate so-called
  plateau potentials. Although not well understood, these potentials appear to involve a series of time- and voltage-dependent

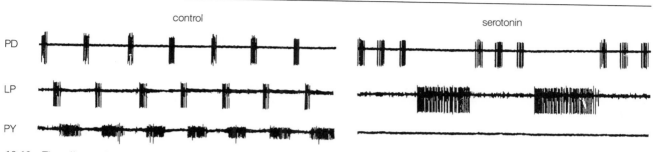

control   serotonin

PD

LP

PY

1 sec

10.12   The effects of serotonin on three of the neurons in the pyloric central pattern generator. Following the application of serotonin, the alternating bursting of activity between the PD and LP neurons is replaced by a pattern in which three PD bursts are interspersed between one long LP burst of activity. Furthermore, the PY neuron ceases to fire when serotonin is present.

conductance changes in the membrane that prolong depolarization of a cell following its excitation.

- *Neuropeptides*   FMRFamide selectively influences the activity of the PY neuron by increasing its firing rate, which may result in the inhibition of the LP neuron. Proctolin, which increases the frequency of the pyloric rhythm, probably enhances activity in LP, AB, and PY neurons.

Other neuroactive substances are known to affect the stomatogastric ganglion and the pyloric system, but the data I have noted so far are sufficient to enable us to draw several conclusions. First, the circuit is affected by a surprisingly large number of substances, each of which acts in a specific way. That is, no two substances so far tested alter the circuit in an identical fashion. Each neuromodulator reconfigures the circuit and creates a different motor pattern from the same network. Also, a single neuromodulator can have multiple effects on a single neuron or different effects on other cells in the system.

What about the mechanisms underlying modulation of neural circuits? Data from the stomatogastric ganglion have been revealing, as have results from other systems. I have emphasized in this chapter that for central pattern generators to function, both circuitry and special neuronal properties are required. Neuromodulators can affect not only intrinsic properties of single cells but also the circuitry of a central pattern generator. An example of the former is the effect of acetylcholine agonists on pyloric neu-

dopamine

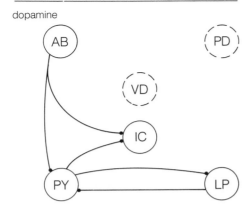

octopamine

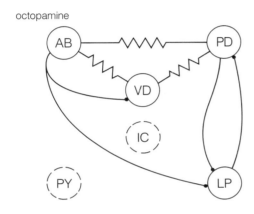

serotonin

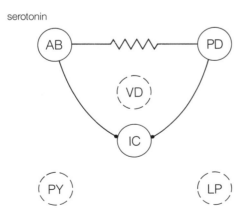

10.13 The major functional synaptic pathways in the pyloric CPG following the application of dopamine, octopamine, and serotonin to the stomatogastric ganglion. These agents increase the strength of some synaptic interactions, weaken others, and leave some unchanged.

rons. After depolarization of the cell, plateau potentials occur in the presence of the acetylcholine analogues, but they are not observed when the analogues are not present. Neuromodulators also can induce bursting activity in isolated pyloric cells that are quiescent, and different modulators induce bursting in different cells. Other intrinsic properties of neurons can be altered by neuromodulators.

Neuroscientists now have good evidence that the strength of synaptic interactions between neurons in central pattern generators can be modified by various neuromodulators. For example, at chemical synapses the release of synaptic transmitter and the responsiveness of the postsynaptic cell to the transmitter can be modified. The extent of coupling at electrical synapses also can be modulated; it even appears that a single modulator can enhance transmission at one synapse while decreasing it at another synapse. Figure 10.13 shows, for example, the major functional synaptic pathways in the pyloric central pattern generator circuit in the presence of dopamine (top), octopamine (middle), and serotonin (bottom). Compare these circuits with the overall circuitry of the generator shown in Figure 10.11.

The main conclusion drawn from these studies is that neuromodulators can redefine the circuitry of a central pattern generator in a variety of ways. Neuromodulators may switch a circuit from one configuration to another so that quite different behaviors are mediated by the circuit, or they may fine-tune the circuit and subtly alter its output in response to slightly different environmental conditions. Changes in the circuit induced by neuromodulatory substances may be relatively short term, on the order of seconds, or they may last for substantial periods of time (minutes or more). In the next chapter I explore how a neuromodulator can alter the strength of synapses in a circuit on short- and long-term bases and how this mechanism provides a model for memory and learning.

Summary

The ability to make purposeful movements is a characteristic feature of animals. Motor systems are made up not only of effectors, such as muscles, but also of the neural circuitry necessary to control the muscles. Much motor activity is repetitive or rhythmic, and the circuits controlling these movements have been studied extensively, particularly in invertebrates.

The neural circuit underlying rhythmic behavior is called a *central pattern generator*. Its neurons interact to provide an output that leads to coordinated muscle activity. Typically, the generator is activated by sensory input, by a command neuron or system, or by both. Its output is synaptic signals onto motor neurons that innervate the relevant muscles. The three motor systems described in this chapter have different degrees of complexity.

The central pattern generator that controls the heartbeat in the leech consists of 7 pairs of interneurons, found in the first 7 ganglia of the ventral nerve cord. The first 4 pairs of interneurons are responsible for regulating the heartbeat rhythm. The first 2 interneuron pairs, termed HN(1) and HN(2) cells, coordinate the activity of the other two pairs, the HN(3) and HN(4) cells, which innervate the motor neurons that drive the heart muscles. All of the synapses between interneurons and between interneurons and motor neurons are inhibitory. When synaptically isolated from one another, all of the neurons, both interneurons and motor neurons, are spontaneously active. Thus, the cells affect the activity of one another strictly by inhibition.

The alternating or oscillatory bursting activity of the HN(3) and HN(4) cells is established by reciprocal inhibition between pairs of cells: when one cell is active, the other is inhibited. Crucial to the rhythmic activity is a special membrane mechanism that permits one member of an interneuron pair to escape from inhibition after a period of time. This cell then becomes active and inhibits the other member of the pair. Thus activity shifts after a set period of time from one member of a pair to the other. The special membrane mechanism is a voltage-sensitive channel that slowly activates upon hyperpolarization of the membrane. That is, during inhibition when the membrane is hyperpolarized, this channel slowly opens and allows mainly Na^+ to flow into the cell. This positive current depolarizes the cell and permits it to resume activity.

Escape swimming in the sea slug, *Tritonia,* is controlled by a central pattern generator made up of 12 interneurons of three types. Two of these types, the DSI and VSI cells, innervate the motor neurons that contact, respectively, the dorsal and ventral flexion muscles. Swimming consists of alternating dorsal and ventral flexion movements. The third cell type, the C2 cell, has connections only with the VSI and DSI cells. The circuit works as follows: Depolarizing input activates the DSI cells, which inhibit the VSI cells but activate the C2 cells. Activation of the C2 cells leads to excitation of the VSI cells but only after a delay of a few seconds. Once the VSI cells are activated, they inhibit the DSI cells and the C2 cells, which stop firing. When the C2 cells cease to be active, the VSI cell is no longer excited and it stops firing. This relieves the inhibition of the DSI cells, which resume activity; thus, DSI and VSI cells alternately burst.

The key to the rhythmic firing of the DSI and VSI cells is the delay in the excitation of the VSI cells by the C2 cells. Two mechanisms effect this delay. VSI-A cells receive a mixed inhibitory-excitatory input from the C2 cells. Initially the synapse is inhibitory, but after 1–2 seconds the synapse becomes excitatory and the VSI-A cells begin to fire. VSI-B cells receive only excitatory input from the C2 cells, but they have a voltage-dependent K^+ channel in their membrane that activates early and opposes the depolarizing excitatory input from the C2 cells. This channel inactivates over the course of 1 or 2 seconds, after which the VSI-B cells are able to fire.

For either of these central pattern generators to function, neural circuitry and special membrane mechanisms are required. *Tritonia* also exhibits defensive withdrawal movements when touched, and both the DSI and VSI neurons are involved in this behavior. In this instance, the DSI and VSI neurons fire in concert, not in an alternating fashion. Under these conditions, the C2 neuron is silent and another interneuron, the I-neuron, is active in the circuit. These observations indicate the neurons can participate in more than one circuit and that circuits can assume different configurations and mediate different behaviors.

Networks can be modulated or reconfigured by synaptic input. Several neuromodulatory substances alter the output of the pyloric central pattern generators found in the stomatogastric ganglion of lobsters and crabs. The pyloric central pattern generator, consisting of 14 neurons of 6 types, controls the muscles of the pyloric region of the stomach. Five of these types are motor neurons that fire rhythmically but out of phase with one another. Different neuromodulators, such as monoamines, muscarinic agonists, and peptides, alter the rhythm and strength of the firing and the phase relations between the cells—each one having a unique effect. Neuromodulators can alter the intrinsic membrane properties of the neurons and change synaptic strengths. Both chemical and electrical synapses can be modulated, and one substance can increase the strength of one synaptic interaction but reduce it at another synapse.

Further Reading

BOOKS

Selverston, A. I., ed. 1985. *Model Neural Networks and Behavior*. New York: Plenum Press. (A superb collection of articles on invertebrate model systems.)

ARTICLES

Calabrese, R. L., J. D. Angstadt, and E. A. Arbas. 1989. A neural oscillator based on reciprocal inhibition. In *Perspectives in Neural Systems and Behavior*, ed. Thomas J. Carew, pp. 35–50. New York: Alan R. Liss.

Getting, P. A., and M. S. Dekin. 1985. *Tritonia* swimming: A model system for integration within rhythmic motor systems. In *Model Neural Networks and Behavior*, ed. A. I. Selverston. pp. 3–20. New York: Plenum Press.

Harris-Warrick, R. M., and E. Marder. 1991. Modulation of neural networks for behavior. *Annual Review of Neuroscience* 14:39–57.

Marder, E., and S. O. Hooper. 1985. Neurotransmitter modulation of the stomatogastric ganglion of decapod crustaceans. In *Model Neural Networks and Behavior*, ed. A. I. Selverston, pp. 319–337. New York: Plenum Press.

Miller, J. P., and A. I. Selverston. 1985. Neural mechanisms for the production of the lobster pyloric motor pattern. In *Model Neural Networks and Behavior*, ed. A. I. Selverston, pp. 37–48. New York: Plenum Press.

Moulins, M., and F. Nagy. 1985. Extrinsic inputs and flexibility in the motor output of the lobster pyloric neural network. In *Model Neural Networks and Behavior*, ed. A. I. Selverston, pp. 49–68. New York: Plenum Press.

ELEVEN

An Invertebrate Model of Memory and Learning

HOW THE BRAIN learns things and remembers them is one of the key questions in neuroscience. Perhaps when we understand memory and learning we will be close to providing an overall theory of brain function. The past two decades have seen substantial progress in our understanding of how neurons are altered biochemically by synaptic input, and these findings have led to important new concepts regarding possible mechanisms underlying memory and learning. A good deal of this work has involved invertebrates, and in this chapter I describe one series of studies on the marine snail, *Aplysia*. Over the past century, however, a variety of hypotheses explaining memory and learning have been advanced, and I begin by summarizing some of these ideas.

Early Theories

In the latter part of the nineteenth century two principal theories of memory and learning—both involving specific cellular mechanisms—were presented. The first proposed that learning was the result of a change in synaptic function. Early ideas of this sort grew out of Ramón y Cajal's classic view of the nervous system as an assemblage of discrete units, neurons, that are connected in specific ways and that communicate at specialized sites (only later were these termed synapses). When a specific pathway in the brain experiences repeated activity, changes occur in the synapses that may be long-lasting or even permanent. Specific synapses might become more efficacious during learning; that is, transmission becomes easier or greater in magnitude as a result of alterations on either the pre- or postsynaptic side of the junction. One early idea was simply that the distance between pre- and postsynaptic elements decreased during learning.

A second nineteenth-century idea about memory was that neurons sprout new processes and make new synapses during the learning process. Over the years these two basic ideas have been elaborated upon as we have learned more about cellular and neuronal mechanisms. For example, in the 1960s investigators hypothesized that synaptic input led to the synthesis of specific proteins in neurons that could affect either neuronal or synaptic function and hence account for memory and learning. In many animals, inhibitors of protein synthesis interfered with learning, and this was taken as support for the hypothesis. At that time a few investigators believed they had isolated specific memory proteins; some claimed that an animal injected with RNA extracted from animals that had learned a task would show evidence of having learned the task as well. There is now good evidence that changes do take place at synapses when those synapses are repetitively activated, and some changes last for weeks, months, or even longer. Yet it is uncertain whether there are specific memory proteins or whether learning is produced or even enhanced by providing RNA or any other substance to an animal or human.

Reverberating Circuits

Another cellular theory of memory and learning, known as the reverberating circuit theory, was proposed by Alexander Forbes at Harvard in the early 1920s. His theory was based on observations of persistent activity in neurons and pathways after stimulation. Forbes suggested that this prolonged activity, called an after-discharge, was sustained by impulses circulating around a loop of neurons. Such activity, he postulated, could represent a "memory." This theory did not require any functional changes in synapses or neurons or any anatomical modifications in the neurons.

Although this theory was attractive to physiologists and psychologists, it is not held in high regard today. Compelling evidence argued against it: an animal trained to perform a task was chilled to the point when all electrical activity in its brain ceased, yet when the animal was revived it could still do the task. Hence, it retained the memory even though any reverberating circuit activity had been stilled.

Aggregate Field Theory

In the 1930s Karl Lashley, a psychologist at Harvard, offered a theory of memory and learning that challenged cellular theories. This became known as the aggregate field theory. Lashley's premise was that specific cellular connections are less important to learning than are the many parts of the brain. What really counts are not circuits but the total amount of brain available. Lashley came to these conclusions by making lesions in the brains of animals and then testing their ability to learn a maze. He found that the number and extent of lesions had a greater effect on their ability to learn the maze than did the placement of the lesion. The whole cerebral cortex—where he placed his lesions—seemed essential for learning a task and remembering it. A memory, he proposed, is not stored in one discrete place in the brain, as cellular theories would dictate, but is distributed widely in the cortex.

Although Lashley's ideas were warmly embraced by some brain scientists, and some still hold to the general ideas, recent work has led to a reinterpretation of Lashley's results and to the accumulation of more evidence in favor of discrete cellular mechanisms underlying memory and learning. For example, many hold that maze learning is not an appropriate paradigm for localizing aspects of brain function because it requires several motor and sensory capabilities. Furthermore, certain functions have now been linked to specific areas of the cerebral cortex, in contradistinction to Lashley's main thesis. Some of the strongest evidence that specific and unique cells and synapses relate to memory and learning has come from studies on invertebrates; so in this chapter I describe short- and long-term changes in a simple reflex in a well-studied invertebrate.

The Gill-Withdrawal Reflex in *Aplysia*

Aplysia californica, a gastropod mollusk or marine snail about 6–8 inches long, dwells along the coast of California, mainly in tidal pools. Floating near the water surface, it eats seaweed. Figure 11.1, a drawing of *Aplysia*, shows that its dorsal surface is covered by two fleshy parapodia that ordinarily cover the mantle, a protective covering which overlies the gill. When the parapodia are separated, the mantle is exposed and the underlying gill is partially revealed.

When the mantle or associated siphon are gently touched, the gill rapidly withdraws into the mantle cavity. The gill-withdrawal reflex gradually decreases in magnitude if the stimulation is repeated; this is called *habituation*. If a strong stimulus is applied

11.1 *Aplysia californica,* in side and dorsal views. The gill lies under the mantle and siphon. Also indicated in the drawing are the approximate positions of the head ganglia (buccal, cerebral, pleural, and pedal) and abdominal ganglia.

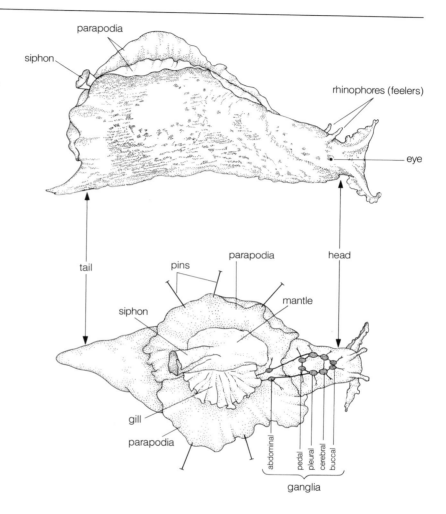

elsewhere on the animal, to the head or tail for example, and then a gentle stimulus is applied to the mantle, the reflex is enhanced. That is, the amount of gill withdrawal is greater than it ordinarily is; this is called *sensitization*. A strong stimulus applied to the head or tail also rapidly reversed habituation of the reflex, a process known as *dishabituation*. Dishabituation and sensitization share many of the same properties in *Aplysia* and the underlying mechanisms may be similar for the two phenomena.

Habituation and sensitization in *Aplysia* reflect significant changes in the efficacy of the reflex response. Depending on how many conditioning stimuli are applied and how long they are delivered, the changes can last from hours to weeks. These changes in the reflex, induced by repeated stimuli, provide a model for an elementary form of memory and learning.

The gill-withdrawal reflex is not the only neurobiological system in which possible memory and learning mechanisms have been studied, but it represents in my view the most satisfactory analysis so far provided. The approaches taken and principles used in the analysis are understandable in terms of what was presented in Part I of this book, and so I shall describe the analysis step by step. It is useful to understand how an analysis is carried out experimentally, the thinking behind the experiments, and the conclusions reached. The work, mainly that of Eric Kandel, James Schwartz, and their co-workers, first at New York University and more recently at Columbia, has had a profound impact on our thinking about the mechanisms responsible for memory and learning.

Habituation

When an *Aplysia* is immobilized in a small aquarium and a gentle stimulus—a jet of sea water from a Water Pik—is applied to the mantle, the gill rapidly withdraws. The amount of gill withdrawal can be quantitatively measured by placing a photocell under the gill. As the gill withdraws, more of the photocell is exposed to light and its electrical output increases. The record of photocell output provides a measure of the extent and duration of gill withdrawal. Figure 11.2A shows an experimental setup for studying changes in the reflex in the whole animal.

In a single training session in which stimuli are presented approximately every 30 seconds to 3 minutes, the extent of gill withdrawal is reduced by about 50 percent after only 4 trials and by 70 percent after 10–15 trials (Figure 11.2B). Further habituation is observed if the stimulation continues for another 30–40 trials. When the stimulation is stopped, recovery begins shortly thereafter, but it may require 2 hours or more to reach completion. If a strong stimulus is presented to the head or neck region of the animal, the response is immediately restored completely; it is dishabituated. The strong stimulus can also create a response that is larger than the initial unhabituated one; the reflex is now sensi-

11.2 **(A)** An experiment set up to study the gill-withdrawal reflex. The animal is restrained with a head clamp and a stabilizing plate is positioned between the gill and mantle. A photocell is placed under the gill and a water pik is used to stimulate the mantle or siphon. **(B)** The gill-withdrawal response rapidly habituates—by the tenth trial, the response is only about 30 percent of the original response. Recovery requires about 2 hours. After the gill-withdrawal response is habituated, it can be quickly restored (dishabituated) with a strong stimulus to the head or tail. If the stimulus is strong enough, the response may be larger than the control response. This is sensitization.

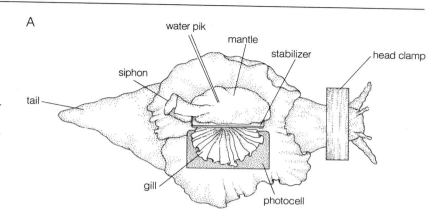

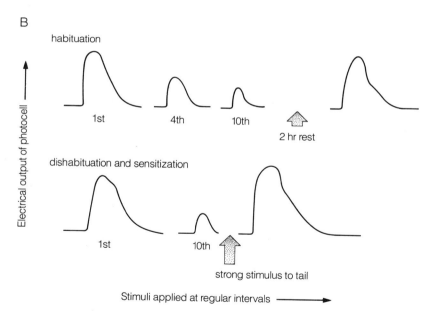

tized, and the responses remain elevated in magnitude for some time thereafter.

A single training session of 10 tactile stimuli leads to a significant habituation of the gill-withdrawal reflex that can last 2 hours or more. If this training session is repeated for 4 consecutive days, the animals continue to show habituation 1 day or even 1 week later. Hence, repeated training sessions result in long-term habituation.

Studying long-term habituation of the gill-withdrawal reflex is difficult because the animal must be restrained for testing. When the siphon is touched, however, not only does the gill withdraw but so does the siphon itself—and this response habituates as the gill-withdrawal reflex does. By simply watching for the reappearance of the siphon in an unrestrained animal, and by timing how long it takes the siphon to reappear after its withdrawal, it is possible to study habituation of this reflex in the same animal for weeks at a time.

If the siphon is stimulated 10 times in a training session and the training sessions are repeated on 4 consecutive days, habituation progressively builds so that by the fourth day, the median duration of the response is only 20 percent of that observed on the first day. Furthermore, the response shows clear evidence of habituation for several weeks thereafter. Records of typical experiments are shown in Figure 11.3. In (A) we see the progressive decrease in response duration per trial during the training and recovery sessions; and in (B) the median duration of the response for the training and recovery sessions plotted as a percent of the control measurement (the median duration of the trials on day 1).

That habituation can be either short-term (lasting hours) or long-term (lasting weeks) is of particular interest in light of the widely held view that there are two forms of memory, short- and long-term memories. In humans and animals, we find both short- and long-term forms, and in all cases short-term memories are much more labile than long-term ones. For instance, a blow to the head or intense mental activity will often cause an individual to forget recent experiences or recently learned material. Brian Boycott and J. Z. Young, working at the Zoology Station in Naples, Italy, showed in the octopus that removal of a specific region of the brain prevented the animal from remembering to avoid a noxious stimulus for more than about 15 minutes; the animal was unable to form long-term memories although short-term memory seemed unaffected. Hence, the animal would remember to avoid the noxious stimulus for up to 15 minutes but not thereafter. Control animals, though, would remember to avoid the noxious stimulus for weeks.

From these and many other observations came the notion that

11.3 Long-term habituation of siphon with-drawal in *Aplysia*. **(A)** Habituation during daily training sessions *(open circles)* and during recovery from long-term habituation *(closed circles)*. Ten stimuli were presented on each of the days indicated and the duration of the withdrawal response measured. During the training, the initial response time was shorter each day and habituation became faster and more extensive. Reversal of these effects was seen during the recovery period. **(B)** Median response during training and recovery periods in the experiments described in **(A)**. Control duration (100 percent) is the median response time of the 10 trials during the first day. After just 4 days of training, the median duration time was less than 20 percent of the control, and even on day 26 of recovery, the response duration was still only about half its original level.

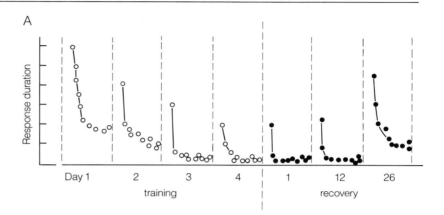

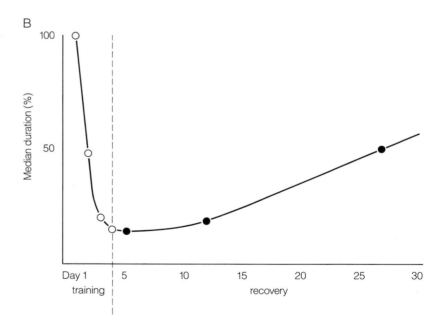

new memories are fragile; only after some time do they become stable. In other words, initial memories may be the effect of active neural processes whereas long-term memories may involve permanent structural changes in the brain. (Donald Hebb, a Canadian neuroscientist, suggested in 1949 that short-term memory could be reverberating circuit activity and long-term memory could be alterations in synaptic or neuronal structure, thus linking the two cellular theories of memory and learning.) That habituation in *Aplysia* can be a short- or a long-term phenomenon provides an opportunity to examine whether these are two separate processes or an extension of the same process, and to see how they interact.

The Neural Circuitry of the Reflex

The head region of *Aplysia* has four prominent ganglia—the buccal, cerebral, pleural, and pedal—that form a ring around the animal's esophagus. Posteriorly (approximately under the anterior margin of the mantle) lies the abdominal ganglion (Figure 11.1). The abdominal ganglion contains neurons that innervate the gill and is, therefore, the ganglion where the gill-withdrawal reflex is mediated. The abdominal ganglion contains about 1,500 neurons. By recording from individual neurons or their axons, stimulating them singly or in pairs, and observing effects on gill withdrawal, Kandel and his colleagues identified 24 sensory neurons coming from the siphon or mantle, 6 principal motor neurons innervating the gill musculature, and 3 interneurons within the ganglion involved in the reflex. The sensory neurons innervate both the interneurons and the motor neurons, and the interneurons innervate the motor neurons. Two of the interneurons are excitatory to the motor neurons, and one is inhibitory to the sensory neurons, the excitatory interneurons, and perhaps one of the motor neurons. A summary of the neural circuitry is presented in Figure 11.4A. For simplicity only one sensory neuron is illustrated and the figure depicts the synapses being made onto cell bodies. In fact, the sensory neuron synapses are made onto dendrites of the motor neurons and interneurons in the ganglion's neuropil.

Except for the inhibitory interneuron, all the other classes of neurons make similar synaptic contacts, and so the basic circuitry of the reflex can be simplified further (Figure 11.4B). The cell bodies of some sensory neurons lie outside the abdominal ganglia but they extend their axons into the ganglion, where they synapse on the dendrites of the interneurons and motor neurons.

Despite the circuit's obvious simplicity, many sites could participate in habituation. Some are peripheral (outside of the ganglion), involving sensory adaptation, muscle fatigue, or the nerve muscle synapse; others are central (within the ganglion), including syn-

11.4 Neural circuitry for the gill-withdrawal reflex in *Aplysia*. **(A)** Six motor neurons and 3 interneurons (2 excitatory and 1 inhibitory) are involved in the reflex. Only 1 of the 24 sensory neurons is shown; the others make similar synaptic contacts. **(B)** The basic circuit simplified further. Only 1 motor neuron and interneuron are shown. Those parts of the circuit within the abdominal ganglion are enclosed within the circle. Note in both **(A)** and **(B)** that excitatory synapses are indicated as open symbols, inhibitory as filled symbols. Since the interneurons make both excitatory and inhibitory synapses on motor neurons, the interneuron synapse in **(B)** is partially filled.

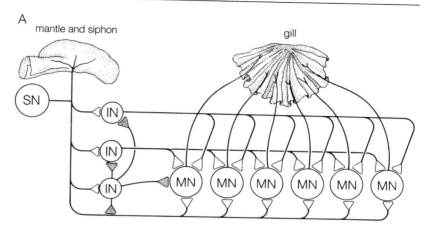

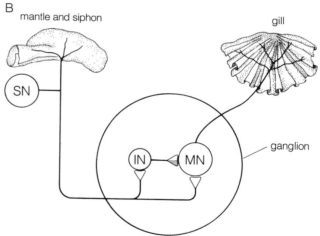

aptic interactions between sensory neurons and motor neurons, sensory neurons and interneurons, or interneurons and motor neurons.

PERIPHERAL MECHANISMS

To test whether sensory receptor adaptation contributes to habituation, Kandel and his colleagues recorded the action potential discharge in the axons of sensory neurons following repeated stimuli to the skin. As we see in Figure 11.5A, the responses recorded

11.5 Experiments showing that peripheral mechanisms do not contribute significantly to habituation. **(A)** Sensory neuron responses do not vary substantially upon repeated stimulation of the mantle. **(B)** Following repeated activation of the motor neuron by intracellular current injection *(I)*, gill contractions (as measured by the electrical output of the photocell) remain constant.

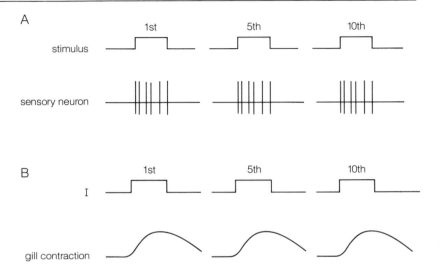

from the sensory neuron axons did not vary significantly from the first to the tenth stimuli. Some adaptation can be seen in the responses; following application of the stimulus, the initial rate of impulse discharge was greater than the maintained rate, but the total number of impulses recorded per stimulus was the same for all trials. These experiments indicate that adaptation of the sensory neuron does not contribute significantly to the habituation process.

The possibility that habituation of the nerve-muscle synapse comes into play—or perhaps even muscle fatigue—was tested also. Motor neurons innervating the gill were impaled with micropipettes and stimulated directly by passing a depolarizing current into the cell. Gill contractions remained constant despite repeated and prolonged stimulation of the motor neurons (Figure 11.5B). In other words, repeated stimulation of the motor neurons, at intervals that cause habituation in the intact animal, did not result in habituated responses.

CENTRAL MECHANISMS

These experiments indicated that habituation comes about as a result of changes in the abdominal ganglion itself and not as a result of a peripheral mechanism. This was tested by recording

11.6 Habituation and dishabituation observed in the discharge of the motor neurons following repeated stimulation of the mantle. The lower records show that the extent of gill contraction is closely related to the number of impulses generated in the motor neuron.

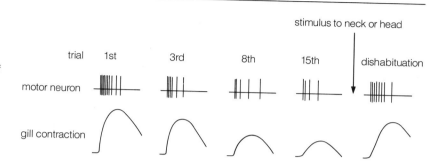

from the motor neurons while stimulating the mantle in the intact animal (Figure 11.6). The first tactile stimulus applied to the mantle elicited a strong burst of impulses in the motor neuron, but with repeated stimulation the number of impulses generated in the motor neuron dropped substantially. Furthermore, the extent of gill contraction followed closely the output of the motor neurons—that is, the number of spikes traveling down the axon. The extent of gill withdrawal, then, reflects motor neuron output. It was also possible to dishabituate the motor neuron at any time by applying a noxious stimulus to the neck or head of the animal.

Following the demonstration that the motor neurons themselves habituate after repeated stimulation of the mantle or siphon, Kandel and his co-workers simplified the experimental setup. They reasoned that since the output of the motor neurons is an indicator of gill contraction, and since no habituation of the sensory neuron response occurs, they might be able to study habituation by working with the ganglion alone, isolated from the rest of the animal. Figure 11.7 shows that this is indeed possible. The siphon nerve that carries sensory neuron axons from the siphon was electrically stimulated with extracellular electrodes, and responses of the motor neurons innervating the gill were recorded intracellularly. The recorded motor neuron was hyperpolarized to prevent spike-firing of the cell and to enhance EPSP amplitude (see Chapter 5). With repeated stimulation of the siphon nerve, the EPSP gradually habituated—it grew smaller in amplitude. Subsequent stimulation of a connective resulted in dishabituation of the response. Hence, the isolated ganglion demonstrated both habituation and dishabituation equivalent to that seen in the intact animal.

11.7 Habituation and dishabituation in an isolated abdominal ganglion preparation. Repeated stimulation of the nerve leading from the siphon resulted in decreased EPSP amplitudes recorded from a motor neuron. Stimulation of one of the connectives leading to the head resulted in dishabituation of the EPSP.

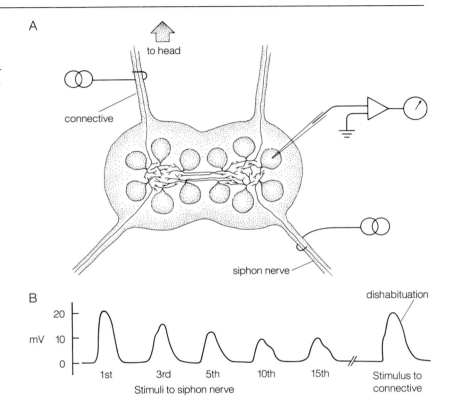

The next stage of the analysis was to investigate the cellular mechanisms at work. For example, the changes in EPSP amplitude during habituation could be caused by a change in synaptic input to the motor neuron or by a change in the properties of the motor neuron itself. If, for example, the membrane resistance of the motor neuron gradually changed after repeated stimulation of the cell, EPSP amplitude would decrease. As we found in Chapter 5, if the resistance of the membrane decreases over time, the voltage developed across the membrane in response to a constant synaptic input would decrease in accord with Ohm's law, $V = IR$. This possibility was tested by measuring input resistance of a motor neuron during habituation of its EPSP. No change in resistance was observed, indicating that the depression of EPSP during habituation was likely to result from altered synaptic input, not altered membrane resistance of the motor neuron.

SYNAPTIC CHANGES

Two changes in synaptic input to the motor neurons might account for habituation. Either there could be a buildup of inhibitory synaptic input to the motor neurons with repeated stimulation, or there could be a progressive decrease in excitatory synaptic drive to the motor neurons or interneurons.

As a test, the membrane potential of the motor neurons was hyperpolarized to a level below that of IPSP reversal potential (more negative than -80 mV). Hence, when evoked, IPSPs would be reversed in polarity. If IPSP buildup accounts for habituation in the motor neuron, the response during repetitive stimulation should be greater, not less. But the tested response of the motor neuron decreased in amplitude, as it did at normal membrane potentials.

Kandel and his co-workers concluded from this experiment, that IPSPs do not decrease motor neuron activity and that habituation occurs when the excitatory synaptic drive to the neurons is decreased. Furthermore, a variety of experiments indicated that the major excitatory input to the motor neurons is directly from the sensory neurons themselves, not from the interneurons. Since EPSP amplitude declines substantially during habituation, it follows that the major locus for habituation is at the sensory-motor neuron synapse.

PRE- OR POSTSYNAPTIC MECHANISMS

EPSP amplitude could decrease as a result of presynaptic changes (for example, a decrease of transmitter release) or postsynaptic changes (a decrease in the sensitivity of the postsynaptic receptors to the transmitter). An ideal way to begin to examine the properties of a synapse is to record the miniature EPSPs generated across the postsynaptic membrane. A lower frequency of min EPSPs means that fewer synaptic vesicles are released per stimulus, whereas a diminished amplitude of min EPSPs means that less transmitter per vesicle is released or that the postsynaptic receptors let fewer ions cross the membrane in response to a vesicle's worth of transmitter.

Individual min EPSPs cannot be recorded from the motor neurons in *Aplysia* as they can from muscle fibers, primarily because the recording site is so remote from the synapses themselves. However, by stimulating a single sensory neuron and recording from a motor neuron under conditions of decreased release of transmitter from the sensory neuron (achieved by lowering Ca^{2+} levels in the bathing medium), one can record tiny EPSPs having substantial fluctuations in amplitude (see Figure 11.8A). These results indicate that transmitter released from the sensory neuron terminals is quantal (so we know the transmitter is released from vesicles) and that estimates of the number and relative size of the quanta released at the synapse can be obtained by quantal analysis.

The analysis was carried out by stimulating the sensory neuron to fire one action potential over and over again and measuring each time the amplitude of the EPSP that was evoked in the motor neuron. Sometimes no EPSP was recorded, but usually a response was observed. By plotting a histogram of the amplitudes recorded, one can identify amplitude peaks that are multiples of one another, as shown in Figure 11.8B. The first and largest of the peaks is at 34 μV, the second and next largest peak at is 68 μV, the next at 102 μV, and so on. The presumption is that the 34 μV response represents the unit (single vesicle) response, whereas the other peaks represent a response to the release of 2 or 3 vesicles of transmitter.

After habituation, the experiment can be repeated and the data plotted on the histogram. Habituation dramatically increases the number of failures (no response recorded) and correspondingly decreases the number of responses observed. The amplitude peaks, though, are comparable in both the control and habituated states; peaks at 34 μV, 68 μV, and perhaps 102 μV are still obvious in the histogram. These results are evidence that EPSP amplitude during habituation is depressed by a reduction in the number of vesicles released from the synaptic terminals and not by a decline in the amount of transmitter per vesicle or by a drop in the sensitivity of the postsynaptic receptors to the released transmitter.

Why are there fewer vesicles releasing transmitter from the presynaptic terminal? The best guess is that Ca^{2+} has something to do with it. As we found in Chapter 6, Ca^{2+} entry into the presynaptic terminal is required to bind synaptic vesicles to the membrane and then to release transmitter. In preparations whose Mg^{2+} levels

11.8 Quantal analysis of transmission at the sensory-motor neuron synapse in *Aplysia*. **(A)** EPSP responses recorded in a motor neuron following stimulation of the sensory neuron under conditions of low levels of transmitter release. The EPSP amplitude varied significantly in response to a constant stimulus, suggesting quantal release of transmitter. Occasionally no response was evoked in the motor neuron following stimulation *(arrows)*. **(B)** Idealized response-amplitude histograms of data generated by experiments similar to the one shown in **(A)** under control, habituated, and sensitized conditions. Under all conditions, similar voltage peaks were observed, at approximately 34, 68, and 102 μV. Following habituation, many more response failures occurred than under control conditions, whereas following sensitization there were fewer response failures. Correspondingly, more responses were recorded following sensitization and fewer following habituation than under control conditions.

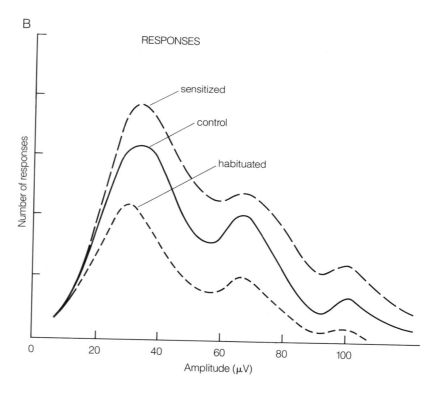

are raised externally, preventing the entry of Ca^{2+} into the terminal, the response does not habituate when the sensory neurons are repeatedly stimulated. This means that Ca^{2+} must enter the presynaptic terminal for habituation to occur. The proposal is that the buildup of Ca^{2+} depresses further entry of Ca^{2+} into the terminal; when the terminal is next depolarized by an action potential, less Ca^{2+} enters, fewer vesicles bind to the presynaptic membrane, and less transmitter is released. The result: a smaller EPSP in the motor neuron.

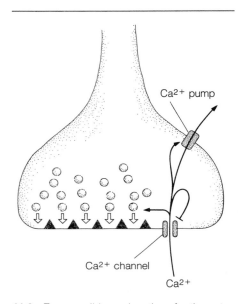

11.9 Two possible explanations for the auto-regulation of Ca²⁺ levels in the sensory neuron's presynaptic terminal. Ca²⁺ admitted into the terminal via the voltage-sensitive channels could decrease the conductance of the channels or enhance the activity of Ca²⁺ pumps. Either effect would decrease the buildup of Ca²⁺ following subsequent depolarization of the terminal.

How internal levels of Ca²⁺ regulate Ca²⁺ entry into the terminals is not fully understood. One possibility is that elevated Ca²⁺ levels within the terminal alter the voltage-sensitive Ca²⁺ channels that open upon membrane depolarization so they admit less Ca²⁺ per unit depolarization. Another possibility is that Ca²⁺ stimulates pumps in the terminal membrane to transport Ca²⁺ out of the neuron, which also would limit Ca²⁺ from accumulating in the terminal after depolarization. These two possibilities are diagrammed in Figure 11.9. Later, I return to more specific mechanisms by which Ca²⁺ could alter channel or pump activity within the terminal.

Sensitization

A strong stimulus applied to the head or tail region of *Aplysia* rapidly enhances both gill- and siphon-withdrawal reflexes, and the responses remain enhanced for some time. Indeed, by presenting a sensitizing stimulus to an animal 4 times a day over a period of 4 days, enhanced responses are recorded for more than a week after the last sensitizing stimulus was presented. If a sensitizing stimulus is presented to an animal whose withdrawal responses are habituated, the reflexes are rapidly dishabituated—and, as noted earlier, sensitization and dishabituation in *Aplysia* are effected by similar mechanisms.

Examination of the circuitry involved in sensitization reveals that sensory neurons from the head or tail (which I will call HSN cells) synapse on interneurons within the ganglion (HIN cells) that make synapses onto the mantle and siphon sensory neuron (MSN) terminals. Hence, sensitization is mediated by a presynaptic mechanism (namely, before the synapse onto the motor neuron). This circuitry is represented in Figure 11.10.

The terminals of the HIN cells contain small, dense-cored vesicles (see Figure 11.11), which suggests the neuroactive substance released from these terminals is a monoamine; a body of evidence indicates that serotonin, an indoleamine, or a substance very similar to serotonin is the neuroactive agent released. Applying serotonin to the ganglion raises the amplitude of EPSPs generated in the motor neuron (MN) when the mantle sensory neurons (MSN) are stimulated. Serotonin antagonists, like LSD, prevent EPSP enhancement when a sensitizing stimulus is presented.

How does serotonin increase EPSP amplitude? An obvious sug-

11.10 Simplified neural circuitry for both habituation and sensitization of the gill-withdrawal reflex. The pathway for habituation of the reflex is the same as shown in Figure 11.4B; for clarity's sake, the sensory neuron from the mantle and siphon is here labeled MSN. The sensitization pathway involves neural input from sensory neurons from the head or tail, here labeled HSN. When the head or tail is prodded, the HSN signals an interneuron (HIN) that synapses onto the MSN terminal. Through this connection, a stimulus to the head or tail affects the input from the mantle sensory neuron (MSN) to the motor neuron (MN).

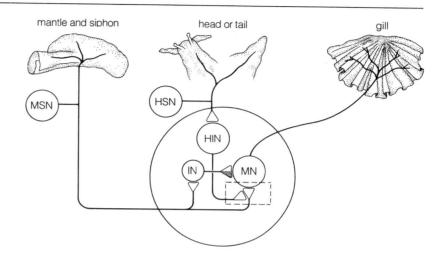

gestion is that more transmitter is released from the MSN terminals after the sensitizing pathway is activated. By repeating the quantal analysis for sensitized ganglia, we find that this is the answer (Figure 11.8). After sensitization, fewer response failures and more responses are recorded following stimulation of the mantle sensory neuron. The amplitude peaks, however, are again in the same range: The first peak is at about 34 μV, the second at 68 μV, and the third at 102 μV. Thus, sensitization is a result of *more* vesicles released per unit depolarization of the terminal membrane!

How might this come about? As noted in Chapter 6, monoamines often interact with membrane receptors linked to G-proteins and second messenger systems. Prolonged stimulation of the nerves from the head leads to increases in cyclic AMP levels within the abdominal ganglion, suggesting that serotonin activates adenylate cyclase in the MSN terminals.

Confirmation that serotonin's effect on these terminals is mediated by cyclic AMP was provided by experiments that applied a membrane-permeable analogue of cyclic AMP to the ganglion. Like serotonin, it enhanced excitatory synaptic transmission between the sensory and motor neurons. Intracellular injection of cyclic AMP into a mantle sensory neuron also enhanced EPSP amplitude in the motor neuron (MN) when that sensory neuron was stimulated.

11.11 (A) A close-up view of the synaptic terminals in the area boxed off in Figure 11.10. Serotonin, released by the HIN cell onto the MSN terminal, promotes the synthesis of the cyclic AMP from ATP. The cyclic AMP activates kinases that phosphorylate K^+ channels. The conductance of these channels is decreased, prolonging the duration of action potentials invading the terminal, as the trace of the membrane potential in (B) shows. Consequently more Ca^{2+} enters the MSN terminals and more transmitter is released.

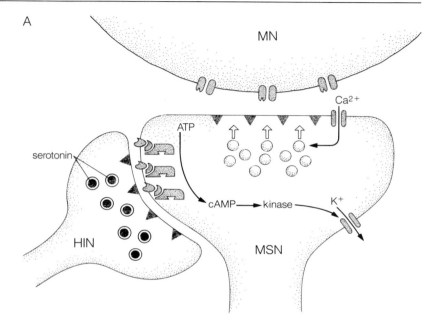

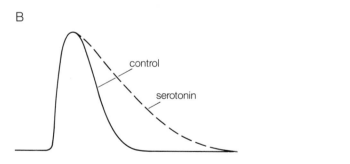

The next question is how cyclic AMP increases the number of vesicles released from the MSN terminal when it is depolarized by a set amount. The obvious answer would seem to be that the Ca^{2+} channels are altered, but this turned out to be incorrect. Rather, serotonin via cyclic AMP affects a specific K^+ channel (called the S-channel) present in the MSN terminal membrane. This is accomplished when cyclic AMP activates a kinase that phosphorylates the K^+ channel and thereby lowers the K^+ conductance of the terminal. Serotonin and cyclic AMP accomplish this by reducing the number of functional S-channels in the terminal membrane.

How does reducing the conductance of a K^+ channel release more transmitter from the terminal? When an action potential invades and depolarizes the MSN terminal, K^+ channels open and allow K^+ to leave. The efflux of K^+ from the terminal contributes to the terminal's repolarization; that is, it helps the membrane potential return to its resting potential. Phosphorylation of the K^+ channel reduces the current flow through these channels and prolongs the terminal's depolarization induced by the invading action potential. This, in turn, allows the voltage-sensitive Ca^{2+} channels in the membrane to remain open longer than is normally the case, and for more Ca^{2+} to enter the terminal. With more Ca^{2+} in the terminal, more synaptic vesicles bind to the presynaptic membrane and more transmitter is released. This process is illustrated in Figure 11.11, along with records of the prolonged action potential following the application of serotonin.

Long-Term Habituation and Sensitization

Short-term habituation and sensitization can be explained by modifications in synaptic transmitter release from the sensory neuron terminals and, further, the modifications can be related to the amount of Ca^{2+} entering the terminal under various conditions. What about long-term habituation and sensitization? Are additional factors at work here?

Evidence of other factors came initially from experiments on synaptic interactions between individual sensory neurons and motor neurons. In control animals, the chance that any one sensory neuron makes a detectable synaptic connection with a motor neuron is about 90 percent. Following long-term habituation, however, the number of detectable connections between any two cells falls to about 30 percent. These results suggest that the number of synapses between sensory and motor neurons is reduced or, at the very least, that the synapses are altered in strength.

To test these possibilities, sensory neurons were injected with horseradish peroxidase so their terminals could be visualized in the electron microscope. The synapses made by the sensory neurons are typically found in terminals that arise along the sensory neuron axons. Several distinctive features were observed at the synapses. First of all, the number of terminals varied in control, long-term habituated, and long-term sensitized animals. There were significantly more terminals in sensitized animals and less in habituated animals as compared with controls. Furthermore, the number of terminals with active zones (presumed sites of transmitter release) also varied significantly. Whereas in control animals 40 percent of the terminals had active zones, only 10 percent of terminals in habituated animals had active zones. In sensitized animals, though, about 65 percent of the terminals contained active zones.

There were other alterations. The size of the active zone grew after long-term sensitization (to 160 percent of the size of control active zones), whereas it decreased after long-term habituation (to 60 percent of control size). Furthermore, the number of vesicles declined in habituated terminals. Even so, the changes in active zone size and vesicle content of the terminals appeared more labile than the changes in terminal number or incidence of active zones. Hence, terminal number and incidence of active zones appeared to conform more to physiological changes than did active zone size or vesicle number.

These changes indicate that during long-term habituation and sensitization, there are structural changes in the neurons and their synapses. Presumably these are not local changes restricted strictly to the terminals themselves, but they are likely to include protein synthesis or protein breakdown related to events elsewhere in the neuron. It is also probable that changes in gene transcription come into play, which would necessarily include the nucleus of the sensory neurons.

A Summary Scheme of Habituation and Sensitization

Figure 11.12 presents a summary scheme of habituation and sensitization that ties together much of what we presently know. During habituation, repetitive stimulation of the mantle or siphon induces a buildup of Ca^{2+} in the presynaptic terminals or MSN neurons, which results in lowered conductance of the Ca^{2+} channels or enhanced Ca^{2+} pump activity in the terminal membrane. In either case, subsequent Ca^{2+} input into the terminal following depolarization is depressed. With less Ca^{2+} available during these later depolarizations, fewer vesicles bind to the presynaptic membrane and less transmitter is released. The decreased postsynaptic response is short-term habituation.

11.12 A summary scheme of the cellular mechanisms underlying short- and long-term habituation and sensitization and classical conditioning in *Aplysia*.

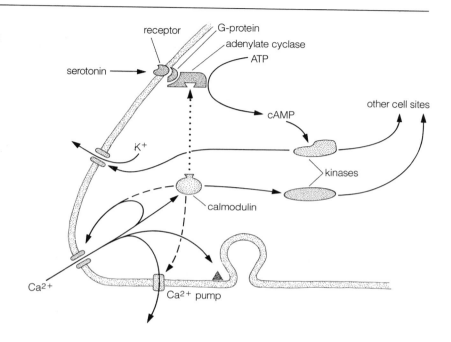

The Ca^{2+} entering the terminal can interact with the calcium binding protein, calmodulin, which activates specific kinases, and the Ca^{2+} channels, or pumps, or both may be modulated via calmodulin and phosphorylation (shown by dashed lines in Figure 11.12). Calmodulin and various kinases can also move from the terminals to other sites in the cell. In this way they mediate long-term changes that occur during long-term habituation.

Sensitization is a result of stimulation of sensory neurons from the head or tail (here called HSN cells). These cells release serotonin onto the mantle or siphon sensory neuron (MSN) terminals. The serotonin interacts with specific receptors in the MSN terminal membrane that are linked to a G-protein and adenylate cyclase. When adenylate cyclase is activated, cyclic AMP accumulates in the MSN terminal and activates a kinase resulting in phosphorylation of a K^+ channel. As a result, K^+ leaves the terminal more slowly after the action potential invasion, depolarization is prolonged, and more Ca^{2+} flows into the terminal, augmenting transmitter release.

Cyclic AMP and the cAMP-dependent kinases, like calmodulin and calmodulin-dependent kinases, can have effects elsewhere in the MSN cells and mediate the long-term changes observed in long-term sensitization. In addition to the structural changes in long-term sensitization, long-term changes are induced in the cAMP-dependent kinase molecule itself. That is, following repeated sensitizing stimuli, the kinase remains persistently active. The activity of the kinase is regulated by an inhibitory subunit of the molecule that, when dissociated from the kinase, leaves an active enzyme.

During long-term sensitization the levels of the regulatory subunit in the cells are reduced, perhaps because of enhanced protein breakdown. Thus cyclic AMP, or perhaps the kinase itself, activates proteases that break down the regulatory subunit. These interesting results suggest a variety of changes in long-term sensitization and habituation.

Classical Conditioning

So far I have described the pathways for habituation and sensitization as separate, and not as interactive. These pathways can interact, however, and their interaction explains a kind of classical conditioning exhibited by the gill-withdrawal reflex.

Classical conditioning was discovered by Ivan Pavlov, the famous Russian biologist, at the turn of the century. He found that if two stimuli are paired closely in time, with one stimulus (the conditioned stimulus) producing a response and the other (unconditioned stimulus) producing no response, eventually the animal responds to the unconditioned stimulus, when it is presented alone, in the same way it responds to the conditioned stimulus. The gill- and siphon-withdrawal reflexes demonstrate a type of classical conditioning—namely, a weak stimulus to the siphon or mantle is made more effective by pairing it with a strong stimulus to the head or tail. That is, if a strong stimulus is presented to the head or tail within 1 second after a gentle stimulus is applied to the mantle or siphon, the resulting sensitization of the withdrawal reflex is larger and more persistent than if the strong stimulus is given independently of the gentle stimulus.

Figure 11.13 depicts an experiment in which a strong stimulus

11.13 A form of classical conditioning exhibited by the gill-withdrawal reflex in *Aplysia*. When strong stimuli to the tail are paired with gentle stimuli to the mantle (**A**), the resulting sensitization (**B**) of the response is larger and longer-lasting than is the sensitization of the response when gentle stimuli to the siphon and strong stimuli to the tail are not paired.

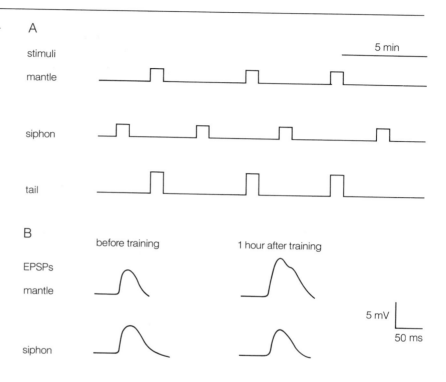

to the tail was paired with a gentle stimulus to the mantle. Gentle stimuli were also presented to the siphon in the experiment, but these stimuli were not applied simultaneously with the strong tail stimulus. EPSPs elicited by mantle or siphon stimulation were recorded before and 1 hour after the training session. The response elicited by the mantle stimulation after 1 hour was considerably larger than the response elicited by siphon stimulation. When the strong tail stimulus was paired with siphon stimulation during the training session and not with mantle stimulation, the opposite result was observed: the response to siphon stimulation 1 hour after the training session was now significantly larger than the response to mantle stimulation.

Although all the details underlying this phenomenon are not clear, calmodulin appears to play an important part. It has been proposed that Ca^{2+} entering the sensory neuron terminal upon mantle or siphon stimulation binds to calmodulin, which interacts with the serotonin-stimulated adenylate cyclase, thereby amplifying and perhaps prolonging its action. For calmodulin to enhance adenylate cyclase activity, adenylate cyclase must be activated; hence the necessity for pairing the stimuli. If the mantle or siphon stimuli are presented independently of the strong tail or head stimulus, adenylate cyclase activity is not enhanced and the resulting sensitization reflects the stimulation of adenylate cyclase by serotonin alone. The possible interaction of calmodulin with adenylate cyclase is represented in Figure 11.12 by the dotted line.

To summarize, the main findings in *Aplysia* are as follows. Short-term modulation of synaptic action (from minutes to hours) is due to an alteration in the amount of transmitter released from the MSN presynaptic terminals. These changes are linked to changes in Ca^{2+} influx into the terminal, which are caused by alterations to the Ca^{2+} channels, to Ca^{2+} pumps, or the K^+ channels. In all cases, the number of synaptic vesicles released per unit depolarization changes and subsequently alters the response of the motor neurons. In long-term modulation, synaptic junctions undergo structural changes and certain enzymes also change. These changes involve biochemical mechanisms probably including enzyme activation, protein synthesis, and gene transcription.

Do the results in *Aplysia* have application to other species? As noted at the beginning of the chapter, the analysis of habituation, sensitization, and classical conditioning is more complete in *Aplysia* than in any other organism, but these phenomena have also been studied in other species. From collective results, we know that the mechanisms are likely to be similar. In crayfish, habituation occurs when presynaptic terminals release less transmitter. Studies of classical conditioning in the locust, the mollusc *Hermissenda,* and the cat reveal that modulation of K^+ channels affects learning in these animals. Certain fruitflies that cannot learn because of genetic mutation (and hence are called dunce mutants) have defects in the cyclic AMP cascade. Other fruitfly mutants that have learning deficits have defects in their serotonin receptors or in their calmodulin-dependent adenylate cyclase. Finally, in Chapter 18, I describe the results of experiments that suggest similar mechanisms are operative in regions of the vertebrate brain linked to long-term memory formation.

Summary

Nearly a century ago it was proposed that specific changes at synaptic sites in the brain affect memory and learning. This cellular theory of memory and learning postulated that synaptic transmission was altered because of learning, or that new synapses were formed during learning and that these new connections result in memories. A second cellular theory, advanced in the 1920s, stated that persistent electrical activity in loops of neurons might also account for memories. The cellular theories were challenged in the 1930s by Karl Lashley, who postulated that specific cellular connections are not important for memory and learning. Today, the pendulum has swung back to the early ideas that synaptic modification underlies learning and memory. Some of the most compelling evidence for this return has come from work on the gill-withdrawal reflex in *Aplysia*.

A gentle stimulus applied to the mantle or siphon of the *Aplysia* causes rapid withdrawal of the gill. With repeated stimulation of the mantle or siphon, the response diminishes, or habituates. If the stimuli are repeated over several training sessions, habituation of the response may be prolonged for weeks (long-term habituation). A strong stimulus to the head or tail of the animal preceding gentle mantle or siphon stimuli sensitizes the reflex; that is, the response is greater than it is in a control (unstimulated) animal. Following strong stimuli to the head or tail of the animal over several training sessions, sensitization can also be made to last for weeks.

The gill-withdrawal reflex is mediated in the abdominal ganglion: sensory neurons receive input from the mantle or siphon, motor neurons innervate the gill, and a few interneurons are interposed between the sensory and motor neurons. The site of habituation has been localized to the synapse between the sensory neuron terminals and the motor neurons. Upon repeated stimulation of the sensory neurons, their terminals release less and less neurotransmitter, which produces smaller responses in the motor neurons and less vigorous gill-withdrawal responses. Sensitization is mediated via interneuronal synapses on the sensory neuron terminals: after the head or tail receives a sensitizing stimulus, action potentials from the sensory neurons in the head or tail activate interneurons that then release serotonin at synapses with the mantle sensory neurons; these cells then release more transmitter onto the motor neurons than is normally the case.

Short-term habituation and sensitization can be explained by variations in Ca^{2+} influx into the terminal upon membrane depolarization. With habituation, Ca^{2+} itself regulates further Ca^{2+} influx into the terminal; with sensitization, serotonin released onto the terminal promotes the synthesis of cyclic AMP by activating the enzyme adenylate cyclase. The cyclic AMP, via a kinase, phosphorylates a K^+ channel in the terminal and prolongs terminal depolarization. More Ca^{2+} is allowed into the terminal, and more transmitter is released from the terminal.

Long-term habituation and sensitization creates structural changes in the sensory neurons. More synaptic terminals per sensory neuron are seen and more terminals make synapses. The size and number of active zones containing vesicles also change during long-term habituation and sensitization. Also, there may be long-term changes in certain enzymes during long-term habituation and sensitization.

Finally, the gill-withdrawal reflex demonstrates evidence of classical conditioning. Pairing a strong sensitizing stimulus to the head or tail with a gentle stimulus to the mantle or siphon induces sensitization that is greater and more persistent than is the case when the two stimuli are presented to the animal independently. This interaction between the two types of stimuli appears to be mediated by the Ca^{2+} binding protein, calmodulin, which amplifies and prolongs the activity of the serotonin-stimulated adenylate cyclase in the terminal of the mantle sensory neuron.

Further Reading

BOOKS

Dudai, Y. 1989. *The Neurobiology of Memory.* Oxford: Oxford University Press. (A critical overview of contemporary research on memory and learning. Discusses both invertebrate and vertebrate model systems.)

Kandel, E. R. 1976. *Cellular Basis of Behavior.* San Francisco: W. H. Freeman. (A detailed account of the research on *Aplysia,* the methods and thinking behind the experiments.)

ARTICLES

Bailey, C. H., and M. Chen. 1983. Morphological basis of long-term habituation and sensitization in *Aplysia*. *Science* 220:91–93.

Carew, T. J., and C. L. Sahley. 1986. Invertebrate learning and memory: From behavior to molecules. *Annual Review of Neuroscience* 9:435–487.

Kandel, E. R., and J. H. Schwartz. 1982. Molecular biology of learning: Modulation of transmitter release. *Science* 218:433–443.

Lashley, K. S. 1950. In search of the engram. In *Physiological Mechanisms in Animal Behavior*, pp. 454–482. Symposia of the Society for Experimental Biology, no. 4. New York: Academic Press. (An account of Lashley's experiments.)

Schwartz, J. H., and S. M. Greenberg. 1987. Molecular mechanisms for memory: Second-messenger induced modifications of protein kinases in nerve cells. *Annual Review of Neuroscience* 10:459–476.

Squire, L. R. 1986. Mechanisms of memory. *Science* 232:1612–19.

The Vertebrate Brain

THE COMPLEXITY of the vertebrate brain is overwhelming, and it is fair to say that we are only beginning to unravel its intricacies. There are those who hold we will never reach one of the major goals of neuroscience, an understanding of the human brain, who say "the brain cannot understand itself." But most neuroscientists believe otherwise, that with time, patience, and hard work we will understand how our brains work.

We are optimistic for two reasons. First, as discussed in Part I of this book, the brain is made of cells, as are other tissues, and brain cells, although highly specialized, operate according to the rules of all cells. There seems to be nothing unique about brain cells. Second, enormous progress has been made in understanding how individual nerve cells work—how they receive stimuli, process information, and transmit it—and how the networks connecting those cells operate. We have many techniques available to study neural function, and more are being developed. Our understanding of brain function should increase dramatically over the next few decades.

The vertebrate brain will be explored in some detail in the remaining chapters of this book. Here I introduce the central nervous system, which consists of the brain and spinal cord, and describe their parts and function. I then take several chapters to cover selected aspects of the operation of the central nervous system: motor control and the processing of visual information. The final chapters of the book are devoted to brain development and the higher brain functions, such as speech and memory, that distinguish humans from other vertebrates.

The Central Nervous System

The vertebrate nervous system is composed of two parts: the central nervous system, made up of the spinal cord and the brain

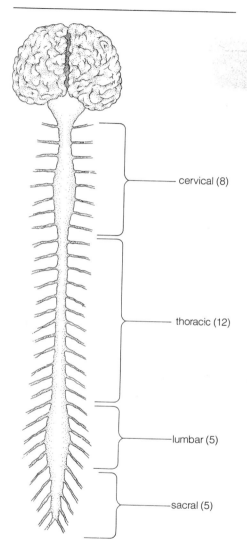

cervical (8)

thoracic (12)

lumbar (5)

sacral (5)

12.1   Schematic diagram of the central nervous system.

proper, and the peripheral nervous system, all the nerves and ganglia that lie outside the spinal cord and brain. Unlike many invertebrate nervous systems, which may be distributed more or less evenly along the length of the animal, the vertebrate nervous system is highly centralized. All higher neural functions—perception, movement initiation, memory, learning, and consciousness—are carried out within the brain proper. The spinal cord, on the other hand, serves three major roles. First, the spinal cord contains the circuitry (central pattern generators) for rhythmic motor behaviors such as scratching and walking; second, many reflexes are mediated within the spinal cord; and third, the cord serves as a conduit for sensory information flow from the periphery of the animal to the brain, and for motor information flow from the brain to the neurons in the cord that innervate the muscles of the limbs and trunk.

Figure 12.1 is a schematic diagram of the vertebrate central nervous system. The spinal cord is housed in the vertebral column, and from the cord 30 pairs of nerves (one member of a pair for each side of the body) exit between the individual vertebrae and extend out to the periphery. Eight exit in the cervical (neck) region; twelve in the thoracic (chest) region, and five each in the lumbar (abdominal) and sacral (lower back) regions. The spinal cord does not run the entire length of the vertebral column but terminates at the beginning of the lumbar region. The nerves that exit between the lumbar and sacral vertebrae run inside the spinal column some distance before exiting.

The sensory information entering the cord is mostly somatosensory; touch, temperature, pressure, and pain. Other sensory information, such as visual, auditory, and olfactory, enters the brain directly via *cranial nerves*. The twelve pairs of cranial nerves are concerned not only with carrying sensory information; they also control eye, tongue, and facial movements, and they innervate the viscera, including the heart, stomach, and diaphragm. Three of the cranial nerves are exclusively sensory (I, II, VIII), three control eye movements (III, IV, VI), four innervate the face, tongue, and jaw and carry sensory and motor information (V, VII, XII), two innervate the viscera and are also sensory and motor (IX, X), and one controls head and neck movement (XI). Table 12.1 lists the cranial nerves and the sensory-motor information that they carry.

Table 12.1   The cranial nerves and their functions

	Nerve	Functions
I	Olfactory	Smell
II	Optic	Vision
III	Oculomotor	Eye movements Pupillary constriction Lens accommodation
IV	Trochlear	Eye movements
V	Trigeminal	Jaw movements
VI	Abducens	Eye movements
VII	Facial	Facial movements Taste
VIII	Vestibulocochlear	Hearing Balance
IX	Glossopharyngeal	Swallowing Taste
X	Vagus	Speech Cardiac and visceral muscle control Throat and visceral sensations
XI	Spinal accessory	Head and neck movements
XII	Hypoglossal	Tongue movements

The Spinal Cord

A cross section of the spinal cord reveals its characteristic organization. The cord is bilaterally symmetrical and is divided into distinct zones of gray and white matter. As we know from Chapter 2, gray matter consists mainly of cell bodies and dendrites, whereas white matter is mainly composed of myelinated axons. The gray matter, centrally located within the spinal cord, has the shape of a butterfly. The wing regions are called horns, and on each side of the cord there is a dorsal and a ventral horn. As one moves down the cord, from cervical to lumbar regions, there is more gray matter relative to white matter; thus, fewer axons are within the

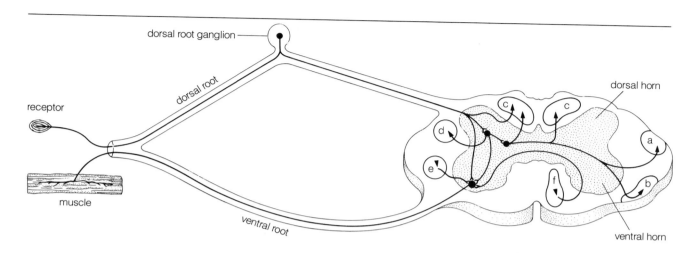

12.2 Cross section of the spinal cord in the lumbar region. Nerves entering and exiting the cord, and the pathways within the cord, are shown for one side of the body only. Sensory information enters the cord via the dorsal root; motor information exits the cord via the ventral root. The cell bodies of sensory neurons reside in the dorsal root ganglion. The interneurons and motor neurons that the sensory axons impinge upon are found, respectively, in the dorsal and ventral horn regions of the gray matter within the cord. Sensory information ascending the cord, or motor information descending the cord, runs in specific tracts labeled a–f. The sensory tracts carry information regarding pain and temperature (a), crude touch (b), light touch and proprioception (c), and proprioception (d). Motor tracks are of two kinds: crossed (e) and uncrossed (f).

cord in its lower part. Figure 12.2 presents a cross section of the cord at approximately the lumbar region.

Sensory information enters the cord dorsally, motor information leaves the cord ventrally. The nerve bundles carrying this information are called the *dorsal* and *ventral roots*. As shown in Figure 12.2, the cell bodies of the sensory neurons are found in ganglia (the dorsal root ganglia) adjacent to the cord. As we learned in Chapter 7, somatosensory neurons are typically monopolar. From a short process coming from the cell body, two myelinated processes arise: one that extends to the periphery and the sensory endings, and another that projects into the spinal cord.

Sensory neuron axons entering the spinal cord have at least three possible fates. They may ascend up the cord to higher cord levels or to the brain itself, they may terminate on interneurons that sit in the dorsal horn region of the cord, or they may pass through the dorsal horn to terminate directly on motor neurons in the ventral horn.

Spinal Cord Reflexes

Sensory neurons that directly innervate motor neurons create a simple reflex response. A reflex is an involuntary movement caused by stimulating receptors that activate motor output. In the knee-jerk reflex, sensory neurons that innervate the quadriceps (thigh) muscle directly innervate motor neurons that contract the muscle. If you are sitting with your legs crossed and someone delivers a sharp tap to the tendon of the top leg just below the kneecap, sensory neurons within the quadriceps muscle are activated by the resulting stretch of the muscle. The sensory neurons (which are described in the next chapter) activate motor neurons that contract the quadriceps and make the leg kick out slightly. This reflex helps keep the knee from collapsing when, for one reason or another, the leg is suddenly bent.

The knee-jerk reflex is *monosynaptic;* that is, only one synapse intervenes between the sensory and motor neurons. Other reflexes may be *polysynaptic;* one or more interneurons are interposed between sensory and motor neurons. The circuitry for both a monosynaptic reflex and a polysynaptic reflex with one interposed interneuron is shown in Figure 12.2.

Reflex pathways may be both excitatory and inhibitory; when one set of muscles in a limb contracts, another set usually must relax. For example, during the knee-jerk reflex, tension decreases in the muscle on the back of the thigh that causes the knee to bend. The interneurons between sensory and motor neurons in a polysynaptic reflex circuit thus may be either excitatory or inhibitory. Reflex action is also influenced by nerve fibers descending from the brain, as well as activity from other regions of the spinal cord. Furthermore, reflex pathways are often used in the generation of complex motor behaviors initiated by higher brain centers, so the neural circuitry of some reflexes can be quite complicated. As noted earlier, the central pattern generators involved in rhythmic motor behaviors are found in the spinal cord, and it is likely that some of this circuitry is involved in certain reflexes. The next chapter discusses in more detail motor control in vertebrates and the role of the spinal cord in mediating motor activity.

Ascending and Descending Pathways

Information traveling to and from the brain within the spinal cord is carefully segregated. Different sensory information ascends along specific pathways, whereas motor information from different parts of the brain descends along specific tracts. For example, fibers carrying pain and temperature information typically ascend the cord in the region marked *a* in Figure 12.2, crude touch in the region marked *b,* and light touch in the region marked *c.*

Figure 12.2 is drawn to show that most of the sensory information ascends along pathways on the side of the spinal cord opposite from where the information entered the cord. This is intentional. Indeed, the right side of the brain controls the left side of the body and vice versa. Thus, sensory information coming from the left side of the body must cross over to the right side of the central nervous system; information regarding pain, temperature, and crude touch crosses over at the level of the cord where it enters. A good deal of light-touch information ascends the cord on the same side that it enters, but then it crosses over to the other side at the level of the brain stem. We do not understand why neural information from one side of the body is dealt with by the other side of the brain, or why one side of the brain controls the other side of the body.

The sensory information ascending the cord that I have described so far eventually reaches the cerebral cortex of the brain. It is conscious information; we are aware of it. But other sensory information coming from our muscles, joints, and limbs—crucial for making coordinated movements—does not reach the cortex concerned with conscious sensation. We may not be aware of this information, but it is used by lower brain structures (particularly the cerebellum) to ensure efficient and coordinated movements. This proprioceptive (muscle, tendon, and joint) and kinesthetic (position and movement) sensory information tells the brain about limb position, limb movement, joint angles, muscle tension, and so forth. It ascends the spinal cord in tracts (areas marked *c* and *d* in Figure 12.2) on the same side that it enters the cord. If this sensory information does not reach beyond the cerebellum, it remains on the same side of the brain where it entered the cord. Proprioceptive muscle receptors will be discussed in some detail in the next chapter.

Descending motor information also travels along the cord in prescribed tracts. Two of these are indicated in Figure 12.2 as *e* and *f*. Most descending motor information that originates in the cortex crosses over to the other side in the brain stem and runs down the cord on the same side that it innervates motor neurons *(e)* but some information descends on the same side of the cord that it arises from and crosses over to the other side at the level where it innervates motor neurons *(f)*.

One consequence of the crossing over of sensory and motor

information at certain levels in the central nervous system is that a lesion of the spinal cord can bring about what appear to be curious deficits. For example, hemisection (destruction of either the right or left sides of the cord) results in an inability to sense crude touch, temperature, and pain on the opposite side of the body from the lesion; partial loss of light-touch, proprioceptive, and kinesthetic information on both sides of the body; and paralysis mainly on the same side of the body.

The Brain

Figure 12.3, a longitudinal section of the primate brain, illustrates the major brain structures. It is convenient, first of all, to divide the brain into three regions: hindbrain, midbrain, and forebrain. The hindbrain emerges from the spinal cord and consists mainly

12.3 A longitudinal section of the primate brain showing its major structures. The hindbrain emerges from the spinal cord and consists of the pons, cerebellum, and medulla. The midbrain encompasses a relatively small part of the primate brain, whereas the forebrain is very prominent. Major structures of the forebrain include the cerebral cortex, corpus callosum, thalamus, and hypothalamus. The pituitary gland is found just below the hypothalamus.

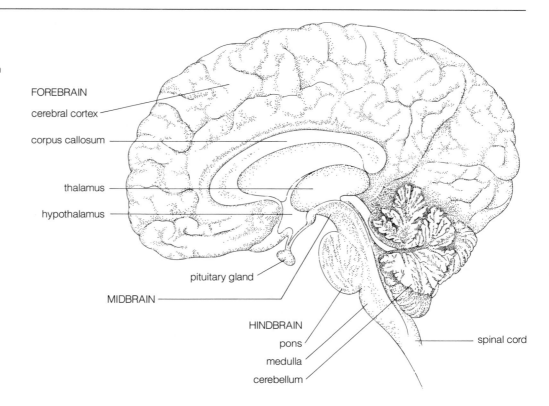

FOREBRAIN
cerebral cortex
corpus callosum
thalamus
hypothalamus
pituitary gland
MIDBRAIN
HINDBRAIN
pons
medulla
cerebellum
spinal cord

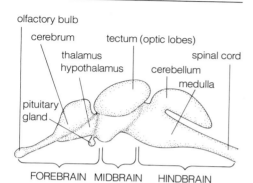

olfactory bulb
cerebrum   tectum (optic lobes)
thalamus   spinal cord
hypothalamus   cerebellum
                         medulla
pituitary
gland

FOREBRAIN MIDBRAIN   HINDBRAIN

12.4 A longitudinal section of the codfish brain. The codfish brain has the same basic structure as the primate brain, but it is much smaller. Furthermore, the forebrain is much smaller relative to the midbrain than it is in the primate. The midbrain, on the other hand, is relatively large and is dominated by the tectum.

of the medulla, cerebellum, and pons. Above the pons lies the midbrain, a small region in the primate brain that sits between the hindbrain and the forebrain. The forebrain consists of two major regions, one that encompasses the thalamus and hypothalamus, and another that includes the basal ganglia and the cerebral cortex (the region termed the cerebrum).

Brains of cold-blooded vertebrates—such as fish, frogs and other amphibia, and turtles and other reptiles—have an especially small cerebrum relative to the rest of the brain. In these animals, the cerebrum is concerned primarily (but not exclusively) with the analysis of smell; the olfactory tract, for example, projects directly into the cerebrum. In mammals, the cerebrum is greatly expanded and most higher neural function is centered in the cerebral cortex. Indeed, the expansion of the forebrain, especially the cerebral cortex, is by far the most significant difference we observe when comparing the brains of vertebrates. The evolution of the vertebrate brain has mainly centered on the forebrain; it has evolved from a relatively small structure concerned primarily with the processing of one sense to, in humans, a much larger organ responsible for consciousness, sensation, memory, and intelligence.

Figure 12.4, a longitudinal section of a codfish brain, illustrates how the brain of a lower vertebrate differs from that of a primate. The entire brain is considerably smaller than the primate's, but particularly striking is the small size of the cerebrum compared with other brain structures. In the codfish, the midbrain is quite prominent and is dominated by a structure called the tectum. The tectum receives a large input from the optic nerve (in regions called the optic lobes), and it also receives other sensory inputs. In addition, tectal cells project to the spinal cord, where they control muscle activity. Thus the tectum is the key structure in nonmammalian species for integrating sensory inputs and for controlling motor outputs.

With the development of the forebrain in the higher vertebrates, the midbrain and tectum have become less important brain structures. The more complex functions—sensory processing, sensory-motor integration, and the initiation of motor activity—are localized in the cerebral cortex. The midbrain and tectum have been gradually reduced in size relative to the rest of the brain; in mammals, much of the tectum eventually is relegated to mediating noncortical visual reflexes, such as pupillary and eye movement responses. In higher mammals, the tectum also helps to coordinate head and eye movements.

Medulla and Pons

Running through the medulla are numerous ascending and descending nerve tracts from the spinal cord to higher brain centers and vice versa. Half (VII–XII) of the cranial nerves enter the brain in the medulla, and within the medulla are numerous nuclei (clusters of neurons) related to the cranial nerves. These nuclei are concerned with the control of the head, face, eyes, and tongue, as well as with vital body functions. Thus, nuclei in the medulla play an important role in the regulation of heart rate, respiration, and gastrointestinal function.

In addition to the neurons found in discrete nuclei within the medulla, small groups of neurons distributed diffusely throughout the medulla and pons make up what is called the *reticular formation*. Many neurons of the reticular formation contain monoamines or peptides (or both), and their axons extend widely throughout the rest of the brain. One reticular formation neuron can extend axonal branches into the spinal cord as well as into the thalamus and hypothalamus, and perhaps even to the cortex. Thus, reticular neurons exert widespread influence on virtually all parts of the brain.

The reticular formation has several functions. It regulates the level of activity of many parts of the brain and is thus important for arousal and controlling levels of awareness. Lesions of parts of the reticular formation cause animals, including humans, to become comatose or to fall into a stupor from which they cannot be aroused. Reticular neurons also help to control respiration and heart function. Some, for example, send axons directly to the spinal cord, where they control the motor neurons innervating the muscles used in respiration. Finally, some reticular neurons modulate pain pathways.

The pons, lying just above the medulla, contains many neurons that relay information from the cerebral cortex to the cerebellum. In addition, some neurons in the pons, like some neurons of the reticular formation of the medulla, participate in the control of respiratory and cardiac function.

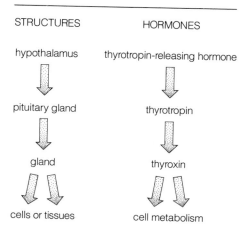

STRUCTURES	HORMONES
hypothalamus	thyrotropin-releasing hormone
pituitary gland	thyrotropin
gland	thyroxin
cells or tissues	cell metabolism

12.5　The structures involved in the regulation of hormone release *(left)* and a specific example involving the thyroid hormone *(right)*. A peptide (TRH) released from the hypothalamus induces the release of a pituitary gland hormone (thyrotropin), which promotes the release of hormone (thyroxin) from the thyroid gland. Thyroxin exerts widespread cellular effects.

Hypothalamus

The hypothalamus, like the medulla, is an important regulatory center. Indeed, the medulla and hypothalamus are often considered the regulatory centers of the brain. The nuclei of the hypothalamus are concerned mainly with the regulation of basic drives and acts. For example, hypothalamic nuclei regulate eating, drinking, sexual activity, body temperature, heart rate, blood pressure, and so forth. The hypothalamus also affects emotional behavior. Stimulating parts of the hypothalamus elicits irritability in an animal, whereas stimulation of other parts induces placidity. Lesions of the hypothalamus can lead to aggressive behavior.

Another important function of the hypothalamus is the regulation of hormone release from the pituitary gland. As indicated in Figure 12.4, the pituitary gland sits just below the hypothalamus, and the hormones it releases are under the control of small peptides released by hypothalamic neurons. Some of these small peptides promote the release of pituitary hormones; others inhibit them. The pituitary hormones in turn often regulate the release of hormones from endocrine glands found elsewhere in the body, such as the thyroid and adrenal glands.

The basic scheme for the regulation of hormone release by the hypothalamus and pituitary gland is shown in Figure 12.5. Specifically, the example shown is the release of thyroxin, the hormone of the thyroid gland that controls levels of metabolic activity in the cells of many tissues. A small, three-amino-acid peptide, called thyrotropin-releasing hormone (TRH), is released by hypothalamic neurons. TRH reaches the pituitary gland via blood vessels that link the two brain regions. In the pituitary, TRH initiates the release of thyrotropin, which, via the general circulation, promotes the release of thyroxin from the thyroid gland. Thyroxin is distributed to its target cells via the circulation.

Some peptide hormones released by the hypothalamic neurons exert direct effects on glands or tissues. They do not mediate the release of substances from the pituitary, but instead enter the general circulation and are distributed directly to their target tissues. Oxytocin is one such peptide; it releases milk from the mammary gland. Under the right hormonal conditions, it also influences uterine contractions. Vasopressin, another example, regulates the water permeability of kidney cells.

Cerebellum

The cerebellum is one of the most extensively studied parts of the central nervous system, partly because it has a highly regular structure. This regularity suggests that all areas of the cerebellum do roughly the same thing and that the circuits in the cerebellum are similar in all parts. The detailed circuitry of the cerebellum is now understood quite well, and studies confirm that one area of the cerebellum is equivalent to other areas.

The main function of the cerebellum is to coordinate and integrate motor activity. It receives input from the cerebral cortex concerning the initiation of movement, as well as sensory input from the spinal cord and other sensory systems, including much proprioceptive and kinesthetic information. The cerebellum compares the inputs, integrates the information they provide, and projects signals to descending motor systems. The cortex gives the command for a movement, but it is the cerebellum that coordinates the motor command with sensory inputs to ensure a smooth, appropriate movement.

Lesions of the cerebellum typically result in jerky movements. A delay in the initiation of a movement often connotes a cerebellar lesion, as do movements that are erratic or oriented in the wrong direction. Sometimes the movement falls short of its intended goal; on other occasions the movement is exaggerated relative to the intended one.

Thalamus

The thalamus consists of numerous prominent nuclei that relay sensory information to the cerebral cortex. Some nuclei also send information to the cortex about motor activity. The individual nuclei receive specific sensory or motor information and relay it to the cortex. An example is the lateral geniculate nucleus, the site of termination of the optic nerves. The lateral geniculate neurons receive direct input from the optic nerve axons, and they project to the primary visual area of the cortex. Virtually all visual information reaching the cortex ascends via this route.

The thalamic nuclei also receive input from the reticular formation and back from the cortex, so it is believed that information is gated in the thalamus. That is, sensory and motor information relayed to the cortex is modified by these inputs to the thalamus.

Basal Ganglia

The basal ganglia are five prominent nuclei that are positioned above, below, and on each side of the thalamus. Figure 12.6 is a vertical cross section through about the middle of the brain. The plane of section, a coronal plane, is indicated by the dashed line of the inset. The positions of the basal ganglia are indicated by the light stippling; the dark-stippled areas show the position and extent of the thalamic nuclei and the light-hatched areas represent the cellular layers of the cortex. The basal ganglia are concerned primarily with movement, its initiation and execution. The basal ganglia receive input primarily from the cortex and their output back to the cortex is via the thalamus. Lesions in the basal ganglia cause movement dysfunctions. Humans with basal ganglia deficits will demonstrate tremors, repetitive movements, and rigidity of limbs. They also have difficulty initiating movements, which are characteristically slow.

One disease affecting the basal ganglia, mentioned briefly in Chapter 6, is Parkinson's disease. This disease, caused by a defi-

12.6 A vertical section cut approximately through the middle of the brain. The plane of section is shown by the dashed line in the inset at the lower right. Note that the cortex is a layer about 2 mm thick covering the hemispheres; note also the approximate positions of the basal ganglia and thalamic nuclei. The corpus callosum, the large tract of axons that connects the two hemispheres, is also indicated.

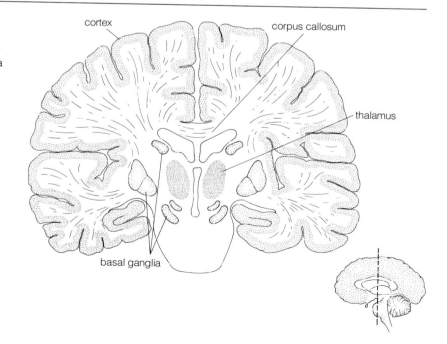

ciency of dopamine in one of the basal ganglia, creates a rhythmic tremor of the limbs at rest, a rigidity of muscle tone, and a slowness in the initiation and execution of movements. Patients can be aided significantly with the drug L-dopa, a precursor in the synthesis of dopamine that raises dopamine levels in the brain. Dopamine itself cannot be given to patients with Parkinson's disease because it will not cross from the blood to the brain (this is known as the blood-brain barrier); L-dopa readily crosses the blood-brain barrier. L-dopa therapy is not a cure for the disease, but it does help many patients. The disease progresses even when L-dopa therapy is in effect, but the drug controls many of the symptoms.

Another disease of the basal ganglia, Huntington's disease, is an inherited affliction that also stems from the loss of specific neuroactive substances from the basal ganglia. Cholinergic and GABA-ergic neurons degenerate in basal ganglia. The initial stages of the disease are characterized by small, uncontrolled movements. Usually beginning in middle age, these symptoms gradually increase in severity until the patient is confined to bed. Mental capability undergoes deterioration and death ensues 15–20 years after the onset of the disease. The loss of cognition by Huntington's disease patients suggests that the basal ganglia contribute to cognitive behavior. On the other hand, there may be a concomitant loss of cortical neurons in Huntington's disease, which could also account for cognitive dysfunction.

The gene for Huntington's disease has recently been localized to the short-arm portion of a specific chromosome in humans (number 4). The disease is inherited dominantly; that is, individuals receiving just one copy of the defective gene from either parent will be affected. This means that if one parent has the disease, there is a 50 percent chance that each offspring will have the disease; if both parents are affected, the risk increases to 75 percent.

Now that the gene has been approximately localized, efforts are under way to isolate it and to discover how gene mutation leads to the disease; that is, how mutation of a single gene leads to degeneration of specific neuronal pathways.

Cerebral Cortex

In humans and other mammals, most higher central nervous system functions are localized in the cerebral cortex. Skilled move-

12.7 The primate cerebral cortex viewed from the left side. The cortex is divided into two hemispheres, right and left, and each hemisphere is subdivided into four lobes: frontal, parietal, temporal, and occipital. Two major sulci, or fissures, are shown—central and lateral—and the primary sensory and motor areas are indicated by stippling.

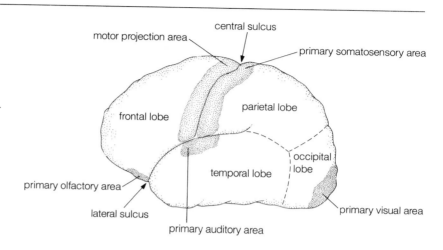

ments originate in the cortex, the seat of sensation, consciousness, memory, and intelligence. The cortex is divided into two hemispheres, a left and a right hemisphere, and each hemisphere is subdivided into four lobes: frontal, parietal, occipital, and temporal (see Figure 12.7). The neurons in the cortex are all close to the surface, forming a layer about 2 mm thick that covers the hemispheres (see Figure 12.6).

Specific neural functions are carried out within the lobes. For example, the frontal lobes are concerned primarily with movement and olfaction, the parietal lobes with somatic sensation, the occipital lobes with vision, and the temporal lobes with audition and memory. The lobes of the cortex are also characteristically infolded or convoluted in higher mammals; the folds increase the surface area of the cortex and the number of cortical cells. Many of the convolutions and the fissures between them are named, but with the exception of the central sulcus, which separates the frontal lobe from the parietal lobe, and the lateral sulcus, which separates the temporal lobe from the parietal and frontal lobes, we shall not be concerned with this detailed anatomy.

Within each of the lobes are areas devoted to the initial cortical analysis of specific sensory information or to the initiation of specific movements. The primary sensory areas receive sensory information directly from the thalamus, whereas the primary motor area initiates fine movements. The primary sensory and motor areas are indicated in Figure 12.7. Note that the central sulcus sits between the primary motor area and the primary somatosensory area.

At specific locations on the somatosensory cortex, information from different parts of the body is analyzed; likewise, specific regions of the cortex in the primary motor area initiate movement of certain parts of the body. Thus, an approximate representation of the body surface exists on the primary somatosensory and motor areas, and a representation of the visual field exists on the primary visual cortex. These representations are not strictly proportional. More cortical area is devoted to parts of the body where sensation is more acute or to those parts involved in precise movements. For example, the face and hand encompass a disproportionate amount of the primary somatosensory and motor areas, and the central foveal region of the retina encompasses a disproportionate amount of the primary visual cortex. Figure 12.8 shows the topographic projection, or map, of the body on both the somatosensory cortex and motor projection areas of the human. Wilder Penfield, a Canadian neurosurgeon, obtained these data in the 1940s and 1950s by stimulating the cortex of patients undergoing brain surgery for epilepsy. (Penfield's techniques and findings will be discussed further in Chapter 18.)

Spreading out from the primary projection areas are secondary or association areas of the cortex. These areas handle the more complex processing of sensory input, the integration of sensory information, and recognition, understanding, and memory. To illustrate, consider the effects of lesions in the occipital cortex, where vision is mediated. The primary visual cortex is called V1; with a massive lesion in V1, a patient has no awareness of sight. Such patients may exhibit visual reflexes mediated by lower (subcortical) brain centers, but they will not be conscious of seeing anything. For example, an object brought swiftly toward the patient will cause the patient to blink (the blink reflex), but the person will be unable to explain why he blinked and will deny that he saw something.

With a lesion in the first visual association area, V2, patients are aware of seeing things but they have trouble recognizing them. A circle and a square drawn on a piece of paper will be seen but not identified by shape. With a lesion in higher visual association

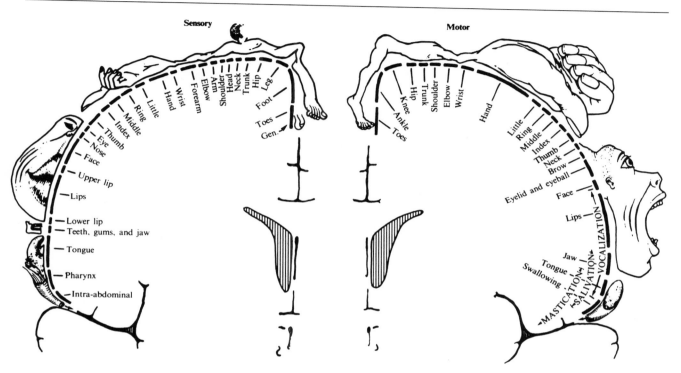

Sensory   **Motor**

12.8 The representation of the body surface on the primary somatosensory cortex *(left)* and primary motor cortex *(right)*. The drawings represent cross sections of the cortex through each area. Note that parts of the body that have acute sensation or fine motor control, such as the face and hands, occupy a disproportionate amount of cortex.

areas, a patient may see an object and recognize it but might have difficulty remembering having seen the object even after a very short time.

Penfield's stimulation experiments of the human cortex have also shed light on the association cortex. When the primary visual cortex (V1) is electrically stimulated, patients report seeing flashes of light, shooting stars, or other simple visual sensations. Stimulation of the visual association areas, on the other hand, may evoke a visual memory from the patient. That is, the patient recalls seeing something at some earlier time. In Chapter 18, I shall discuss in more detail localization of function in the cerebral cortex.

Summary

The vertebrate central nervous system consists of the spinal cord and the brain. Higher neural processing is carried out within the brain proper; the spinal cord contains the circuitry for rhythmic motor activity, mediates many reflexes, and serves as a conduit for information flow from the brain to the periphery of the body and vice versa. Most of the sensory information entering the spinal cord is somatosensory in nature; other sensory information (from the eye and ear, for example) enters the brain directly via the cranial nerves. The cranial nerves also carry motor information that controls eye, face, and tongue movements and visceral function. Motor information controlling the rest of the body passes down the spinal cord and out to the periphery via motor nerves.

Sensory information enters the spinal cord via the dorsal roots, whereas motor information exits the cord via the ventral roots. Sensory fibers entering the cord may ascend the cord to the brain, innervate interneurons found within the dorsal horn region of the cord, or synapse on motor neurons within the ventral horn of the cord. The latter arrangement is the basic circuitry of a simple monosynaptic reflex, where sensory stimulation leads directly to a motor output, such as the knee-jerk reflex. Reflex pathways are often more complex and may involve interneurons (polysynaptic reflexes) that may be either excitatory or inhibitory.

Sensory and motor information runs along the spinal cord in specific tracts depending on the type of information or its origin. Neural information from the left side of the body is processed by the right side of the brain and vice versa. The crossover can occur within the spinal cord—close to where information enters or exits the cord—or in the brain stem.

The brain is conveniently divided into three regions: the hindbrain, midbrain, and forebrain. The forebrain is far more developed in mammals and especially in primates than it is in nonmammalian species. The forebrain in mammals is the site of more complex sensory processing, sensory-motor integration, and motor initiation, whereas in nonmammalian species these activities are centered in the midbrain. The hindbrain consists of the medulla, pons, and cerebellum. The medulla contains numerous axon tracts that carry information between higher brain centers and the spinal cord. Some of the cranial nerves enter the medulla, and nuclei related to the cranial nerves are found in this region of the brain. These nuclei control the head, face, eyes, and tongue, and they regulate heart rate, respiration, and gastrointestinal function.

Within the medulla are also small clusters of neurons that make up the reticular formation. These neurons innervate wide regions of the brain and regulate levels of brain activity. Reticular neurons also regulate respiration, heart function, and pain. The pons contains neurons that relay information from the cortex to the cerebellum, and the cerebellum serves to coordinate and integrate motor activity. It receives information from all sensory systems, as well as input from the cortex concerning the body's movement. It compares and integrates the inputs and provides an output to the descending motor systems that ensures a smooth, appropriate movement.

The hypothalamus, like the medulla, is a regulatory center. The nuclei in the hypothalamus are concerned with basic drives and acts, such as eating, drinking, sexual behavior, body temperature, and so forth. Certain hypothalamic neurons release small peptides into the circulation that act as hormones, either by regulating the release of other hormones from the pituitary gland or by acting directly on target tissues. The thalamus, like the hypothalamus, is part of the forebrain. Its nuclei relay mainly sensory information to the cerebral cortex, but also information about ongoing motor activity. The thalamus receives some input back from the cortex and also from the reticular formation that is used to gate (modify) the information sent by the thalamic neurons to the cortex. The basal ganglia, also part of the forebrain, are concerned primarily with the initiation and execution of movement. They receive input from the cortex and feed information back to the cortex via the thalamus. Two diseases of the basal ganglia, Parkinson's and Huntington's disease, are characterized by limb tremors and spontaneous movements.

The cerebral cortex is the seat of sensation, consciousness, memory, and intelligence in humans. Skilled movement also originates in the cortex. The cortex is divided into four lobes, each concerned with specific functions. The frontal lobe deals with movement and smell, the parietal lobe with somatic sensation, the occipital lobe with vision, and the temporal lobe with audition and memory. Within each lobe are primary sensory or motor areas, and spreading beyond these are association areas. The primary sensory or motor areas are devoted to the initial processing of sensory information in the cortex or to the initiation of movement, whereas the association areas are concerned with more complex processing, with the integration of sensory information, and with recognition, understanding, and memory.

Further Reading

BOOKS

Angevine, J. B. Jr., and C. W. Cotman. 1981. *Principles of Neuroanatomy.* Oxford: Oxford University Press. (A readable introduction to neuroanatomy.)

Heimer, L. 1983. *The Human Brain and Spinal Cord: Functional Neuroanatomy and Dissection Guide.* New York: Springer-Verlag.

Kandel, E. R., J. H. Schwartz, and T. M. Jessel, eds. 1991. *Principles of Neural Science.* 3d ed. New York: Elsevier. (A medical school textbook emphasizing the mammalian brain.)

Sarnat, H. B., and M. G. Netsky. 1974. *Evolution of the Nervous System.* New York: Oxford University Press.

ARTICLES

Guillemin, R., and R. Burgus. 1972. The hormones of the hypothalamus. *Scientific American* 227(5):24–33

Llinas, R. R. 1975. The cortex of the cerebellum. *Scientific American* 232(1):56–71.

Motor Control in Vertebrates

VERTEBRATES have two motor systems. One, the autonomic nervous system, regulates the internal organs of an animal, including the heart, digestive tract, lungs, bladder, blood vessels, various glands, and the pupil of the eye. The other, the voluntary motor system, controls the muscles of the limbs, body, and head. The name "autonomic nervous system" suggests that the control exerted by this system is involuntary and autonomous. This is generally correct but not always the case. On occasion it is possible to exert voluntary control over the autonomic nervous system; moreover, the system is not always strictly autonomous. For example, the autonomic nervous system controls the blood flow to the muscles, a function that is crucial for sustained and vigorous voluntary movement. Thus, important interactions occur between the autonomic and voluntary motor systems.

In this chapter, I describe first the autonomic nervous system in the vertebrate and focus mainly on the control of the heart. I then turn to voluntary movement, especially the mechanisms underlying walking and the initiation of skilled movement in primates.

The Autonomic Nervous System

The autonomic nervous system's two main divisions, the *sympathetic* and *parasympathetic* systems, are organized similarly in all vertebrates and generally they have opposing effects. The effects of sympathetic innervation are to stimulate the heart and enhance skeletal muscle activity, but they decrease activity in the digestive system and other internal organs. The parasympathetic system, by contrast, decreases the heart rate and the strength of heart muscle contraction and stimulates activity in the digestive system and in other internal organs. An easily observed example of the opposing actions of the sympathetic and parasympathetic systems is the pupil of the eye, which dilates by sympathetic activity and constricts by parasympathetic innervation.

The sympathetic system, often referred to as the "fight-or-flight" system, prepares an animal for strenuous activity. When we are frightened, for example, its effects are obvious: our heart rate increases and the strength of cardiac contraction is enhanced (which is why we feel our heart pounding), our pupils dilate, we sweat because of sympathetic stimulation of sweat glands but feel parched because the glands releasing saliva are inhibited. The parasympathetic system has the opposite function. It prepares the body for relaxation or rest; heart rate and blood pressure decrease and the digestive system becomes more active.

The anatomical organization of the sympathetic and parasympathetic systems is special and the systems demonstrate distinct differences. In the sympathetic system, spinal cord neurons innervate ganglia that lie on either side of the vertebral column or nearby. Postsynaptic motor neurons in the sympathetic ganglia then innervate the internal organs and glands. In the parasympathetic system, the ganglia innervating an organ are located in the organ itself. These parasympathetic ganglia are innervated by presynaptic neurons found in the brain stem or in the sacral region of the spinal cord. Thus many of the preganglionic parasympathetic fibers run out from the brain stem in the cranial nerves, particularly the vagus nerve (cranial nerve X). The basic organization of the autonomic nervous system is shown in Figure 13.1.

An important consequence of this anatomy is that the sympathetic motor neuron axons running to the internal organs are long and innervate relatively large areas, whereas the parasympathetic motor neuron axons are shorter and are much more restricted in their innervation patterns. Thus, sympathetic nervous activity tends to be much more widespread than is parasympathetic activity, which may be restricted to an individual organ or gland.

The autonomic nervous system is controlled principally by nuclei in the hypothalamus. Neurons in the hypothalamus extend axons to the preganglionic sympathetic and parasympathetic neurons located in the brain stem and spinal cord. The hypothalamic neurons innervating the preganglionic motor neurons receive input from many brain areas, including the cortex and reticular formation.

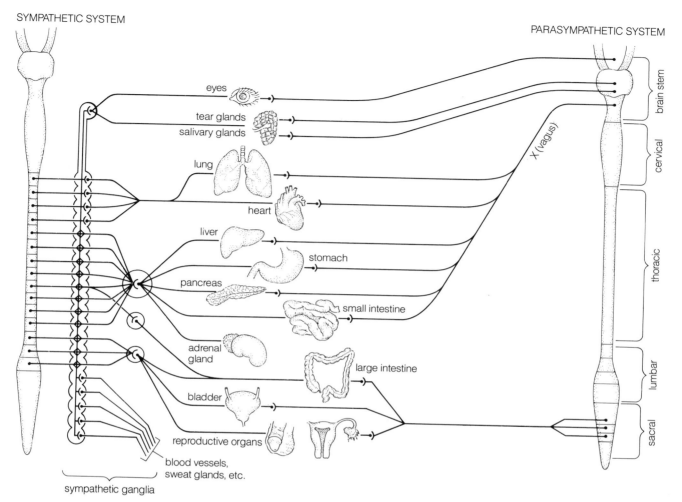

SYMPATHETIC SYSTEM

PARASYMPATHETIC SYSTEM

brain stem

cervical

thoracic

lumbar

sacral

eyes

tear glands
salivary glands

lung

heart

liver

stomach

pancreas

small intestine

adrenal gland

large intestine

bladder

reproductive organs

blood vessels, sweat glands, etc.

sympathetic ganglia

X (vagus)

13.1 The organization of the autonomic nervous system. In the sympathetic system, spinal cord neurons innervate a series of ganglia that lie on either side of the spinal cord (lateral ganglia) or nearby (collateral ganglia). Neurons from these ganglia innervate the internal organs, blood vessels, and glands. In the parasympathetic system, motor neurons from the brain stem and sacral region of the spinal cord innervate ganglia found in the organs themselves. Thus, the postsynaptic neurons in the parasympathetic system are much shorter than are the postsynaptic neurons in the sympathetic system, and they innervate a more restricted area. The parasympathetic motor neurons from the brain stem run to the internal organs via cranial nerves.

Ganglion Organization

In the sympathetic and parasympathetic ganglia, the preganglionic neurons release acetylcholine. One interesting aspect of postganglionic neurons is that they release acetylcholine in the parasympathetic system but norepinephrine in the sympathetic system. In addition, some peptides are in the ganglia, either coexisting in the pre- or postganglionic neurons or in separate sensory neurons that arise in the internal organs and terminate in the ganglia. So, for example, in the sympathetic system, enkephalin has been found coexisting with acetylcholine in preganglionic neurons and with somatostatin in some postganglionic neurons. In addition, vasoactive intestinal peptide has been identified in certain sensory neurons in the ganglia.

The synaptic interactions that occur in the autonomic nervous system ganglia are complex and not completely understood. Certainly the ganglia are not simple relay stations where information is passed on unmodified from pre- to postsynaptic neurons. Indeed, the ganglia are intricate integration centers, and hence they are being studied intensively. Sympathetic ganglia in the frog and rabbit have been favored preparations for investigation and much has been learned from them. For example, upon stimulation of preganglionic fibers, a sequence of synaptic potentials appear in the postsynaptic neurons. These include three distinct depolarizing potentials, or EPSPs, and, in the rabbit, one hyperpolarizing IPSP. The potentials have different time courses, are mediated by different neuroactive substances, and are generated by different mechanisms. In them we see a range of synaptic interactions that can occur in vertebrate neural circuits.

The earliest potential recorded is a fast EPSP, which lasts for 10–20 ms. It is mediated by acetylcholine acting directly on membrane channels, a nicotinic action of acetylcholine. This response is similar to the postsynaptic potential at the neuromuscular junction. The next potential is a slow IPSP, which lasts several hundred milliseconds. It is believed that in the rabbit this response is mediated by a catecholamine, probably dopamine, released by an interneuron in the ganglion that is activated by the preganglionic axons. This hyperpolarizing response results from a decrease in the conductance of a channel that allows mainly Na^+ to cross the membrane.

The third potential is a slow EPSP; it lasts several seconds and is mediated by acetylcholine. The receptors that mediate this response are muscarinic, however, which indicates that a second messenger participates in the generation of the potential. The response also appears to be caused by decreased membrane conductance—this time a change in K^+ conductance. The last potential is an even slower EPSP, called the late, slow EPSP, and it lasts for several minutes. The potential is mediated by a neuropeptide very similar to luteinizing hormone–releasing hormone (LHRH) found in the hypothalamus, and it also results from a decreased K^+ conductance.

Figure 13.2 is a composite drawing of the synaptically mediated

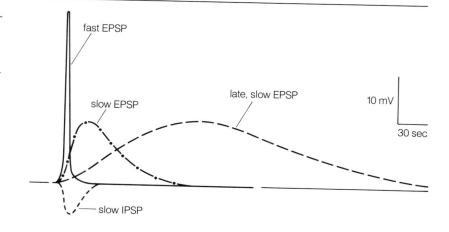

13.2   A composite drawing showing the variety of synaptically mediated potentials that can be recorded from sympathetic ganglion cells. The fast and slow EPSPs are both mediated by acetylcholine, the slow IPSP by dopamine, and the late, slow EPSP by a neuropeptide similar to luteinizing hormone–releasing hormone (LHRH).

potentials on a common time and voltage scale. Not only do the potentials have different durations, but they vary in sign and amplitude. The fast EPSP is by far the largest of the potentials, often 20 mV or more in amplitude; the other potentials are usually less than about 10 mV. Presumably, initial spike generation is mediated mainly by the fast EPSPs; the slow EPSPs and the IPSP modulate the level of membrane potential and hence the pattern and rate of action potential firing by the ganglion cells.

Control of the Heartbeat

The vertebrate heart is spontaneously active; it beats rhythmically when isolated. It is richly innervated by both sympathetic and parasympathetic neurons, however, and this innervation controls the rate and strength of cardiac contraction. To understand how the sympathetic and parasympathetic systems control the heart, one must first understand how the heart's rhythmic activity is established.

The heart is driven by a group of specialized cells located in the right upper chamber, or atrium, of the heart in the so-called sino-atrial node. These cells are rhythmically active and the impulses they generate spread along the muscles of the atrium via electrical junctions to a second pacemaker site between the upper (atrial) and lower (ventricular) heart chambers. Other specialized cells (Purkinje fibers) then spread the excitation via electrical junctions

13.3 The generation of rhythmic action potential firing in pacemaker cells of the heart. At resting membrane voltage (−60 mV), the membrane is leaky to Na⁺ and K⁺ $(I_{Na,K})$. This causes the membrane gradually to depolarize to action potential threshold, and a Ca²⁺ spike (I_{Ca}) is fired. K⁺ channels (I_K) open as the Ca²⁺ current inactivates, repolarizing the cell to −60 mV.

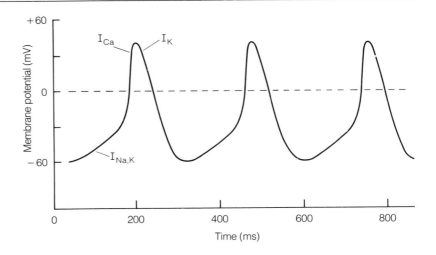

to the rest of the heart. The Purkinje fibers are large and conduct action potentials faster than do the other cardiac cells. Here I focus on the primary pacemaker cells in the sino-atrial node.

Figure 13.3 illustrates the rhythmic firing pattern of one of these pacemaker cells; the underlying membrane conductance changes that give rise to the cell's voltage changes are indicated on the drawing. At resting membrane voltage (−60 mV), the membrane is leaky to Na⁺ and K⁺; that is, a channel in the membrane is activated at this membrane potential. This channel causes the membrane to depolarize. Remember, more Na⁺ enters the cell than K⁺ leaves through such a channel when membrane voltage is more negative than −20 mV (see Chapter 5). The slow membrane depolarization induced by this current $(I_{Na,K})$ opens voltage-sensitive Ca²⁺ channels in the membrane. When the Ca²⁺ channels are sufficiently activated, a Ca²⁺ action potential rapidly drives the membrane potential to about +25 mV (I_{Ca}). The recovery of membrane potential comes about as a result of voltage-sensitive K⁺ channels opening (I_K), allowing K⁺ to flow out of the cell. This efflux of K⁺ brings the membrane potential down to about −60 mV, which activates the Na⁺ and K⁺ channels and starts the cycle over again.

Sympathetic innervation of the heart comes from ganglia that lie along the upper part of the thoracic region of the vertebral

norepinephrine   acetylcholine

G_S G_I

+   −

AC

cyclic AMP

protein phosphorylation

Ca^{2+} channels

neural, metabolic activity

13.4   A scheme showing how norepinephrine and acetylcholine, through different G-proteins, affect adenylate cyclase (AC) activity. Norepinephrine enhances adenylate cyclase activity via a stimulatory G-protein, G_s; acetylcholine inhibits it via an inhibitory G-protein, G_I. The subsequent effects on cyclic AMP, protein phosphorylation, Ca^{2+} channels, and neural and metabolic activity relate to the activation or inhibition of adenylate cyclase.

column. As noted above, sympathetic ganglionic neurons release norepinephrine and stimulate the heart. Parasympathetic innervation, by contrast, comes via the vagus nerve to parasympathetic ganglia located in the heart itself, and it is the neurons in the ganglia that innervate the heart muscle. In the parasympathetic system, both the pre- and postganglionic fibers release acetylcholine; yet, the preganglionic fibers excite neurons in the ganglion, whereas the heart's activity is depressed by the acetylcholine released onto it by the ganglionic neurons. Thus, acetylcholine has both excitatory and inhibitory effects in this system; the excitatory effects are mediated by nicotinic receptors, the inhibitory effects by muscarinic receptors (see Chapter 6).

How do norepinephrine and acetylcholine affect the heart rate and the strength of cardiac muscle contraction? Both neuroactive substances stimulate receptors, found on all heart cells, that are linked to G-proteins and adenylate cyclase. The G-proteins stimulated by the norepinephrine receptors activate adenylate cyclase, while the G-proteins linked to the acetylcholine receptors inhibit adenylate cyclase. Norepinephrine thus raises levels of cyclic AMP in the heart and acetylcholine lowers cyclic AMP levels. In addition, the effects of cyclic AMP are mediated by protein phosphorylation; and so sympathetic innervation increases phosphorylation of cell proteins, whereas parasympathetic innervation depresses protein phosphorylation. This convergent scheme is presented in Figure 13.4.

The principal targets of the cyclic AMP–dependent phosphorylation in the heart cells are voltage-dependent Ca^{2+} channels. Phosphorylation of these channels extends the time they are open during depolarization and it also makes it more probable that Ca^{2+} channels will open during depolarization; that is, it recruits Ca^{2+} channels. These effects result in a greater Ca^{2+} current for a set level of membrane depolarization and hence they have several consequences. First, the heart rate speeds up because the Ca^{2+} spikes are generated more quickly and the rise of the Ca^{2+} spike is faster. Second, because Ca^{2+} is important in muscle contraction, the added Ca^{2+} current into heart cells due to sympathetic innervation enhances the force of muscle contraction.

The pacemaker and other heart cells are also affected by the increase in Ca^{2+} levels within the cells. K^+ channel currents are enhanced, and this helps speed membrane potential recovery following the Ca^{2+} spike and accelerates the heart rate. In addition, enzymes involved in energy metabolism of the cells are activated and thereby raise ATP levels, the energy source for muscular contraction. Thus a variety of effects are induced in the heart cells by sympathetic stimulation, most of which are mediated by modulation of the voltage-sensitive Ca^{2+} channels.

Parasympathetic stimulation, on the contrary, reduces cyclic AMP levels and the phosphorylation of Ca^{2+} channels, thus reversing the effects of sympathetic stimulation. The Ca^{2+} channels are open for a shorter time and fewer are opened by depolarization. K^+ channel currents are reduced and metabolism of the heart cells slows.

Voluntary Movement

We can explain the effects of the autonomic nervous system on the heart down to the molecular level (as in Figure 13.4). Our understanding of most of the voluntary nervous system is much more rudimentary. Indeed, we know little in detail even of the circuitry involved in most voluntary movements, never mind the synaptic mechanisms involved. Nevertheless, we can identify the locus of some of the circuits and the main pathways of information flow controlling motor activity. The rest of this chapter is concerned with these matters.

Locomotion

Walking is a rhythmic motor behavior involving a stereotyped pattern of limb movement. Interest in understanding the neural mechanisms of rhythmic movements such as walking goes back to the early part of this century. Two British scientists, Charles Scott Sherrington and T. Graham Brown, were pioneers in much of the early work on mammals. Sherrington showed, for example, that rhythmic movements could be elicited from the hind legs of dogs and cats for weeks after the spinal cord had been severed. By tickling the animal's skin or lightly pulling on its hair, a rhythmic scratch reflex could be elicited in animals whose spinal cord was deprived of input from the brain. It was a revelation to learn that higher levels of the nervous system are not needed for the generation of rhythmic limb movements; spinal cord circuits are sufficient.

In 1911, Graham Brown carried these observations forward by showing that transection of the spinal cord initiated rhythmic movements of the limbs that closely resembled the movements of walking. The movements persisted for a minute or two before dying away. From these experiments Brown proposed that the basic neural circuitry needed for walking is located within the spinal cord.

Beginning in the 1960s, three Russian scientists, M. K. Shik, F. V. Severin, and G. N. Orlovsky, and a Swedish scientist, Sten Grillner, significantly extended the work of Brown by showing that when the brain stem of a cat is cut across its middle, the animal is capable of standing and can be induced to walk. The lower brain stem and spinal cord remain intact, but the animal lacks any input from higher neural centers. If a specific region of the brain stem, called the mesencephalic locomotor region, is stimulated, the animal will walk in a controlled manner on a treadmill. The setup for such an experiment is shown in Figure 13.5, along with a drawing of the spinal cord and brain indicating the level of transection and the approximate region of the locomotor region. If muscle activity is recorded while the animal is walking during the experiment, one observes alternating bursting of activity in the flexor and extensor muscles in conjunction with the walking. A record is shown in Figure 13.5.

Strengthening the electrical stimulation of the brain stem enhances the force of the walking movements but does not speed the walking. Rather, with more stimulation strength, the animal behaves as though it were walking uphill or against a load. Speeding up the treadmill, though, accelerates the walking. If the treadmill is made to go fast enough, the animal will begin to trot and eventually it will gallop at high treadmill speeds.

The general conclusion is that Graham Brown's early proposal is correct: the spinal cord has central pattern generators for walking. Indeed, each limb appears to have separate pattern generators that are nevertheless highly coordinated. These central pattern generators provide the motor output necessary to initiate and control coordinated walking movements. To activate the generator programs, descending stimulation—a command signal—from the brain stem is required. It appears, then, that rhythmic motor behavior in vertebrates is organized in a fashion similar to rhythmic motor behaviors in invertebrates. But at the present time much

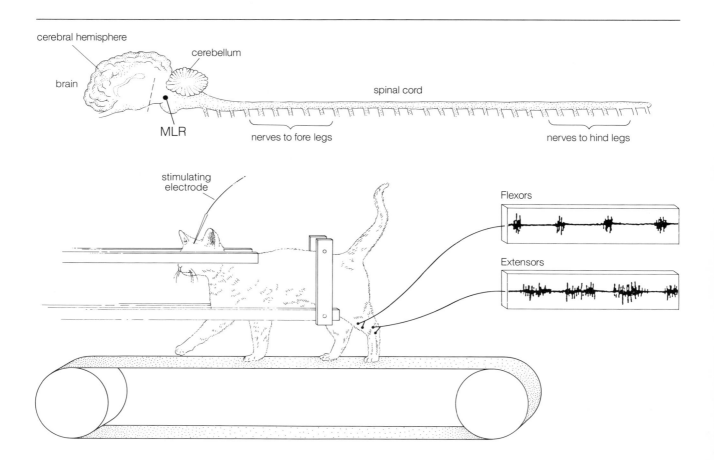

cerebral hemisphere

cerebellum

brain

spinal cord

MLR

nerves to fore legs

nerves to hind legs

stimulating electrode

Flexors

Extensors

13.5   A cat whose brain stem is cut walks when a specific region of the brain stem, the mesencephalic locomotor region (MLR), is stimulated. At the top of the figure is shown a drawing of the brain and spinal cord locating the transection *(dotted line)* and the MLR. Muscle activity can be recorded during such experiments (right side of figure), and alternating bursting activity is observed in the extensor and flexor muscles in conjunction with the walking.

less is known of the detailed circuitry of the spinal cord central pattern generators for walking than of the central pattern generators in invertebrates (Chapter 10).

Sensory Feedback and Locomotion

If we cut the dorsal roots carrying sensory information into a cat's spinal cord, and the cat's brain stem has been transected, the animal will still walk when placed on a treadmill and stimulated appropriately. If the treadmill changes speed, however, the animal no longer alters its rate of walking or gait; rather, the animal walks slowly and at a constant speed. Sensory feedback is essential,

13.6 The sequence of leg movements in walking, trotting, and galloping in a four-legged mammal, and the relative lengths of time that the feet are on *(black areas)* or off *(white areas)* the ground.

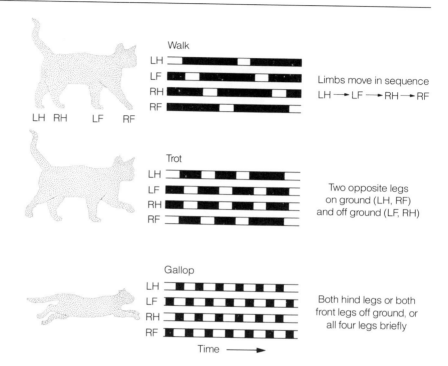

then, for modulating the central pattern generators and allowing the animal to change walking speed or its gait to a trotting or galloping mode. Presumably, sensory input not only modulates the output frequency of the central pattern generators but also reconfigures them so the firing pattern of the output neurons is appropriate for the animal's gait. In essence, a different pattern of motor output is required for walking, trotting, and galloping. This is made clear by comparing the stepping patterns of the limbs in walking, trotting, and galloping animals. Figure 13.6 illustrates the sequence of leg movement and the time that a foot is off the ground by these three gaits. While the animal is walking, the limbs move in sequence and only one leg is off the ground at a time. While it is trotting, the front and back legs on opposite sides of the body are off the ground at the same time, and while it is galloping both hind legs or both front legs are off the ground simultaneously. In addition to sensory modulation of the central pattern generators, there must be descending inputs that can similarly modify these circuits. In other words, the animal can voluntarily change its speed and its gait.

What is the nature of the sensory input from the limbs that modulates the central pattern generators in the spinal cord? There are a variety of sensory receptors associated with the limbs and muscular activity, including sensory receptors within the joints, tendons, and muscles themselves. Muscle receptors are of special interest because they signal information about the elongation and tension of the muscles. This sensory information, generated within an organism itself, was accordingly called "proprioceptive information," from the Latin word, *proprius,* meaning "one's own," by Sherrington. As noted in Chapter 12, we are not aware of most proprioceptive information; it doesn't reach those parts of the cortex associated with conscious sensation. Rather, its major functions are to regulate the motor neurons themselves and to provide information to the central pattern generators within the spinal cord and to brain centers, such as the cerebellum, that have to do with the coordination of movement.

Two main types of muscle proprioceptors innervate a group of modified muscle fibers enclosed in a capsule and found among the regular muscle fibers. This structure, depicted in Figure 13.7, resembles the spindle used in spinning wool and was therefore called a spindle by the early histologists who discovered it. The two types of sensory neurons innervating a spindle are differentiated on the basis of their size and response to stretch of the spindle. The sensory endings from the larger axons, called group I afferents, are sensitive mainly to the rate of stretching of the spindle; that is, they respond most actively during elongation of the muscle surrounding the spindle and their discharge rate subsides while the stretch is maintained. The endings from smaller axons, termed group II afferents, respond less vigorously during muscle elongation, but their steady firing rate reflects the static tension of the surrounding muscle. These differences are shown also in Figure 13.7. Both the group I and group II sensory endings are mechanoreceptors; they respond directly to deformation of their terminals (see Chapter 7).

An interesting and important feature of the muscle spindles is that they are also innervated by small motor neurons (γ-fibers). This means that the modified muscle fibers that make up the

13.7 A muscle spindle, which is made up of modified muscle fibers, is innervated by two types of sensory neurons (group I and group II afferents) and one type of motor neuron (γ-efferents). The spindle sits among the regular muscle fibers and senses both the static tension of the surrounding muscle fibers and the rate of stretching of the fibers. The group I afferents respond most actively during muscle elongation, whereas the steady discharge of the group II afferents reflects the static tension of the muscle. Typical records from group I and group II afferent fibers are shown below.

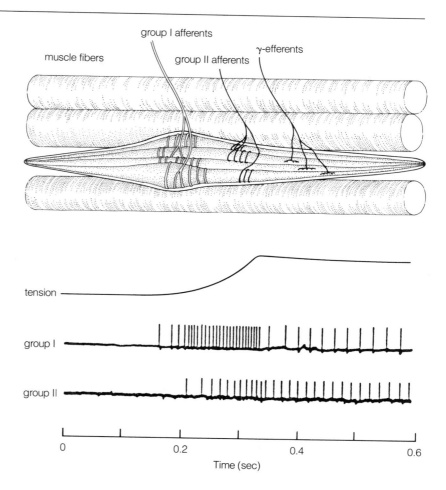

spindle are themselves under motor control and they contract when the γ-fibers fire. The motor neuron innervation of the spindles regulates the tension on the spindle muscle fibers, enabling the afferent fibers to function over a wide range of muscle length. For example, when a muscle contracts, tension on the spindle decreases, and both group I and group II afferents would stop firing if the spindles themselves did not change length. The motor neuron innervation of the spindles causes the spindle to contract when the muscle contracts. Because of the restored tension on the sensory endings, the afferents can then signal further changes in elongation and tension of the muscle surrounding the spindle.

Feedback and Motor Control: An Overall Scheme

Figure 13.8 shows the general organization of many motor systems in vertebrates. This scheme is quite similar to the scheme I described in Chapter 10 for motor control in invertebrates (see especially Figure 10.1). The spinal cord contains the neural circuitry required to generate coordinated and rhythmic outputs of the motor neurons—the central pattern generators. These circuits are activated and controlled by commands from higher motor centers in the brain. Later in the chapter I discuss a hierarchy of higher motor centers. The output of the central pattern generators controls the muscles of the body and limbs and thus the interactions of the animal with its environment.

Several types of feedback interactions regulate the central pattern generators and the higher motor centers. Proprioceptive information feeds back from muscles, joints, and tendons to the central pattern generators within the spinal cord and also to higher brain centers. Information about the output from the central pattern generators is also fed back to the higher brain centers. This

13.8 Organization of vertebrate motor systems. Central pattern generators, located in the spinal cord, are activated by commands from higher motor centers. The output of the central pattern generators controls rhythmic limb and body movements. There is feedback from the muscles to the central pattern generators and to higher brain centers. The output of the central pattern generators also feeds back to higher brain centers. In addition, sensory input from the environment impinges on both the central pattern generator and higher brain centers.

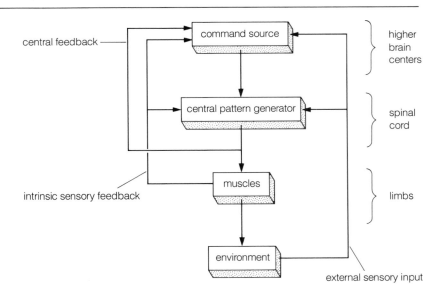

feedback information tells the higher brain centers about the output from the central pattern generators and how the central pattern generators are responding to central commands and to intrinsic and extrinsic sensory inputs. Finally, sensory input from the environment impinges on the central pattern generators as well as onto the higher brain centers. During movement there is a constant interplay between sensory input, central pattern generators, and higher brain centers.

The Initiation of Movements

Electrical stimulation of the primary motor cortex (described in Chapter 12) generates restricted movement of a part of the body, a limb, or even a digit. For this reason the primary motor region of the cortex has long been believed to be critical for the initiation of movements. Some output from the primary motor cortex extends into the spinal cord, and in some species even to the motor neurons themselves. In 1874 the Russian anatomist Vladimir Betz observed very large neurons in this part of the cortex in primates; subsequently, researchers found that the axons of these giant cells, which had come to be known as the Betz cells, terminate on motor neurons that participate in limb movements and speech. These cells have been studied intensively in recent years, particularly by Edward Evarts and his colleagues at the National Institutes of Health.

By recording Betz cells in awake, behaving animals trained to carry out a task requiring wrist flexion or extension, Evarts and his colleagues showed that some Betz cells responded to wrist flexion and others to wrist extension. What they discovered is that the neurons typically discharged before the onset of muscle contraction. This provided evidence that the neuronal activity observed was initiating the movement, not a corollary of the movement. They also found that a cell's discharge frequency was related to the force of the movement, not to the kind of movement made.

The conclusion from these experiments is that motor cortex neurons participate directly in the initiation of skilled, voluntary movements. These neurons are also known to receive feedback sensory information about the movement they initiated and its consequences; that is, they receive both proprioceptive sensory information and other sensory input related to the movement produced. The same sensory pathways may also participate in

13.9 Stretch reflex pathways function in parallel with sensory-motor pathways involving the motor cortex. Stretch of a muscle can directly activate a motor neuron via a sensory neuron. The sensory information also ascends to the cortex and it may impinge eventually on cortical neurons (Betz cells) that innervate the same motor neurons.

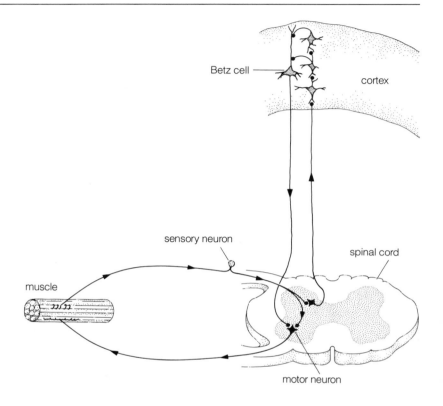

mediating spinal cord reflexes. Indeed, it is likely that sensory-motor pathways in the motor cortex function in parallel with the simple stretch reflexes mediated in the spinal cord. The two pathways are illustrated schematically in Figure 13.9. An imposed stretch of a muscle can lead to direct activation of the motor neurons via sensory neurons; this is the "short," spinal cord loop. But sensory information also ascends to the cortex and ultimately impinges on Betz cells that directly innervate and drive the motor neurons (the "long," cortical loop). The cortical loop could even assist and refine the stretch reflexive movements. Presumably the sensory neurons and motor neurons are also part of, or at least contribute to, the central pattern generators that control more complex and rhythmic movements.

It is important to note, finally, that not all movements are controlled by the primary motor cortex. The primary motor cortex mainly produces skilled and precise movements. Relatively automatic (reflexive) or rhythmic movements are controlled by other motor systems both higher and lower in the central nervous system. Walking, for example, is initiated by stimulating the brain stem, which receives input from several higher motor centers, but the neural circuitry controlling walking is located within the spinal cord.

Motor Programs

A natural question to ask is how and where in the cortex complex motor activities are planned and programmed. Three regions—the supplementary motor area and premotor cortex in the frontal lobe, and the posterior parietal cortex in the parietal lobe—have been implicated in the generation of such motor programs. The location of these premotor areas is shown in Figure 13.10.

The supplementary motor area and premotor cortex receive substantial input from the posterior parietal cortex. As might be expected, because of its location adjacent to the somatosensory cortex, the posterior parietal cortex provides the frontal lobe motor areas with sensory information, particularly with spatial information necessary for making a correct movement toward a target. People with lesions in this area will reach for an object but miss it. They have good limb and muscle control, but they have

13.10 The left hemisphere of the primate cerebral cortex showing the areas of the cortex involved with planning and programming movements. Two of the areas are in the frontal lobes (the supplementary motor area and premotor cortex) and one is in the parietal lobe (the posterior parietal cortex).

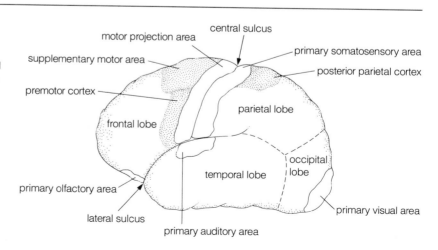

difficulty orienting their limbs to objects they wish to touch or grasp. The conclusion drawn from both experimental and clinical evidence is that this cortical area processes sensory information necessary for purposeful movement.

The premotor cortex is also primarily involved in the sensory guidance of limb movements. Lesions in this area cause deficits quite similar to lesions in the posterior parietal cortex. Neurons in this cortical area project axons to brain stem regions that contribute to motor activity as well to the spinal cord itself. In addition, the premotor cortex provides substantial input to the primary motor cortex, so its influence is widespread. This cortical area may be particularly important in the orientation of the body or limb toward a target. This has yet to be determined with certainty.

The supplementary motor area appears to be essential in programming complex movements. Compared with the primary motor cortex, it seems to play a more indirect and preparatory role in motor activity. A dramatic illustration of the activity of the supplementary motor cortex was provided by Per Roland and his co-workers in Denmark. They used the technique of PET (positron emission tomography) scanning to measure cerebral metabolic activity and provide an index of neural activity. Roland and his colleagues asked human subjects to move their fingers in a complex sequence and observed increased metabolic activity in both the supplementary motor area and motor cortex. When the subjects were asked to rehearse mentally the sequence of finger movements, but not actually move their fingers, increased metabolic activity was observed only in the supplementary motor area and not in the motor cortex. This suggests that the supplementary motor area is likely to be involved in motor sequence programming. I discuss the technique of PET scanning further in Chapter 18.

Figure 13.11 is a simplified diagram of the principal structures and pathways of motor control. The three major structures for motor activity are the primary motor cortex, the brain stem, and the spinal cord. The motor cortex issues motor commands to the brain stem and spinal cord, and it also sends motor information to the cerebellum and basal ganglia. The brain stem contains nuclei that integrate commands from higher levels with sensory input; it is particularly important for the control of posture and for initiating integrative and rhythmic movements of body and limbs, such as walking. The spinal cord mediates various reflexes and contains

13.11 The principal structures and pathways involved in motor control in the vertebrate.

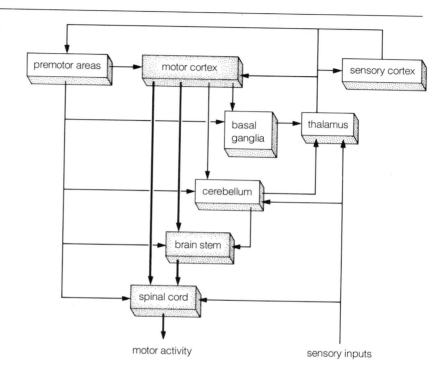

motor activity

sensory inputs

the central pattern generators underlying rhythmic behaviors, such as scratching and walking. It also contains the motor neurons that drive the muscles and are, in Sherrington's words, the "final common pathway."

Control of these structures is extensive: premotor areas plan and program movements. They also integrate the motor programs with spatial information so that targets in space can be accurately approached. Their major output is to the primary motor cortex, but they also send information to the brain stem, spinal cord, cerebellum, and basal ganglia. The cerebellum coordinates motor activity by comparing descending motor commands with sensory signals. Its main output is back up to the motor cortex via the thalamus or to brain stem nuclei. The basal ganglia are concerned more with the initiation of movement; they receive input from many parts of the cortex and feed information back primarily to the premotor cortical areas via the thalamus. The basal ganglia appear to provide tight feedback control of most voluntary movements, and thus damage to the basal ganglia often results in uncontrolled oscillatory movements, presumably related to an upset in the feedback circuits.

Summary

Two motor systems, the autonomic nervous system and the voluntary motor system, are recognized in vertebrates. The autonomic nervous system controls the internal organs whereas the voluntary motor system controls muscles of the limbs, body, and head. The autonomic nervous system is organized into two opposing systems: the sympathetic system prepares an animal for strenuous activity—cardiac and other muscular activity heightens whereas digestive tract activity diminishes; the parasympathetic system has opposite effects—the heart slows and beats less strongly, whereas digestive tract activity increases. The effects of both systems are mediated by neurons found in ganglia that lie along the spinal cord or nearby (sympathetic ganglia) or that are present in the organs themselves (parasympathetic ganglia). These ganglia are innervated by motor neurons in the brain stem and spinal cord, all of which are cholinergic. The ganglionic neurons, by contrast, release either norepinephrine (sympathetic neurons) or acetylcholine (parasympathetic neurons). Complex synaptic interactions within the ganglia bring into play numerous neuroactive substances that generate synaptic potentials of various sizes, polarities, and duration.

Sympathetic neurons via norepinephrine speed the heart rate and the strength of cardiac contraction; parasympathetic neurons via acetylcholine have opposite effects. Both neuroactive substances activate receptors on heart cells linked to G-proteins and adenylate cyclase. Norepinephrine, however, enhances adenylate cyclase activity whereas acetylcholine depresses it. Levels of cyclic AMP increase upon sympathetic stimulation but decrease upon parasympathetic stimulation. The cyclic AMP exerts its effects via protein phosphorylation, and the primary targets of this phosphorylation are voltage-sensitive Ca^{2+} channels in the membranes of heart pacemaker and muscle cells. These channels are important in generating the heartbeat: upon sympathetic stimulation, the amount of Ca^{2+} they admit into the cells is enhanced upon phosphorylation and the heart rate is speeded, cardiac contraction becomes stronger, and other changes are induced—in K^+ channel activity and cell metabolism—that enhance cardiac performance. Parasympathetic stimulation, conversely, lowers Ca^{2+} channel activity and opposite effects are observed.

Walking is a rhythmic motor behavior mediated by central pattern generators located in the spinal cord. A cat whose brain stem has been transected can be induced to walk on a treadmill if a specific region of its lower brain stem is stimulated. More brain stem stimulation increases the force of walking, whereas speeding up the treadmill accelerates the rate of walking and may change the animal's gait. If sensory input into the spinal cord is blocked, the animal walks at a constant speed; no longer will it respond to changes in treadmill speed. This indicates that sensory feedback is essential for modulating the central pattern generators that participate in locomotion.

Various kinds of sensory information come from joints, tendons, and the muscles themselves during motor activity. Muscle receptors signal both the rate of elongation of muscles and the steady tension the muscles maintain. This is accomplished by two types of mechanoreceptors that innervate encapsulated groups of specialized muscle cells called spindles. The muscle cells in a spindle are also innervated by small motor neurons that adjust the spindle's length in response to how much the surrounding muscle mass is contracting or being stretched. Thus the elongation and tension receptors innervating the spindle respond over a wide range of muscle length.

Sensory feedback affects not only the central pattern generators in the spinal cord but also higher motor centers. The outputs of the central pattern generators are fed back to higher motor centers as well. Both kinds of information are critical for smooth and coordinated movements.

Electrical stimulation of the primary motor cortex results in body, limb, or digit movements. Some neurons in the primary motor cortex, called Betz cells, extend directly into the spinal cord, and in primates these neurons terminate on motor neurons. Betz cells participate in the initiation of skilled voluntary movements. They receive sensory information about the movements they initiate via sensory neurons that also participate in simple spinal reflexes. It is believed that sensory-motor pathways involving the motor cortex function in parallel with simple reflexes mediated at the level of the spinal cord. The more complex cortical loop may assist and refine the simpler reflexive movements. Not all movements are under the control of the primary motor cortex. Automatic (reflexive) and rhythmic movements are controlled by spinal cord circuits that are activated by sensory input or commands from the brain stem and higher motor centers.

Motor programs are planned and initiated by areas in the frontal and parietal lobes. The parietal cortex is concerned with the processing of spatial information crucial for making target-directed movements. The premotor cortex in the frontal lobe is involved with the sensory guidance of limb movements and with body and limb orientation. Its output goes to the primary motor cortex, brain stem, and spinal cord as well as to the cerebellum and basal ganglia. The supplementary motor cortex, also in the frontal lobe, helps to program complex movements. Its output is mainly to the primary motor cortex.

The three major structures that participate in motor activity are the primary motor cortex, the brain stem, and the spinal cord—and each is under extensive control. Premotor areas plan and program movements, and they integrate spatial sensory and motor information. Their output is mainly, but not exclusively, to the motor cortex. The cerebellum coordinates motor activity by comparing motor commands with incoming sensory inputs. Its output is primarily fed back to the motor cortex or to brain stem nuclei. Finally, the basal ganglia are concerned with the initiation of movement; they receive input from many areas of the cortex and feed information back primarily to premotor areas.

Further Reading

BOOKS

Evarts, E. V., S. P. Wise, and D. Bousfield, eds. 1985. *The Motor System in Neurobiology.* Amsterdam: Elsevier Biomedical Press. (A collection of short reviews from *Trends in Neuroscience.* Some are easily readable, others not.)

Matthews, P. B. C. 1972. *Mammalian Muscle Receptors and Their Central Action.* London: Edward Arnold.

Sherrington, C. S. 1906. *The Integrative System of the Nervous System.* New Haven: Yale University Press; 2d ed., 1947. (Sherrington's classic Silliman Lectures at Yale.)

ARTICLES

Evarts, E. V. 1979. Brain mechanisms of movement. *Scientific American* 241(3):98–106.

Grillner, S. 1985. Neurobiological bases of rhythmic motor acts in vertebrates. *Science* 228:143–149.

Kuffler, S. W., and T. J. Sejnowski. 1983. Peptidergic and muscarinic excitation at amphibian sympathetic synapses. *Journal of Physiology* 341:257–278.

Noble, D. 1985. Ionic mechanisms in rhythmic firing of heart and nerve. *Trends in Neuroscience* 8:499–504.

Pearson, K. G. 1976. The control of walking. *Scientific American* 235:72–86.

Tsien, R. W. 1987. Calcium currents in heart cells and neurons. In *Neuromodulation: The Biochemical Control of Neuronal Excitability,* ed. L. K. Kaczmarek and I. B. Levitan, pp. 206–242. New York: Oxford University Press.

Retinal Processing of Visual Information

IN THIS CHAPTER, we begin to explore central nervous system mechanisms in more detail. The focus will be on the visual system, for two principal reasons. First, we probably know more about the visual system than any other system in the vertebrate brain; furthermore, it is the most important sensory system for us. It has been estimated that over 40 percent of our sensory input is visual, and those individuals who have lost their sight are severely disabled.

The second reason the visual system is particularly attractive for analysis is that its basic organization and anatomy are well understood. For instance, it is possible to examine transformations in a visual message over at least 6 synapses in mammals. Two stages of neural processing take place within the retina, a part of the central nervous system displaced into the eye during development. (How this happens is explained later.) A third synaptic transformation is in the lateral geniculate nucleus of the thalamus, and finally in the cortex we can distinguish at least three levels of processing.

Figure 14.1 depicts the visual pathways from the eye to the cortex in a mammal and also the cellular organization of the visual system from receptors to cortex. From the eye, the optic nerve carries visual information to the lateral geniculate nucleus in the thalamus. And from there the nerve fibers that make up the optic radiation carry the visual signal to the primary visual cortex.

From a cellular point of view, the visual system has two important characteristics of neural processing—convergence and divergence of information. Although there are about 100 million photoreceptors in the human eye, there are only about 1 million optic nerve axons. Thus, there is convergence of approximately 100 to 1 between photoreceptors and optic nerve axons. This convergence is not constant across the retina; the central part of the retina in virtually all species has a higher visual acuity than do peripheral

14.1 Schematic drawings of (**A**) the pathways from the eyes to the visual cortex in the mammal and (**B**) the cellular organization of the mammalian visual system. (**A**) From the retinas in the two eyes, the optic nerves carry visual information to the lateral geniculate nuclei (LGN). From there the optic radiation carries the visual signal to area V1 of the cortex. Visual information passes from area V1 to area V2 and eventually to other visual areas (V3, V4, and so on). (**B**) In the retina, there is an overall convergence of information: 100 million receptors converge on about 1 million optic nerve fibers. In the cortex, on the other hand, there is a substantial divergence of information. There are about 1,000 times more cells in the visual cortex than there are optic nerve axons. White circles indicate projection (Golgi type I) neurons; stippled circles are intrinsic (Golgi type II) neurons.

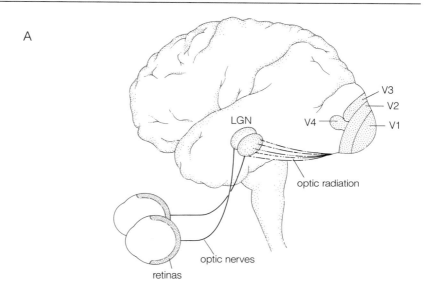

A

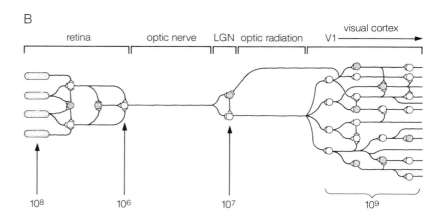

B

retinal regions, and correspondingly there is less convergence in the central retina. In primates such as ourselves, which have a specialized foveal region for high acuity, there is even some divergence from receptors to ganglion cells. That is, for the 20,000 cones that make up the central foveal region in the human eye, there are over 40,000 optic nerve fibers. Hence, for each foveal photoreceptor (and they are all cones), there are more than two third-order ganglion cells whose axons make up the optic nerve.

A more dramatic example of divergence is between optic nerve and cortex. Even though the optic nerve contains approximately 1 million optic nerve fibers, there may be as many as 1 billion neurons in that part of the cortex concerned with vision. So for every optic nerve axon, there are 100 to 1,000 visual cells in the cortex. This enables the cortex to carry out an enormous amount of processing of visual information, as I describe in the following chapters.

The Retina

The reasons for carefully examining the retina are many. First of all, the initial stages of visual processing take place within the retina. The retina is probably the most accessible part of the vertebrate brain, sitting as it does in the eye. Furthermore, it is highly ordered anatomically, such that virtually all of its synapses, between cells of only a few types, are in two synaptic or plexiform layers. Distinct kinds of processing occur in the two synaptic layers. In many vertebrate retinas, the cells are rather large, so intracellular recordings can be made routinely. Thus, the retina in many species can be analyzed like an invertebrate ganglion; it is relatively isolated and is concerned with a specific and defined function. Its circuitry is simple and the responses of each of its cell types can be recorded intracellularly. Yet the retina is a part of the vertebrate central nervous system, and so retinal mechanisms can serve as a model for neural function throughout the vertebrate brain.

Cellular Organization

Because it has especially large cells, the mudpuppy (an amphibian) is an excellent experimental model for studies of cellular organization. Figure 14.2 shows the major cell types observed in its retina, their location and approximate form, and the distribution of their processes within the plexiform layers. The drawing is based on retinal preparations stained by the method of Golgi. Three cellular and two synaptic layers are observed in all vertebrate retinas, as here. The most distal cellular layer, called the outer nuclear layer (ONL), contains the cell bodies of the photorecep-

14.2 A sketch of a mudpuppy and the major cell types found in the vertebrate retina as viewed in vertical section. On the left is a portion of a light micrograph of the mudpuppy retina. *ONL*, outer nuclear layer; *OPL*, outer plexiform layer; *INL*, inner nuclear layer; *IPL*, inner plexiform layer; *GCL*, ganglion cell layer. On the right is a drawing of the major cell types in this retina as revealed in Golgi-stained preparations. *R*, receptors; *H*, horizontal cells; *B*, bipolar cells; *A*, amacrine cells; *G*, ganglion cells.

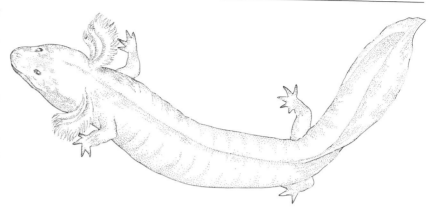

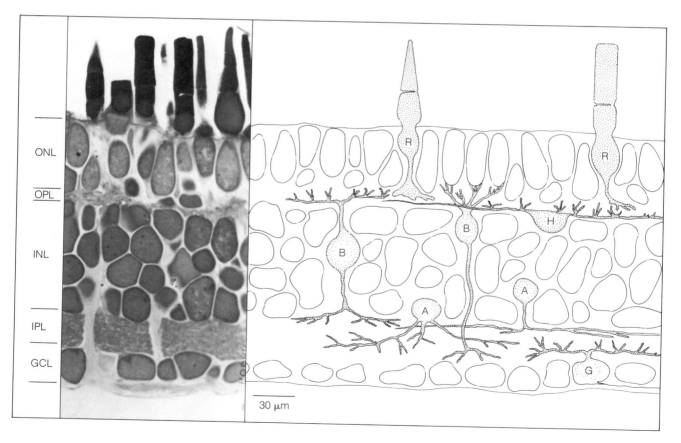

A   neural tube

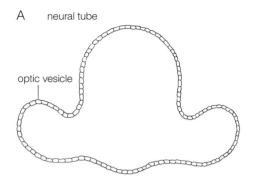

optic vesicle

B   optic cup

outer wall = pigment epithelium

inner wall = retina

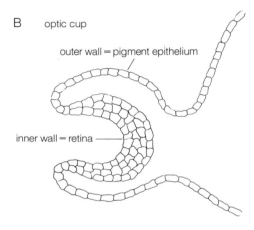

C   retina

GCL   OPL   pigment
    IPL INL   ONL   epithelium

        photoreceptors

light

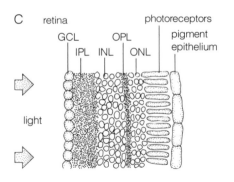

14.3 Schematic diagrams of the development of the retina from the neural tube in the
embryo. **(A)** In the head region of the embryo,
optic vesicles form from outpocketings of the
neural tube. **(B)** The optic vesicles invaginate
to form an optic cup, and the retina develops
from its inner wall. **(C)** The retinal layers in the
adult.

tors. The middle layer, or inner nuclear layer (INL), contains the cell bodies of three major classes of cells—horizontal, bipolar, and amacrine. And the innermost cellular layer contains the ganglion cell bodies (GCL). Between the cellular layers are the synaptic layers, the outer and inner plexiform layers (OPL and IPL).

The retina, like other regions of the vertebrate brain, derives from the neural tube (see Chapter 17). Early in embryonic life the neural tube evaginates to form two optic vesicles; the vesicles then invaginate to form optic cups (Figure 14.3). The cells on the inner wall of the optic cups eventually become the retina. Both walls of the optic cups are initially one cell thick; but the cells of the inner wall divide to form a multicellular layer that becomes the retina. Since the photoreceptors develop along the inner (ventricle) side of the multicellular layer, light must pass through the entire thickness of the retina to strike the photoreceptors. The retina thus appears at first glance to be aligned along the back of the eye in a backward fashion, but this results from the way the retina develops.

In the high-acuity foveal region of the primate eye, the inner layers of the retina are pushed aside so that light can fall directly on the photoreceptors. Thus the retina's foveal region is indented. Other specializations in the fovea also improve visual acuity; it has only cones, which are the thinnest and longest photoreceptors in the retina.

The retina's basic cellular organization is as follows (see Figure 14.2): photoreceptor terminals *(R)* provide the input to the outer plexiform layer; horizontal cells *(H)*, which extend processes widely in the outer plexiform layer, mediate lateral interactions within this first synaptic zone; and bipolar cells *(B)*, the output neurons for the layer, carry the visual signal to the inner plexiform layer. The ganglion cells are the output neurons for the inner plexiform layer. Indeed, the ganglion cells are the output neurons for the entire retina. Their axons run along the margin of the retina and collect at the optic disk to form the optic nerve, which carries all visual information from the eye to higher visual centers. Amacrine cells *(A)*, like the horizontal cells in the outer plexiform layer, extend processes widely in the inner plexiform layer and mediate interactions within it.

To summarize, all visual information passes across at least two synapses, one in the outer plexiform layer and another in the inner plexiform layer, before it leaves the eye. The situation is much more complex than this because each plexiform layer analyzes and processes the visual signal. In virtually all instances, therefore, visual signals go through many more than two synapses before they leave the eye. Thus, the retina can be viewed as the equivalent of two brain nuclei: one processing visual information in the outer plexiform layer, the other processing information in the inner plexiform layer.

The Initial Stages of Visual Processing

What kinds of processing occur within the retina? The ganglion cells sit along the retina's inner side (see Figure 14.2) and are accessible for recording with extracellular microelectrodes. The long optic nerve in many species also affords an opportunity to record action potential discharges extracellularly from single nerve fibers. This was first accomplished in the late 1930s by Keffer Hartline, who dissected out individual nerve fibers in the frog retina. With extracellular microelectrode techniques, one may record the spike output of ganglion cells and their axons in many species. The results reveal two types of processing in all retinas, one reflecting the neural interactions in the outer plexiform layer and the other reflecting interactions carried out in the inner plexiform layer. Ganglion cells convey the information to higher brain centers.

Ganglion cells typically respond to illumination of a restricted yet relatively large region of the retina; that is, the cell's discharge rate alters when this area is illuminated. This region, the cell's *receptive field,* is typically about a millimeter or so in diameter. Thus, many receptors contribute to the responses of a singe ganglion cell, as do inner nuclear layer neurons. (Although we speak of the activation of different cells upon illumination of the retina, it is important to remember that only the *photoreceptors* respond directly to a light stimulus—all other cells, from the retina to the brain, are activated by signals that begin with rod and cone cells.) The receptive fields of adjacent ganglion cells overlap considerably, but not completely; so any one region of the retina is covered by many different ganglion cells, each related to a slightly different

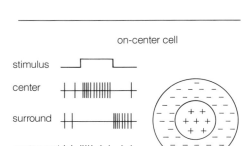

on-center cell

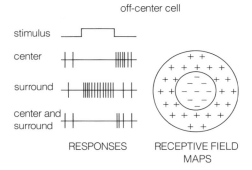

off-center cell

RESPONSES   RECEPTIVE FIELD
                        MAPS

14.4   Idealized responses and receptive field maps for on-center and off-center contrast-sensitive ganglion cells. The drawings on the left represent hypothetical responses of the cell to a spot of light presented to the retina in the center of the receptive field, in the surround of the receptive field, or in both the center and surround regions of the receptive field. A + symbol on the receptive field map indicates an increase in the firing rate of the cell—that is, excitation; a − symbol indicates a decrease in the firing rate—that is, inhibition.

part of the visual field. This arrangement means that any one receptor and inner nuclear layer cell contributes to the responses of many ganglion cells.

The first type of ganglion cell receptive field I describe shows evidence of *spatial* processing of visual information. Ganglion cells of this type, often called contrast-sensitive cells, are subdivided into two mirror-image classes: on-center cells and off-center cells. As shown first by Stephen Kuffler, then at Johns Hopkins University, contrast-sensitive cells have receptive fields that are organized into two concentric regions that are antagonistic to one another. Figure 14.4 shows idealized responses and receptive field maps for on- and off-center ganglion cells.

For an on-center cell, illuminating the center of the field with a spot of light causes the cell to fire a sustained burst of impulses as long as the light is on, the frequency of which depends on light intensity. Illumination within the antagonistic surround of the field inhibits the cell while the light is on, as can be seen by a decrease in the cell's activity followed by a vigorous burst of impulses when the light turns off. If both center and surround are illuminated simultaneously, the two regions antagonize each other and create only a weak response, characteristic of the center. Thus, these cells respond vigorously when illumination is confined to a certain region of the receptive field; they are clearly responsive to the spatial distribution of light on the receptor mosaic. Furthermore, because center and surround regions are antagonistic, these cells respond even better when there is maximum contrast between center and surround stimuli—when a bright center spot is surrounded by a dark ring or annulus.

The receptive field of the off-center cell is the mirror image of that of the on-center cells. Illuminating the center inhibits the cell's activity, and illuminating the surround increases it. Again the cell responds most dramatically when center and surround regions have maximum contrast.

The second basic type of receptive field for ganglion cells evidences *temporal* processing of visual information. These ganglion cells are highly sensitive to movement and, as shown first by Horace Barlow and his colleagues at the University of California, many also respond selectively to the direction of movement. They can respond vigorously to a spot of light moving through a receptive field in a certain direction, but they respond not at all or are

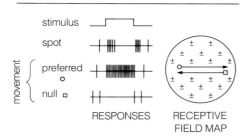

14.5 Idealized responses and a receptive field map for a direction-sensitive ganglion cell. Such cells respond with a burst of impulses at both the onset and termination of a spot of light presented anywhere in the cell's receptive field. This response is indicated by ± symbols all over the map. Movement of a spot of light through the receptive field in the preferred direction elicits firing from the cell that lasts for as long as the spot is within the field. Movement of a spot of light in the opposite (null) direction causes inhibition of the cell's maintained activity for as long as the spot is within the receptive field.

inhibited by the same spot of light moving in the opposite direction.

Figure 14.5 presents idealized responses and a receptive field map of a direction-sensitive cell. When the cell's receptive field is probed with static spots of light, they respond with a burst of impulses at the onset and cessation of illumination. For this reason the cells are commonly referred to as on-off ganglion cells as well as movement- or direction-sensitive ganglion cells. Because the on-off responses do not usually depend on the location of the stimulating spot within the receptive field, these cells are not especially concerned with the spatial aspects of illumination. Furthermore, they can usually respond equally well to a dark spot moving on a bright background as to a bright spot moving on a dark field. Because they are strongly activated by changes in illumination—its onset, cessation or movement—they are clearly more responsive to the temporal aspects of the stimulus.

All retinas appear to have ganglion cells of both types, though some animals have retinas with more ganglion cells of one type than another. Cats and monkeys have many sustained on- and off-center ganglion cells but few on-off direction-sensitive ganglion cells. By contrast, many cold-blooded vertebrates have many transient on-off ganglion cells and few sustained on- and off-center cells. (Why this is so is explained in the beginning of the next chapter.) Also, other types of receptive fields show a mix of spatial and temporal processing of visual information in the ganglion cell. For example, in mammalian retinas some receptive fields are organized in a concentric on- or off-center fashion, but light responses are much more transient than are those shown by the on- and off-center cells just described. Hence, the receptive fields of these ganglion cells are spatially organized but much more sensitive to movement than are the sustained contrast-sensitive cells. Receptive field properties vary in other ways, too. In many retinas, including the primate retina, the on- or off-center responses of many ganglion cells have a different color sensitivity than do the antagonistic surround responses. The cones feeding into the field's central region are spectrally different from the cones driving the antagonistic surround region, and hence the center of the field might reflect input from only red cones whereas the surround might have substantial input from green cones or vice versa. Such ganglion cells are called *color-opponent cells*.

The sizes of receptive fields vary, particularly in the receptive field center, in different parts of the retina. The receptive fields of ganglion cells in the central retina typically have smaller receptive field centers, as one might expect because of the higher visual acuity there. The smallest known receptive field centers are those of primate ganglion cells that receive input from foveal cones. Their centers are extremely tiny and probably reflect input from just a single cone. Earlier I noted that there are at least two ganglion cells per foveal cone, and the best guess is that every foveal cone drives at least two ganglion cells; one that is an on-center cell, the other an off-center cell.

The Significance of Retinal Processing

Why is it important that the initial stages of visual processing are carried out in the retina? What does this arrangement tell us about the processing of visual information in the brain? First we may ask why there are both on-center and off-center cells in the retina. Working with a drug (a synaptic antagonist) that blocks on-center cell activity, investigators have examined an animal's response to illumination when the flow of information from the on-center cells, called the on-pathway, is blocked. What has been found is that an animal whose on-pathways have been blocked cannot detect increases in retinal illumination, but it *can* detect decreases in illumination. Put in other words, these results indicate there is a division of visual information in the retina such that some neurons (on-center cells) are charged with signaling light increments to the rest of the brain, whereas other neurons (off-center cells) signal light decrements. As we shall see, this is a common theme in visual processing—that cells and pathways are concerned with one or another specific aspect of the visual image.

Another question we need to answer: what is the purpose of center-surround antagonism found in ganglion and other cells in the retina? I have already suggested that these cells analyze the visual image spatially. They respond briskly either to central illumination or to surround illumination, but only poorly to diffuse illumination. Thus, they tell the rest of the brain about the distribution of light on the retina, particularly with regard to discontinuities in illumination in their receptive fields. They are particularly effective, then, in signaling edges and borders that pass through the receptive field.

14.6   The effect of surround illumination on our perception of intensity. Both small squares are the same shade of gray, but the square looks brighter against a dark background than a bright background.

But center-surround antagonism provides another important advantage to the visual system. Visual systems respond over a wide range of light intensities, yet we see an object as black or white whether we are in dim light or in bright light. It is not the absolute intensity of light that comes from an object that makes it appear light or dark to us, but the intensity of light coming from the object relative to the intensity of light coming from surrounding objects. So, our visual system is designed to compare light intensities regardless of overall illumination, and center-surround receptive fields help us to do this. Figure 14.6 is a convincing demonstration that our judgment of intensity depends on surrounding illumination. The center spot in both images is identical in intensity, yet the spot looks brighter against a dark surround than a light one.

Finally, what is the function of movement- and direction-sensitive cells in the visual system? I return to this subject in the next chapter, when I review cortical processing of visual information; here I describe briefly some interesting experimental evidence from frogs and toads. These animals do not respond to stationary objects, but they do respond vigorously to moving objects. Frogs and toads have numerous movement- and direction-sensitive ganglion cells in their retinas. Indeed, there is a remarkable parallelism between these cells' discharge rates and the animal's prey-catching. That is, the stimulus parameters that optimally drive certain of the ganglion cells (such as object size and contrast, direction and velocity of object movement) are also the parameters that best activate prey-catching behavior. Hence, their behavior is clearly related to the properties of the retinal ganglion cells and their receptive fields. Some investigators have suggested that the ganglion cells that relate to prey-catching in the frog or toad be called "fly" or "bug" detectors. But, as might be expected, a complex behavior like prey-catching cannot be explained solely by the action of a single type of retinal ganglion cell.

Cellular Mechanisms

Synapses in the Retina

Synaptic interactions within the retina are responsible for on- and off-center receptive fields, center-surround antagonism, and movement and direction sensitivity. The cellular organization of the retina, described earlier in this chapter (see Figure 14.2), offers some clues about these synaptic connections. For example, the photoreceptors can contact bipolar and horizontal cells, but they cannot contact amacrine or ganglion cells because the photoreceptors are not close to these cells. The general pathways of information flow through the retina can be deduced from light microscopic studies, but to work out the exact circuitry between the various retinal cells we must identify the synapses made by the retinal cell types and study their distribution. Only electron microscopy has sufficient resolution to detect the synaptic junctions between neurons.

As noted in Chapter 2, photoreceptor cells make ribbon-type synapses. Figure 14.7A is a drawing of a photoreceptor terminal and its ribbon synapse. It also shows a horizontal cell making a more conventional synapse on a bipolar cell. That is, the horizontal cell synapse is marked by a cluster of vesicles close to the synaptic site, whereas the photoreceptor synapse is marked by a synaptic ribbon surrounded by vesicles. A characteristic of photoreceptor terminal synapses is that they are invaginated—the postsynaptic processes from bipolar and horizontal cells extend into recesses in the terminal membrane. Often three processes are observed in a single invagination. The lateral processes (stippled) come from horizontal cells whereas the central process is typically a bipolar cell dendrite. It is assumed that all three processes receive input

14.7 Drawings of retinal synapses *(left)* and a retinal wiring diagram *(right)*. **(A)** Photoreceptors *(R)* and horizontal cell *(H)* synapses. Bipolar cell *(B)* dendrites and horizontal cell processes extend into invaginations of the receptor terminals. Above the invagination is a synaptic ribbon, marking the synapse. Both horizontal and bipolar cell processes receive input from the photoreceptors at these synapses. Horizontal cells make conventional synapses onto bipolar cell dendrites. **(B)** Bipolar (ribbon) and amacrine cell (conventional) synapses. Bipolar cell terminals may synapse directly onto ganglion cell *(G)* dendrites *(left)* or only onto amacrine cell processes *(A)*, which in turn synapse on the ganglion cell dendrites *(right)*. Amacrine cells make feedback junctions onto bipolar terminals, feed-forward synapses onto ganglion cell dendrites, and lateral synapses on other amacrine cell processes. **(C)** A simplified wiring diagram of the retina. Photoreceptors *(R)* synapse on bipolar *(B)* and horizontal *(H)* cells, whereas horizontal cells make synapses onto the bipolar cells. Two types of ganglion cells (in terms of their synaptic input) are indicated. G_1 cells receive direct input from bipolar cells, whereas G_2 cells receive input from amacrine cells. Amacrine cells *(A)* receive input from bipolar cells or other amacrine cells.

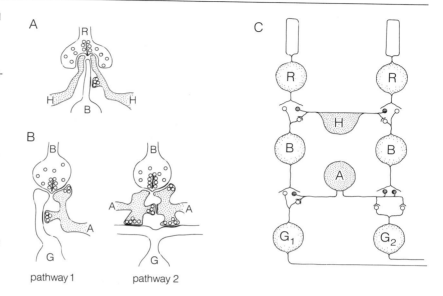

pathway 1   pathway 2

from the photoreceptor, but the reason the photoreceptor synapses are invaginated is not understood.

In the drawing a horizontal cell process synapses on the bipolar dendrite; synapses of these cells are seen in many retinas. There is also evidence that horizontal cells make synapses back onto the photoreceptor terminals, but, for reasons that are not clear, these junctions are not usually seen. The synapses in Figure 14.7A are the principal synaptic junctions observed in the outer plexiform layer.

In the inner plexiform layer, two main synaptic pathways have been observed, as depicted in Figure 14.7B. Bipolar cell terminals, like photoreceptor terminals, make ribbon synapses, whereas amacrine cells, like horizontal cells, make conventional synaptic contacts. One kind of synaptic pathway in the inner plexiform layer consists of a bipolar cell providing input at its ribbon synapse to a ganglion cell dendrite and an amacrine cell process (shaded). Bipolar cells almost always synapse on two processes, as shown here, and these junctions are not invaginated (see Figure 2.14). The amacrine cells can make synapses back onto the bipolar terminal or onto the ganglion cell dendrite, and both kinds of synapses are often seen. This first pathway is similar in many ways to the receptor-bipolar-horizontal cell pathway.

The other pathway seen in the inner plexiform layer is more complex. At the ribbon synapses of the bipolar cells, the two postsynaptic elements are both amacrine cell processes. These amacrine cell processes can make synapses onto each other, back onto the bipolar terminal, or onto ganglion cell dendrites. In other words, in this pathway the ganglion cells do not receive direct input from the bipolar cell terminals; instead, amacrine cell processes are interposed between the bipolar and ganglion cells. Visual information traveling this route (call it pathway 2) must pass across at least 3 synapses before reaching the ganglion cells (as opposed to 2 synapses in the simpler pathway, pathway 1), and the ganglion cells will be more influenced by the amacrine cells in pathway 2 than they are in pathway 1. Some ganglion cells appear to receive much of their input directly from the bipolar cells (via pathway 1), whereas other ganglion cells appear to receive most if not all of their input from the amacrine cells (via pathway 2).

A simplified scheme of the retina's synaptic organization is presented in Figure 14.7C. The photoreceptors provide synaptic input to bipolar and horizontal cells, and the horizontal cells also make synapses onto the bipolar cells. Since horizontal cells extend their processes widely in the outer plexiform layer, a bipolar cell may, via a horizontal cell, receive input from a distant receptor. More proximal photoreceptors, by contrast, contact the bipolar cells directly.

Two types of ganglion cells are indicated in the drawing. One (G_1) receives direct input from bipolar cells, whereas the other (G_2) receives its input from amacrine cells. What might be the significance of the two inner plexiform layer synaptic pathways and the two ganglion cell types indicated in Figure 14.7C? Comparative anatomical studies provide a hint. Cat and monkey retinas have mainly ganglion cells of the G_1 type, whereas frogs, turtles, and other cold-blooded vertebrates have mainly G_2 type ganglion cells. As noted earlier, the cold-blooded vertebrate retinas have many more movement- and direction-sensitive cells as compared with cats and monkey retinas. This suggests that amacrine cells are likely to be involved in movement detection and in fashioning

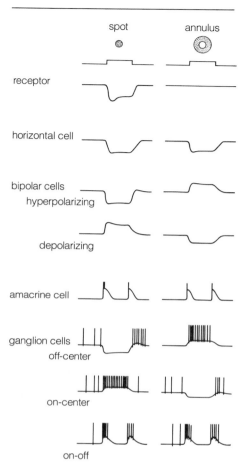

14.8 Intracellular responses and receptive field properties of neurons in the mudpuppy retina. Two types of stimuli were used to evoke the responses; spot and annular (ring) illumination. The distal retinal neurons—the receptor, horizontal, and bipolar cells—respond to illumination with sustained graded potentials, whereas the proximal retinal neurons—the amacrine and ganglion cells—show both sustained and transient potentials and action potentials. The receptor, bipolar, and on- and off-center ganglion cells respond differentially to center (spot) and surround (annular) illumination. Horizontal, amacrine, and on-off ganglion cells respond similarly to spot and annular illumination.

direction-selective responses in the retina. As we shall see, this is indeed the case.

Neuronal Responses

I have noted that it is possible to record intracellularly from the various types of retinal neurons in many species. By stimulating the retina with spot and annular light sources, one can also map the receptive field organization of the cells. Such information, coupled with our knowledge of the circuitry of the retina, permits us to understand where in the retina and by what mechanisms the receptive fields of the neurons are fashioned. Figure 14.8 shows what recordings from various retinal neurons look like when elicited with a spot or annulus of light centered over the recorded cell. These traces are based on recordings from the mudpuppy (see Figure 14.2).

Before describing the receptive fields of the individual neuronal types, I want first to point out some interesting features of the cell responses. For one thing, the distal retinal neurons—the receptor, horizontal, and bipolar cells—respond to light with sustained, graded membrane potential changes. They do not generate action potentials: all of their responses are either receptor or synaptic potentials. This may be the case because these neurons have short processes and do not need to transmit information over long distances: passive spread of potential along the cell membrane is sufficient to transmit information from one end of the cell to the other. A second possible reason the distal retinal cells function with graded potentials is that such potentials are capable of discriminating a wider range of signals than can all-or-none action potentials. The first action potentials observed in the retina are those generated by amacrine cells, and the ganglion cells also generate action potentials.

Another unusual feature of the distal retinal neurons is that most of them respond to light with hyperpolarizing potentials; upon illumination, the cell's membrane potential becomes more negative. This response was explained in Chapter 7, where we noted that darkness appears to be the stimulus for the vertebrate photoreceptors and that light turns them off. The same holds true for horizontal cells and many bipolar cells. Transmitter released from photoreceptors in the dark (when they are depolarized) depolarizes horizontal cells and many bipolar cells. When the photoreceptors hyperpolarize in the light, transmitter flow from the photoreceptors is decreased, and horizontal and many bipolar cells hyperpolarize. Again, matters become more conventional when we reach the inner plexiform layer. There, depolarizing responses with superimposed action potentials are recorded from many amacrine and ganglion cells when the retina is stimulated with light.

Receptive Field Organization

Photoreceptors have the smallest receptive fields of any of the retinal neurons. As is shown in Figure 14.8, the photoreceptor gives a large response to spot illumination but no response to annular illumination. In essence, the photoreceptors respond autonomously with little cross-talk between them. Horizontal cells, by contrast, have much larger receptive fields, as evidenced by the fact that the cells respond with large hyperpolarizing potentials to the spot and annulus (Figure 14.8). There are two reasons for the large receptive fields of horizontal cells. First, horizontal cells extend processes widely in the outer plexiform layer; second, many horizontal cells are electrically coupled together, which serves to extend the size of their receptive fields.

Bipolar cell recordings have two important features. First, there are two types of bipolar cell responses. About half of the bipolar cells depolarize in response to spot illumination whereas the other half hyperpolarize. The existence of both on- and off-center bipolar cells indicates that on- and off-pathways divide at the photoreceptor-bipolar cell synapse. Indeed, there is good evidence that individual photoreceptor terminals activate simultaneously both on- and off-center bipolar cells.

The second important feature of bipolar cell responses is that they have an antagonistic center-surround receptive field organization. Thus, center-surround receptive fields are first generated in the retina's outer plexiform layer as a result of interactions between photoreceptor, bipolar, and horizontal cells.

As already noted, amacrine cell responses are quite different from the responses of the more distal retinal neurons. Amacrine cells generate action potentials, and many respond to light with transient depolarizing potentials at the onset and cessation of illumination. In brief, they are on-off neurons. Amacrine cells typically fire only 1 or 2 spikes at the onset or cessation of illumination and so they behave in many ways like transition cells. Though capable of firing action potentials, the underlying graded (synaptic) potentials are probably more important for signal transmission. Many amacrine cells also respond with on-off potentials to spot and annular stimuli. Thus, their receptive fields do not usually show a center-surround organization. That many amacrine cell responses are on-off and transient, dying away even though illumination or darkness persists, means that these cells are highly responsive to moving stimuli. Amacrine cells respond vigorously to both increases or decreases in retinal illumination and to moving stimuli.

Intracellular recordings from ganglion cells reveal two types of receptive field organizations: the classic center-surround receptive field, which closely resembles the bipolar cell receptive field organization; and the pattern of on-off responses to spot and annular illumination similar to that of the on-off amacrine cells.

Functional Organization of the Retina

Figure 14.9 depicts in a simplified way how some potentials and receptive fields of the retinal neurons are produced by synaptic interactions within the retina. The drawing correlates the connections of the retinal cells with intracellular responses recorded from the mudpuppy.

The photoreceptors, with their small receptive fields, respond well to spots of light centered over the receptor but poorly to light that does not directly strike the cell (the surround). Retinal anatomy indicates that the bipolar and horizontal cells are activated by photoreceptors, and Figure 14.9 shows that both cells respond with sustained graded potentials similar to the photoreceptor potential. The figure shows further that the horizontal cells interact with bipolar cells, and that this synapse is opposite in sign to the receptor-bipolar synapse. In Figure 14.9, the receptor causes the bipolar cell to hyperpolarize whereas the horizontal cell causes the bipolar cells to depolarize. Since the horizontal cells have a much larger lateral extent than do bipolar cells, an antagonistic center-surround receptive field is established in the bipolar cell. Hence, the central bipolar cell response is mediated by the direct photoreceptor-bipolar cell interaction, whereas the antagonistic sur-

14.9 Summary diagram correlating the synaptic organization of the vertebrate retina with some of the intracellularly recorded responses from the mudpuppy retina. This figure attempts to show how the receptive field organizations of the hyperpolarizing bipolar cells, off-center ganglion cells, and on-off ganglion cells are established. Responses occurring in the various neurons upon illumination of the left receptor (R) are shown. The hyperpolarizing bipolar cells (B) and off-center ganglion cells (G₁) respond to direct central illumination *(left side)* by hyperpolarizing; to indirect *(surround)* illumination *(right side)* by depolarizing. Note that the switch from hyperpolarizing to depolarizing potentials along the surround-illumination pathway occurs at the junction of horizontal (H) and bipolar cells. The on-off ganglion cell (G₂) receives strong inhibitory input from amacrine cells (A); the figure suggests that these cells receive their excitatory input (+) from both amacrine and bipolar cells. Inhibitory feedback (−) synapses from amacrine cells onto the bipolar terminals are also indicated.

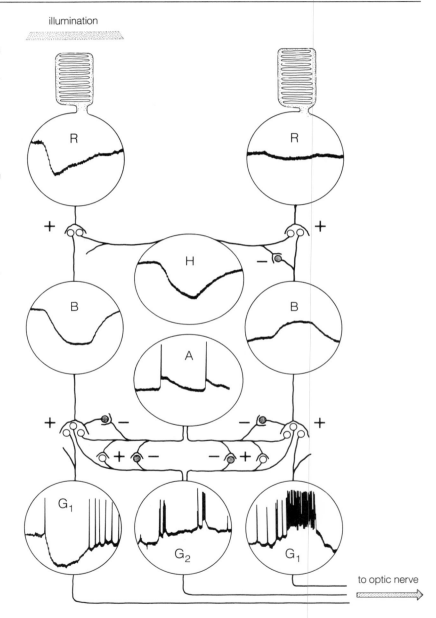

round response of bipolar cells reflects input from the horizontal cells.

Figure 14.9 illustrates center-hyperpolarizing, surround-depolarizing bipolar cells. The synapse is marked as excitatory (+) between the receptor and bipolar cells, and inhibitory (−) between the horizontal and bipolar cells. To generate a center-depolarizing, surround hyperpolarizing bipolar cell, the polarities of the synapses must be reversed. That is, the receptor-bipolar synapse would be inhibitory (−), whereas the horizontal bipolar would be excitatory (+).

The responses of the two types of ganglion cells (G_1 and G_2) illustrated in Figure 14.9 relate to the responses of the neurons providing the bulk of their synaptic input. The off-center, on-surround responses of the G_1 cell, resembling bipolar cell responses, reflect primarily the processing in the retina's outer plexiform layer. In contrast, the on-off responses of the ganglion cell (G_2) resemble on-off amacrine cell responses, and there is good evidence that these cells receive most of their input from amacrine cells. The on-off ganglion cells reflect more the processing that occurs in the inner plexiform layer—these cells are highly movement-sensitive and often direction-sensitive.

To summarize, the outer plexiform layer of the retina is concerned mainly with the static and *spatial* aspects of illumination. The neurons contributing processes to this layer respond primarily with sustained, graded potentials and the neuronal interactions there form an antagonistic center-surround receptive field for the bipolar cells. The on- and off-center ganglion cells, receiving much of their input directly from center-depolarizing or center-hyperpolarizing bipolar cells, reflect the center-surround receptive field organization established in the outer plexiform layer.

The inner plexiform layer is concerned more with the transient or *temporal* aspects of illumination. On-off amacrine and ganglion cells respond vigorously to changes in retinal illumination, particularly to moving stimuli. Interactions in the inner plexiform layer, particularly between amacrine cells, underlie direction-sensitive responses. The receptive fields of movement- and direction-sensitive ganglion cells are formed by the synaptic interactions occurring primarily in the inner plexiform layer.

As was the case for the invertebrate neural circuits discussed in Chapter 10, two factors weigh heavily in determining the properties of the output responses from the two plexiform layers of the retina. So far as the sustained center-surround responses of the bipolar and ganglion cells are concerned, the wiring of the neurons and differences in the polarities of synapses account satisfactorily for the responses and receptive field properties of the cells. But anatomy alone does not explain the response of movement- and direction-sensitive ganglion cells. The special properties of amacrine and ganglion cells—their transient response to retinal illumination (at the onset and offset of the light)—are crucial. We don't yet know precisely what mechanisms are responsible for these responses to light, but it is likely that the answer lies in the special membrane properties of the cells as well as their neuronal connections.

Modulation of Retinal Synapses

The retina, like other regions of the brain, contains a surprisingly large number of neuroactive substances. It is believed that at least fifteen different neuroactive substances are released from retinal neurons during activity. Relatively few of these substances appear to serve as neurotransmitters. L-glutamate is the major excitatory neurotransmitter in the retina and is released by both the photoreceptors and bipolar cells. Acetylcholine also mediates excitation—to on-off ganglion cells in the inner plexiform layer—and is released by certain amacrine cells. GABA and glycine are the main inhibitory neurotransmitters in the retina and are released by many horizontal and amacrine cells. However, the vast majority of neuroactive substances released from retinal neurons appear to be neuromodulatory in nature, although little is known about the action of most of these substances.

Much of what we know about the action of neuromodulators in the retina has come from the study of dopamine in the fish retina. In fishes, dopamine is present in a special amacrine-like cell (see Figure 1.9A), termed the interplexiform cell because it extends processes in both plexiform layers. The synaptic output of the interplexiform cells is mainly on the horizontal cells in the outer plexiform layer. Two effects of dopamine on these cells have been observed; a loss of light responsiveness (that is, light responses are reduced in amplitude after dopamine application to the retina), and a decrease in electrical coupling between horizontal cells.

14.10   A scheme of the modulation of horizontal cells by dopamine in fish. Dopamine *(DA)* interacts with receptors *(R)* linked to a G-protein *(G)* and adenylate cyclase *(AC)*. Adenylate cyclase catalyzes the formation of cyclic AMP (cAMP) from ATP. Cyclic AMP activates kinases *(K)* that phosphorylate *(P)* gap-junctional channels *(GJ)* and channels activated by glutamate *(glut)*. Phosphorylation of the channels modifies ion flow across them.

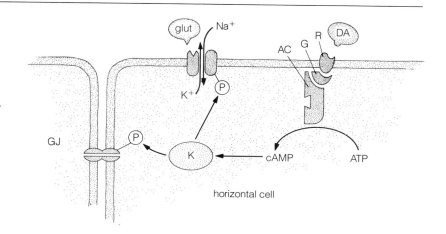

Dopamine does not exert these effects by acting directly on horizontal cell membrane channels; rather, it interacts with membrane receptors linked to adenylate cyclase.

The cyclic AMP thus produced activates kinases that appear to phosphorylate both glutamate channels (the channels activated by the photoreceptor transmitter) and gap-junctional channels (the channels that mediate electrical coupling between horizontal cells—see Figure 14.10). The phosphorylation of these channels serves to alter their properties. In the case of the gap-junctional channels, phosphorylation decreases the time that the channels remain open, thereby decreasing the flow of current that passes across the junction. On the other hand, phosphorylation of the glutamate channels appears to modify the frequency of opening of the channels, which again modulates ion flow across the membrane. The action of dopamine on horizontal cells is not to initiate activity, but to modify the cell's response to the photoreceptor transmitter and the interactions between the cells. Furthermore, these effects are slow; they take many seconds to develop and they last for minutes.

Overall the effect of dopamine is to decrease effectiveness of the horizontal cells in mediating lateral inhibitory effects in the outer plexiform layer. Decreasing the light responsiveness of the cell and shrinking its receptive field size are effective ways of lessening its influence. As noted earlier, horizontal cells drive the antagonistic surround responses of bipolar cells, and thus a decrease in bipolar cell surround responses is observed following application of dopamine to the retina.

What is the significance of the modulation of lateral inhibition and surround antagonism by dopamine and the interplexiform cells in the retina? It has long been known that following prolonged periods of time in the dark, the antagonistic surround responses of ganglion cells are reduced in strength or even eliminated. An obvious speculation is that interplexiform cells and dopamine mediate this effect: they regulate the strength of lateral inhibition and center-surround antagonism in the retina as a function of adaptive state. Evidence in favor of this hypothesis has been found in fish. Following periods of prolonged darkness, horizontal cell receptive field size and light responsiveness are substantially decreased, suggesting that dopamine is released from the interplexiform cells in prolonged darkness. We conclude, therefore, that in vertebrates as well as in invertebrates (see Chapter 10), neuromodulators can modify the strength of synaptic interactions and thus alter the behavior of a neural circuit in response to different environmental conditions.

Summary

More is probably known about the visual system than any other system in the brain. Its basic organization and anatomy are well understood, and the processing of visual information in mammals can be examined over at least six stages—two in the retina, one in the lateral geniculate nucleus, and three in the cortex. Both convergence and divergence of information occurs along the visual pathway. Between photoreceptors and optic nerve axons, there is an average convergence of about 100 to 1; from optic nerve to cortex there is a divergence of 1 to 1,000. The convergence and divergence is not even across the retina or visual field. Less convergence and more divergence is observed in the central retina, where acuity is highest.

Visual processing begins in the retina, which consists of three cellular and two plexiform (synaptic) layers. In each plexiform layer, the processes of three major cell types interact: photoreceptor, bipolar, and horizontal cells in the outer plexiform layer; bipolar, amacrine, and ganglion cells in the inner plexiform layer.

All visual information passes across at least two synapses, one in each plexiform layer, before it leaves the eye. Extracellular recordings from the ganglion cells reveal two types of processing within the retina. One ganglion cell type shows evidence of spatial processing; its receptive field is organized into concentric and antagonistic zones. Both on-center, off-surround cells and off-center, on-surround cells are found in every retina. The other basic ganglion cell type shows evidence of temporal processing; it responds vigorously to moving stimuli, but to static spots of light it gives only transient responses at the onset and cessation of illumination. Mixes of spatial and temporal processing are seen in other ganglion cells, and some ganglion cells show color specificity.

On-center ganglion cells signal increments of illumination to the rest of the brain, whereas off-center cells signal decrements. The center-surround receptive field organization of ganglion cells provides not only spatial information about light distribution on the retina, but also information about light intensity from an object as compared with surround objects. Objects appear light or dark depending on the relative intensity—not the absolute intensity—of light coming from them.

In certain instances, the responses of movement-sensitive ganglion cells can be correlated with the behavior of an animal. In frogs and toads, stimuli that elicit prey-catching behavior are also those that optimally drive one type of movement-sensitive ganglion cells.

The retina's circuitry as revealed by electron microscopy includes two types of synapses: photoreceptors and bipolar terminals make ribbon synapses; horizontal and amacrine cells make conventional synapses. In the outer plexiform layer, the photoreceptors synapse on bipolar and horizontal cells; the horizontal cells synapse onto the bipolar cells. In the inner plexiform layer, two synaptic pathways are observed. In one, the bipolar cell terminals synapse directly onto the ganglion cells; in the other, the bipolar cells synapse onto amacrine cells, which in turn drive the ganglion cells.

Intracellular recordings from the retinal neurons show that the distal neurons—the photoreceptor, bipolar, and horizontal cells—respond with sustained, graded potentials that are usually hyperpolarizing in polarity. Only the amacrine and ganglion cells generate action potentials in the retina. Photoreceptors have small receptive fields as compared with horizontal cells. Two types of bipolar cell responses are recorded, center depolarizing and center hyperpolarizing. Both types also show an antagonistic surround response of opposite polarity. Many amacrine cells respond with on-off responses and are movement-sensitive. The responses of the two types of ganglion cells found in the vertebrate retina resemble either bipolar or amacrine cell responses and their receptive field organizations. The center response of the bipolar cell results from direct photoreceptor-bipolar cell interaction, the surround response from photoreceptor-horizontal-bipolar cell interaction. The ganglion cells that receive substantial input directly from the bipolar cells are the on-center, off-surround cells or the off-center, on-surround cells. The ganglion cells that receive substantial amacrine cell input give on-off responses to spots of light; they are usually highly movement-sensitive and some are direction-sensitive. The outer plexiform layer processes information concerning the spatial aspects of a visual stimulus, while the inner plexiform layer is concerned more with the temporal features of the visual image.

The retina, like other parts of the brain, utilizes a large number of neuroactive substances. Most of these substances appear to be neuromodulatory in nature. An example is dopamine, which in the fish retina modifies the properties of both glutamate and gap-junctional channels in horizontal cells. This modulation affects the strength of lateral inhibition in the outer retina, and it appears to relate to the adaptive state of the retina. The conclusion drawn is that neural circuits such as those found in the retina are not invariant; they can be modified by the action of neuromodulators.

Further Reading

BOOKS

Dowling, J. E. 1987. *The Retina: An Approachable Part of the Brain*. Cambridge, MA: Harvard University Press.

ARTICLES

Barlow, H. H., R. M. Hill, and W. R. Levick. 1964. Retinal ganglion cells responding selectively to direction and speed of image motion in the rabbit. *Journal of Physiology* 173:377–407.

Boycott, B. B., and J. E. Dowling. 1969. Organization of primate retina: Light microscopy. *Philosophical Transactions of the Royal Society of London* B255:109–184.

Dowling, J. E. 1989. Neuromodulation in the retina: The role of dopamine. *Seminars in the Neurosciences* 1:35–43.

Dowling, J. E., and B. B. Boycott. 1966. Organization of the primate retina: Electron microscopy. *Proceedings of the Royal Society of London* B166:80–111.

Kuffler, S. W. 1953. Discharge patterns and functional organization of the mammalian retina. *Journal of Neurophysiology* 16:37–68. (The first description of center-surround receptive field organization in ganglion cells.)

Werblin, F. S., and J. E. Dowling. 1969. Organization of the retina of the mudpuppy, *Necturus maculosus*. II. Intracellular recording. *Journal of Neurophysiology* 32:339–355.

FIFTEEN

Cortical Processing of Visual Information

THE PROCESSING of visual information beyond the retina is understood in detail only in a few higher mammals. The species that have been most extensively studied are the cat and monkey, but our information even about them is still quite fragmentary. For example, virtually all the physiological information we have concerning the neurons in the visual cortex has been derived from extracellular recordings. We therefore know about the pattern of action potential discharge and the receptive field organization of many of the cells in the visual cortex, but we are largely ignorant of the mechanisms underlying activation of the cells or the formation of the cells' receptive fields.

The synaptic circuitry of the visual cortex is not understood at all well, and intracellular recordings from cortical neurons are difficult to make. So, for example, if there are cells in the visual cortex that generate only sustained, graded potentials, like many of the retinal cells, they remain undetected. This is not to say that the extracellular studies have not told us an enormous amount. Indeed, they have yielded much information about higher visual processing and even significant insights to perceptual phenomena.

In nonmammalian species, we still know little about the receptive field properties of cells receiving visual input beyond the retina. Part of the difficulty is that the visual system is organized quite differently in mammalian and nonmammalian species. In mammals, higher visual processing occurs mainly in the cortex, but in nonmammalian species the optic nerve terminates mostly in the midbrain, specifically in the tectum, which receives sensory input of several types and is also the site where movement is initiated. As might be expected, many tectal cells respond in complex ways to sensory stimuli, and so it has not been possible to define the receptive properties of the cells in as satisfactory a way as has been possible in the retina or in the primary visual cortex of mammals, structures devoted exclusively to vision.

Another consequence of the fact that most visual information projects directly to the tectum in nonmammalian species is that more complex processing of visual information tends to take place in the retinas of these animals. As noted in the last chapter, many of the ganglion cells in nonmammalian species have complex receptive field properties, such as movement- and direction-sensitivity. In the higher mammalian and primate visual systems, though, cells with direction-selective properties are not found prominently until one reaches the cortex. Stated differently, more complex visual processing is delayed in the mammalian visual system until the cortex, but in nonmammalian species more complex processing occurs already in the retina.

Many third-order cells in the frog retina (namely, ganglion cells) have properties similar to fifth- or sixth-order neurons in the mammalian visual cortex. This is not without cost, however. The large number of cells in the visual cortex means that visual information can be analyzed in much more detail than is possible in the retina and tectum of nonmammalian species. That frogs and toads are not responsive to stationary objects—dead flies, for example, that would make a perfectly good meal—illustrates the limits of the visual world of nonmammalian species. Other evidence is that the prey-catching behavior of these animals appears constrained by the properties of the ganglion cells; that is, stimuli that best drive certain of the ganglion cells are also those that best activate prey-catching behavior (see Chapter 14).

To summarize: the visual pathways in mammalian and nonmammalian species are different. In the mammal, the retina projects mainly to the lateral geniculate nucleus, and from there information travels to the visual cortex. From the visual cortex, information passes to other cortical regions, where eventually it influences the motor cortex and the behavior of the animal. In the nonmammalian vertebrate, the retina projects to the tectum, and it is from the tectum that the animal's movements and behavior are initiated.

The majority of ganglion cells in the higher mammals (cat and monkey) are contrast-sensitive cells, and so much of the visual information processed in the lateral geniculate nucleus and visual

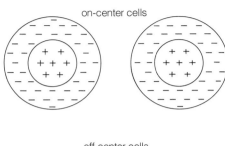

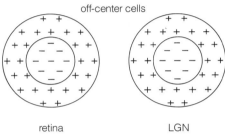

15.1 Receptive field maps of ganglion cells in the retina and lateral geniculate neurons. Both ganglion cells and lateral geniculate neurons have a center-surround receptive field organization and both on- and off-center cells.

cortex originates from the contrast-sensitive or G_1 cells of Figures 14.7C and 14.9. Most of the ganglion cells in nonmammalian species that project to the tectum, on the other hand, are movement- or direction-sensitive (G_2) cells.

Some smaller mammals—such as the ground squirrel or rabbit, whose cerebral cortex is not as developed as that of the cat or monkey—have a number of direction-sensitive ganglion cells in their retinas. These ganglion cells project mainly to the midbrain of the animal and to tectal nuclei.

The Lateral Geniculate Nucleus

The first stop for visual information from the retina is the lateral geniculate nucleus in the thalamus. Extracellular recordings from the lateral geniculate neurons show that they have a receptive field organization very similar to that of the ganglion cells (see Figure 15.1). That is, the receptive fields of the lateral geniculate neurons are concentrically organized, and both on-center, off-surround cells and off-center, on-surround cells are present. But there are also differences between the geniculate neurons and the retinal ganglion cells—for example, in their responses to diffuse illumination. Whereas ganglion cells give a weak response to diffuse illumination, reflective of the center response of the cell, lateral geniculate neurons either give no response or a weak response reflective of the surround. This is not a large difference, however, and so it is generally agreed that lateral geniculate neurons and retinal ganglion cells have a similar receptive field organization.

For what reason, then, does visual information stop in the lateral geniculate nucleus? Of what use is this relay? There is good evidence that the lateral geniculate nucleus receives input from the reticular formation and also from the visual cortex, as well as from the retina. These inputs are received by interneurons of the lateral geniculate nucleus that are inhibitory to the neurons that project to the visual cortex. These interneurons also receive input from the retinal axons as well as from axon collaterals of the lateral geniculate nucleus neurons themselves. As a result of this circuitry, shown in Figure 15.2, the information passed on to the visual cortex from the lateral geniculate nucleus can be modified, or gated, by a variety of influences. The nature of the modifications in the visual signal passed on to the cortex are not well understood.

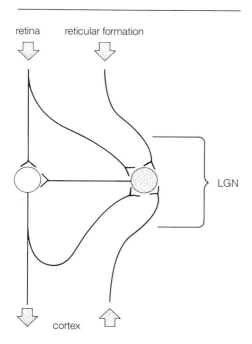

15.2 Axons from the retina synapse on both the projection neurons that extend to the cortex *(open circle)* and interneurons *(stippled circle)* in the lateral geniculate nucleus (LGN). In addition, the interneurons receive input from the reticular formation, back from the cortex, and from axon collaterals of the projection neurons.

One reason that these effects are not better known is undoubtedly that virtually all of the information we have obtained from lateral geniculate neurons has come from anesthetized animals. It is likely that anesthesia significantly modifies input from the reticular formation and other inputs that gate information passing on to the cortex. This is a serious drawback for the study of information processing in the brains of higher mammals, whose behavioral state almost certainly affects neural activity in higher brain centers. It is possible to record from neurons in awake and behaving animals, as I noted in Chapter 13, but the experiments are difficult to set up and great care must be exercised to ensure that the animal experiences no discomfort.

The Visual Cortex

Extracellular recordings from most neurons in the primary visual cortex show clear evidence of further processing of the visual signal. A hierarchy of receptive field organizations is observed in the visual cortex that relates to the stimuli needed to best drive the cells. Some cells require relatively simple stimuli to drive them but other cells require rather complex stimuli to fully activate them. In monkeys (but not cats) some cells with an antagonistic center-surround receptive field organization are found close to the termination of the lateral geniculate neuron axons in the primary visual cortex. Away from this area, however, the receptive field organization is significantly altered; the farther away a recorded cell is from the lateral geniculate nucleus input, the more complex generally is its receptive field organization. (Once again, keep in mind that although we are discussing cells in the visual cortex, stimuli are always presented to the retina. Only photoreceptors are sensitive to light, and all responses throughout the visual pathway are initiated by the receptor cells.)

Two major classes of cortical cells are distinguished on the basis of their receptive field properties—*simple* and *complex* cells. This classification was originally made by David Hubel and Torsten Wiesel, working initially at Johns Hopkins University and later at Harvard Medical School. Indeed, much of what we know concerning the functional organization of the visual cortex has come from the laboratory of these two workers. For their enormous

15.3 Receptive field maps of simple cortical cells. In all cases a correspondingly oriented bar or edge is the optimal stimulus. In (**A**) the central excitatory zone is symmetrically positioned in the receptive field. In (**B**) the central zone (inhibitory in this case) is eccentrically positioned and in (**C**) the excitatory zone is on one side of the field and the inhibitory zone on the other.

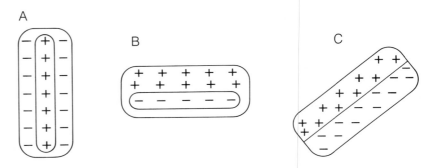

contributions to our understanding of cortical mechanisms, Hubel and Wiesel were awarded the Nobel Prize in 1981.

Simple Cells

Simple cells can be activated with stationary spots of light projected onto the retina, but a map of the field of a simple cell shows that the excitatory and inhibitory zones are characteristically elongated. Thus, the optimal stimulus for a simple cell is not a circular spot or annulus, as is the case for a retinal or lateral geniculate neuron, but an elongated bar. In many simple cells, the central elongated excitatory or inhibitory zone is surrounded by a symmetrical region that gives a response of opposite polarity; that is, the surround is antagonistic to the center zone. A typical simple cell of this type is shown in Figure 15.3A.

Because the cell's receptive field is elongated, proper orientation of the stimulating bar of light is critical for optimal activation of the cell. The cell fires maximally when the bar's orientation matches that of the receptive field, but as the orientation changes, the cell fires less and less well. The reasons for this are twofold; first, less of the excitatory receptive field area is illuminated; second, more of the adjacent inhibitory zone is stimulated. Simple cells do not usually respond at all to diffuse illumination. The responses of a simple cell to various stimuli are shown in Figure 15.4.

In both the cat and the monkey, every simple cell has a preferred orientation of its receptive field, but all orientations are represented

15.4   Responses of a simple cell to a bar of light. When the bar covers the excitatory zone of the receptive field, the cell fires vigorously *(top)*. When the bar stimulates the inhibitory zone, a strong off-response is recorded *(middle)*. When the orientation of the bar is dissimilar to that of the cell, the cell fires very weakly or not at all *(bottom)*.

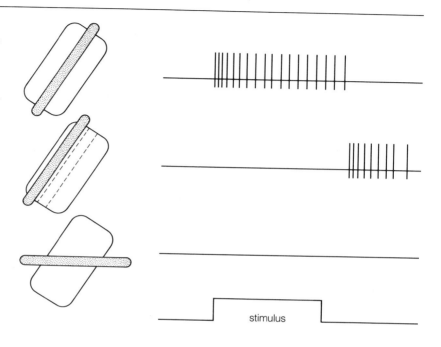

stimulus

in the population of cells. Hence, any one simple cell responds maximally to a bar of a certain orientation, but if enough cells are recorded, every possible orientation is found to drive one or another cell optimally. How fine a discrimination can a simple cell make with regard to orientation? A change of orientation of about 10° can be detected by a typical simple cell, so there are roughly 18–20 simple cells differing with regard to orientation selectivity. That is, as a stimulus is rotated from one orientation back to the same orientation (changed in orientation by 180°), 18–20 different cells will be optimally stimulated.

There are simple cells whose receptive field maps are different from the one illustrated in Figure 15.3A. In Figure 15.3B, for example, the central excitatory zone is eccentrically placed in the receptive field, and in Figure 15.3C the excitatory zone is on one side of the field and the inhibitory zone on the other side. Both of these cells can be activated (or inhibited) with spots of light projected onto the retina, but they are optimally stimulated with precisely positioned and oriented bars or light-dark edges of light. Orientation of the light stimuli away from optimal orientation always diminishes the responses of the cells.

Complex Cells

Simple cells are generally closer to the cortical layers that receive lateral geniculate neuron input than are complex cells. Thus, complex cells are believed to represent the next stage of visual analysis. Furthermore, a hierarchy of complex cells exists with respect to their properties or stimulus requirements, by which several different kinds of complex cells can be distinguished. Complex cells are the most common ones in the visual cortex; and they are located not only in V1 but also in V2 and other visual areas of the cortex. Simple cells are confined pretty much to area V1.

Unlike simple cells, complex cells cannot be activated with small stationary spots of light projected onto the retina, nor are their receptive fields organized into discrete excitatory or inhibitory zones. Rather, to activate vigorously a complex cell requires an oriented bar or slit of light moving briskly across the retina, through its receptive field. Simply turning on and off a slit that is properly oriented within the receptive field will often produce a weak, transient response. To generate a sustained discharge the cell requires movement of the bar, however. The bar must move at right angles (90°) to its orientation, and changes either in bar orientation or direction of movement reduces the response severely. Figure 15.5 illustrates some responses of a complex cell.

DIRECTION-SELECTIVITY AND END-STOPPING

Some complex cells show direction-selectivity: the cells respond vigorously to movement of a bar of light in one direction, but they give no response to movement of the bar in the other direction. In many ways, these cells respond like the direction-sensitive cells found in the retinas of nonmammalian vertebrates. But the cortical cells require a moving bar or slit of light to drive them optimally, whereas the retinal cells are strongly activated by spots of light passing through the receptive fields of the cells in the appropriate direction.

Another elaboration observed in certain cortical cells is end-stopping. Most cortical cells respond best to a slit of light that is

15.5 Responses of a complex cell to a bar of light moving through the receptive field of the cell. For maximal activation of the cell, the bar must be properly oriented and moved at right angles to the optimal orientation. Deviations of bar orientation or direction of movement reduce the cell's response. Some complex cells (as shown here) are direction-selective. Movement of the bar in one direction yields a vigorous response; movement in the other direction evokes no response.

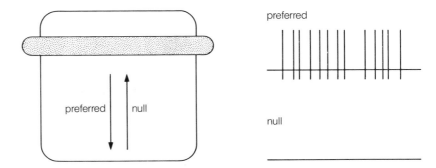

as long as the receptive field of the cell. With shorter slits, the cell responds less well. This means that the cells show summation—the longer the slit, the more vigorous the response. Summation also occurs in retinal and geniculate cells. Large spots or annuli that cover more of the center or surround regions of the receptive fields produce more vigorous responses than do small spots or thin annuli that cover only a portion of the receptive field's center or surround.

With end-stopped cells, a different phenomenon is observed (Figure 15.6). Increasing the length of a properly oriented slit or bar increases the response up to a point, but extending the length of the bar farther in one or in both directions decreases the response! These cells have activating and inactivating regions in their receptive fields, such that when the stimulus is confined to the activating regions, vigorous responses are observed, but when the inactivating regions are stimulated, we get a decrease or loss of responsiveness. Interestingly, the stimulus that best activates the activating region is also the stimulus that best activates the inactivating or inhibitory region, which means that the inactivating zone requires a stimulus with the same orientation and movement direction as the activating zone.

Both simple and complex cells can show end-stopping, although end-stopped complex cells appear to represent a higher-order level of processing in the cortex. For example, more complex cells recorded in area V2 show end-stopping than do complex cells recorded in V1.

15.6 For optimal activation of an end-stopped complex cell, the stimulus must be restricted. Extending the bar on one or both sides reduces the cell's response significantly.

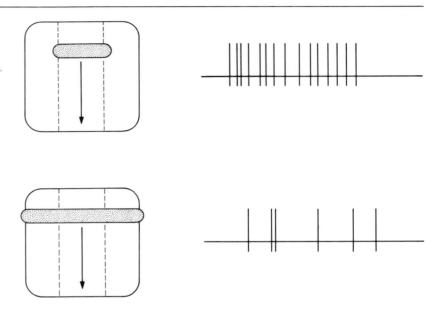

The Significance of Cortical Receptive Fields

At this point, we can draw some conclusions about the analysis of visual information by the cortex. For example, the farther along the visual system, the more specific must be the stimulus if the cell is to be activated strongly. In the retina and lateral geniculate nucleus, neurons respond maximally to appropriately positioned spots or annuli on the retina. In the cortex, however, simple cells require an appropriately positioned and oriented bar of light on the retina. Complex cells require not only a correctly oriented bar or slit of light but also movement of the bar to drive the cell vigorously. More specialized complex cells require movement of the stimulus in a specific direction; certain cortical neurons are best driven by bars or slits of light restricted in length. Activating a specialized complex cell, therefore, requires a precisely oriented slit of light of specified length moving in one direction. As information ascends to higher centers, then, the visual image is broken into components that are encoded in individual cells. Hence, any one neuron responds only to a portion of an image—the portion whose features match the cell's receptive field requirements. As

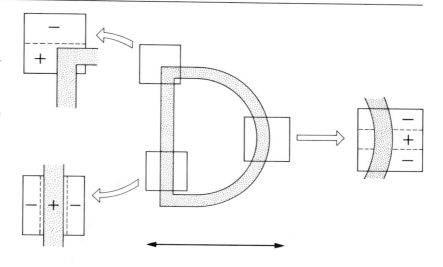

15.7 How a simple figure is detected by cortical neurons. If the figure is stationary, simple cells whose orientation preference is appropriate for a part of the figure will be optimally activated *(bottom left)*. If the figure is moving, complex cells will be activated. End-stopped cells will be selectively activated by the corners *(upper left)* and curved portions of the figure *(right)*.

information ascends to higher centers, the visual image is abstracted in individual neurons.

Another way to think about how the cortex analyzes visual information is to consider the activity of a group of neurons in the cortex when a simple figure such as the one in Figure 15.7 is projected onto the retina within the neurons' receptive fields. First, few neurons will be activated at any one time by the figure because cortical cells do not respond to diffuse stimuli; therefore, only cells whose orientation matches that of one or another part of the figure will be activated. Furthermore, if the figure is stationary, only simple cells will be activated, but if it moves, complex cells will now respond vigorously, and different complex cells will be activated depending on the direction of movement. The corners of the figure will optimally activate complex cells that are end-stopped on one end, whereas the curved portion of the figure will activate cells end-stopped on both ends. Figure 15.7 shows the receptive field maps of some of the cortical cells that will be activated vigorously by the figure when it is stationary or moving in a certain direction.

As I noted, if the image is stationary on the retina, mainly simple cells are activated. This is true, but it is also the case that if the image is made perfectly stationary on the retina, it fades away in

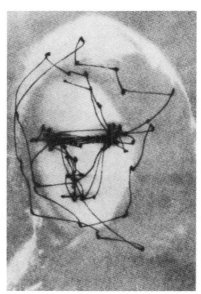

15.8 Eye movements of an observer scanning a face. The eyes jump around and stop momentarily on areas of interest. These pictures were originally published by A. Yarbus.

a few seconds and we no longer perceive it. For us to see objects continuously, the image of the object must be moved periodically on the retina. If the image itself is moving, this is no problem. If the object is stationary, then the eye must move and, indeed, our eyes are in constant motion. The movements of our eyes tend to be jerky, and as we look at an object, our eyes rapidly jump from one position to another, ensuring movement of the image on the retina. These jumps are called *saccades,* and both large and small saccades happen. The large saccades enable us to scan a scene whereas the small ones, microsaccades, keep images moving on the retina.

Records of eye movements were first made by a Russian physiologist, Alfred Yarbus, in the 1950s. Figure 15.8 illustrates one of his experiments—the eye movements of an observer looking at a photograph of a face. The eyes jump around, continuously bringing the focus of gaze to different parts of the image and stopping momentarily on areas of interest. During saccades, our eyes move so rapidly that we are not aware of the movement. Why this is so is not entirely clear, but there is evidence that our vision is turned off momentarily during a saccade.

The conclusion we can draw from these findings is that movement is crucial for the perception of objects and scenes. That most of the cells in area V2 are complex cells, responding selectively to movement, accords with this conclusion. And, as already noted, the decreasing proportion of simple cells, not concerned with movement, in higher visual centers emphasizes the key role of movement in visual perception.

The Formation of Cortical Receptive Fields
We have, as yet, little detailed anatomical information on the synaptic organization of the cortex, and most of the physiological information has been derived from extracellular recordings. But even though little is known about how the receptive fields of cortical neurons are formed, simple schemes can be suggested that might explain how a simple cell's receptive field is produced from geniculate inputs or how a complex cell's receptive field is formed from simple cell inputs. Figure 15.9 depicts wiring that could explain a simple cell's receptive field: it postulates simple excitatory input to the neuron from lateral geniculate neurons. The excitatory centers of the receptive fields of the lateral geniculate

15.9 How a simple cortical cell's receptive field could be fashioned by excitatory inputs from lateral geniculate neurons. The receptive fields of the input neurons (on-center cells) overlap and lie along a straight line on the retina (left). The receptive field of a cortical neuron receiving such input will consist of an elongated central excitatory zone surrounded by an inhibitory zone (right).

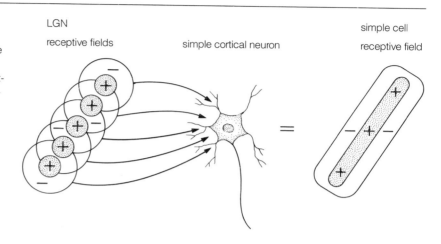

input neurons overlap slightly and lie along a straight line on the retina. The receptive field of the neuron receiving such input will consist of a long, narrow, central excitatory zone surrounded by flanking inhibitory zones—the receptive field of a typical simple cortical cell.

The central excitatory zone of a simple cell, like the one pictured in Figure 15.9, has a width comparable to the diameter of the centers of the receptive fields of the lateral geniculate neurons that provide input to that part of the cortex. As we noted earlier, ganglion cell receptive fields, and particularly their centers, vary in size between the center and periphery of the retina; the same holds for simple cells in the cortex. Simple cell receptive fields that come from the periphery of the visual field are larger and have wider center zones (excitatory or inhibitory) as compared with simple cells that receive input from the central retina.

It is more difficult to postulate how a simple cell whose receptive field is like the one illustrated in Figure 15.3C is wired, but with such edge-sensitive cells it is straightforward to postulate how a directionally selective complex cell is formed. I show this in Figure 15.10. A series of edge-sensitive simple cells with aligned receptive fields provide excitatory input to a complex neuron; with this arrangement, the complex cell should respond vigorously to downward movement, but poorly or not at all to upward movement. The reason is simple: If a slit of light first enters the excitatory

15.10 How a directionally selective complex receptive field could be formed from edge-selective simple cells. If the receptive fields of the input neurons are aligned on the retina as shown on the left, downward movement of a bar will elicit vigorous activity in the complex cell, whereas upward movement will result in no response.

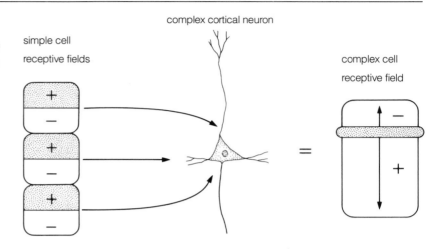

simple cell
receptive fields

complex cortical neuron

complex cell
receptive field

zone of the receptive field of one of the input neurons, the neuron will be activated and will drive the complex cell. As the slit moves into the inhibitory zone it will silence the input cell, but subsequent activation of the excitatory zone of the adjacent simple cell will further activate the complex cell, and so forth. Moving the slit in the opposite direction will initially activate the simple cell's inhibitory zone, which temporarily depresses the cell's excitability. As the slit moves from the inhibitory to the excitatory zone, therefore, the simple cell's output is inhibited and the complex cell will be only poorly activated, if at all.

Complex cells have larger receptive fields than do simple cells, suggesting that they result from inputs from simple cells, in the manner depicted in Figure 15.10. In contrast, experiments in which GABA antagonists have been applied to the cortex have revealed disruptions and alterations of cortical neurons' receptive field properties. These results indicate that other factors besides excitatory pathways also help to fashion cortical receptive fields. Indeed, the formation of receptive fields in the cortex is undoubtedly much more complicated than the schemes proposed in Figures 15.9 and 15.10.

Binocular Interaction and Depth Perception

So far, I have examined the mechanics of activating cortical neurons by presenting specific stimuli to one eye. But cats, monkeys, and we humans receive visual information through two eyes that have extensively overlapping visual fields. Do cortical neurons receive input from both eyes? If so, how do the cortical cells respond to input from the two eyes?

Many neurons in the cortex of cats and monkeys are binocular; they receive input from both eyes. The prominent exceptions to this rule are the concentrically organized cells in the monkey that are close to the input of the lateral geniculate neuron axons. These cells are consistently monocular; they receive input only from either the right or left eye. Most other cortical neurons, both simple and complex cells, are binocular—though most cortical neurons receive more input from one eye or the other. Hence, they show the property of *ocular dominance;* one eye drives the cell better than the other does.

Figure 15.11 presents ocular dominance histograms for the cat and monkey. The data for this figure were collected in the following way. Neurons were recorded in the visual cortex of one hemisphere or the other, and the binocularity of input was estimated for each cell. If the cell received input from only one eye, and that eye was on the side opposite to the cortical recording (the contralateral eye), it was listed as a group 1 cell. The cell was placed in group 7 if it received monocular input from the eye on the same side as the recording (the ipsilateral eye). Cells in group 2 and 6 received some input from the ipsi- or contralateral eyes, but they preferred the other eye. Cells in group 3 and 5 received more equal input from the two eyes, whereas cells in group 4 were perfectly binocular; they had no eye preference.

The data reveal that more cells in cats receive equal input from the two eyes (in other words, are in group 4), but in both cats and monkeys there is a roughly equal number of cells driven better by one eye or the other. Because there is not a dramatic skewing in the histograms, we can conclude that a cortical cell is about as likely to prefer input from the contralateral eye as from the ipsilateral eye.

When cortical cells are binocular, do both eyes have to be

15.11 Ocular dominance histograms for cortical cells in the cat *(left)* and the monkey *(right)*. Cells in group 1 and 7 are monocular (driven by only one eye), whereas cells in group 4 are equally binocular (driven as well by both eyes). Cells in groups 2, 3, 5, and 6 are binocular, but driven better by one eye or the other. In both cats and monkeys, most cells are binocular, but cats have more cells that are equally binocular than do monkeys (that is, cats have relatively more group 4 cells).

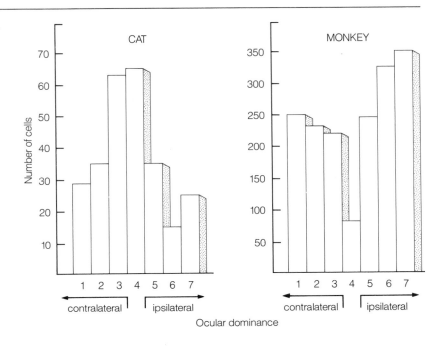

stimulated in the same way? The answer is yes. Whether a cortical neuron is a simple or a complex cell, exactly the same parameters are required to activate it maximally, regardless of which eye is stimulated. The cell's receptive field has about the same position in the visual field of the two eyes, it requires the same orientation of the stimulus, the same direction of movement, and so forth.

As noted, one or the other eye often drives the cell more vigorously; that is, the cell has ocular dominance. Yet the cells often give the strongest response when both eyes are stimulated simultaneously; the effect of stimulating both eyes is synergistic. In a few cells, stimulation of either eye alone causes only a weak or even no response; conversely, stimulation of both eyes produces a vigorous response.

Although binocular cortical neurons have their receptive fields in the same area of the visual fields of both eyes, most of them respond well even if the stimuli presented to the two eyes are not exactly aligned. Some complex cells, though, require precisely aligned stimuli to be presented to the two eyes simultaneously for

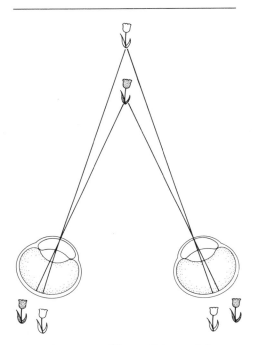

15.12 Images at different distances fall on slightly different parts of the retinas in the two eyes.

them to be activated. These complex cells are believed to play a role in depth perception.

Because the two eyes of a cat, monkey, or human are separated, an image falls on a slightly different part of the visual field depending on the image's distance from each eye, as shown in Figure 15.12. Ordinarily, complex cortical neurons are indifferent to where in the receptive field a moving slit is presented so long as the orientation of the slit is correct and the direction of movement is appropriate. Stimuli presented to the two eyes can fall on somewhat different regions of the receptive fields and still vigorously drive the cells. These cells respond well to both near and far visual stimuli.

Other complex cells require that the two stimuli be presented simultaneously to certain positions within their receptive fields. Some of these cells require alignment appropriate for viewing near objects, others for viewing more distant objects. These cells, termed *disparity-tuned cells*, respond poorly to monocular stimulation. Otherwise, they have the properties of higher-order complex cells: they are found in area V1 but are more common in area V2. It seems likely that the cortex contains disparity-tuned cells that respond to several distances.

Summary

Extracellular recordings from neurons in the visual cortex in the cat and monkey have provided important insights to higher visual processing in mammals. In nonmammalian species, processing of visual information beyond the retina takes place mainly in the tectum, and much less information has been gleaned. More peripheral processing occurs in the retinas of nonmammalian species, so many of the ganglion cells in these species have receptive field properties similar to those of cortical cells in the cat and monkey, such as movement- and direction-sensitivity.

Neurons in the lateral geniculate nucleus of mammals, the relay nucleus between the retina and cortex, have receptive fields organized quite similarly to those of the retinal ganglion cells. They are concentrically organized and have both on-center, off-surround cells and off-center, on-surround cells. Inputs to the lateral geniculate nucleus from the reticular formation and back from the cortex are believed to modify or gate visual information flow from the lateral geniculate nucleus to the cortex, although the nature of the these modifications is not well understood.

Recordings from most neurons in visual area 1 (V1) of the cortex show evidence of further processing of visual information. Two types of cells are distinguished, simple and complex cells, on the basis of their receptive field properties. Simple cells respond to spots of light projected onto the retina, and their receptive fields can be mapped into excitatory and inhibitory zones. The receptive fields of simple cells are typically elongated; hence, properly oriented slits, bars, or edges of light are optimal stimuli for these cells. In a population of simple cells, all orientations are represented, and a rotating stimulus will pass through the receptive fields of 18–20 simple cells as it sweeps from one orientation back to the same orientation, a rotation of 180°. Complex cells do not respond to spots of light; instead, they require for optimal activation well-oriented slits or bars of light that move through the receptive field at right angles to the orientation of the bar or slit. Some complex cells are direction-selective; the cell responds only when the bar moves in one direction. Other cells are end-stopped: the bar of light must be restricted in its length to maximally activate the cell. Thus, one or both ends of the receptive fields of these cells are inhibitory.

Experimental results indicate that the farther along the visual system a neuron is from the photoreceptor, the more specific must the stimulus be to activate it optimally. Specific aspects of an image activate certain cells, suggesting that in the cortex the visual image is broken into components that are encoded in individual neurons. Most neurons in the visual cortex are complex cells, which indicate that movement is crucial for visual perception. Indeed, our eyes are in continual motion, ensuring that images are constantly moving on the retina.

Little is presently known about how cortical receptive fields are established by the synaptic interactions within the cortex. It is possible to suggest how the receptive fields of simple cells can be elaborated by simple excitatory inputs from lateral geniculate neurons and how the receptive fields of complex cells can be formed by excitatory inputs from simple cells, but these schemes are undoubtedly oversimplified. Indeed, we have good evidence that inhibitory mechanisms intrinsic to the cortex are important in establishing the receptive fields of cortical neurons.

Most cells in the cortex are binocular, receiving input from both eyes. The stimuli required to drive the cell optimally by one eye are also the stimuli required to drive the cell optimally by the other eye. Often the cells respond most vigorously when both eyes are stimulated simultaneously. Certain complex cells require precisely aligned stimuli presented to the two eyes to drive the cells. These cells, called disparity-tuned cells, can play a role in depth perception. Because the two eyes of a cat or monkey are separated by some distance, near or far images fall on slightly different parts of the receptive fields in the two eyes. Disparity-tuned cells detect these differences and respond selectively to stimuli that correspond to a certain depth.

Further Reading

BOOKS

Hubel, D. H. 1988. *Eye, Brain and Vision*. Scientific American Library. New York: W. H. Freeman.

Yarbus, A. L. 1967. *Movements and Vision*, trans. B. Haigh. New York: Plenum Press.

ARTICLES

Hubel, D. H., and T. N. Wiesel. 1959. Receptive fields of single neurons in the cat's striate cortex. *Journal of Physiology* 148:574–591.

———— 1962. Receptive fields, binocular interaction and functional architecture in the cat's visual cortex. *Journal of Physiology* 160:106–154.

Pettigrew, J. D. 1972. The neurophysiology of binocular vision. *Scientific American* 227(2):84–95.

The Architecture of the Visual Cortex

ALTHOUGH detailed information concerning the synaptic circuitry of higher visual centers and the cerebral cortex is scanty, we know much about the cellular organization of the primary visual area (V1) in monkeys and cats. In this chapter, I concentrate on the primate visual system and cortex. I also consider how visual deprivation affects the structure and function of the primary visual cortex.

Figure 16.1 is a diagram of the visual pathways in a primate as viewed from the underside of its brain. Visual information from the eyes passes along the optic nerve to the lateral geniculate nucleus, and from there it proceeds by way of the optic radiation to the visual cortex. In primates, the eyes face forward and the visual fields of the two eyes overlap considerably. Information from the right visual field is received by the left half of each retina, information from the left visual field is received by the right half of each retina. In a manner analogous to other sensory systems (see Chapter 12), right visual field information is processed by the left side of the brain and left visual field information by the right side of the brain. The sorting out of ganglion cell axons to project to either the right or left side of the brain takes place at the optic chiasm. There axons from the right and left sides of each retina are routed to the appropriate lateral geniculate nucleus. From the lateral geniculate nucleus, visual information passes on to the corresponding cortical hemisphere.

In Chapter 12, we noted that a topographic representation of the visual field exists on area V1 of the cortex. There is also a topographic representation of the visual fields on the lateral geniculate nuclei, as shown schematically in Figure 16.2. On the left are the retinal fields; the left side of each retina receives input from the right visual field and vice versa. Each retinal half projects to the corresponding lateral geniculate nucleus. The lateral geniculate nuclei in turn project to the corresponding hemispheres of the

16.1 Diagram of the visual pathways in primates viewed from the underside of the brain. Visual information from the eyes passes via the optic nerve to the lateral geniculate nucleus, and from there, via the optic radiation, to the visual cortex. In primates the eyes face forward and the visual fields of each eye overlap. Information from the right visual fields is received by the left half of each retina and vice versa. At the optic chiasm, the fibers from the right and left sides of the two retinas are sorted out so that information from the right visual field projects to the left side of the brain and information from the left visual field projects to the right side of the brain.

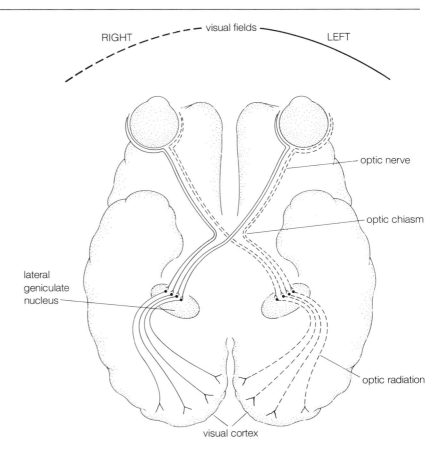

cortex. There is, therefore, a hemi-visual field map on each lateral geniculate nucleus and on the primary visual areas of each cortical hemisphere (area V1). These maps are systematic and consistent; a certain point on the retina projects to a corresponding point in one lateral geniculate nucleus and to an appropriate point in area V1 of the cortex.

An important caveat when comparing retinal and cortical fields is that much more area of the cortex is devoted to foveal and central vision than to peripheral vision. How does this unequal distribution come about? As noted in Chapter 15, the receptive fields of neurons dealing with central visual field information are

16.2 Schematic representation of the projections from the retina to the lateral geniculate nucleus and to area V1 of the visual cortex. The retinal fields are shown on the left and are divided for convenience into four quadrants. Right visual field information is received by the left half of each retina and vice versa. The left half of each retina projects that information to the left lateral geniculate nucleus and the left hemisphere of the cortex. The arrangement corresponds to a hemi-visual field map on each geniculate nucleus and each cortical hemisphere. As shown in the figure, information from points on the retina are projected to corresponding points on the lateral geniculate nuclei and cortical hemispheres.

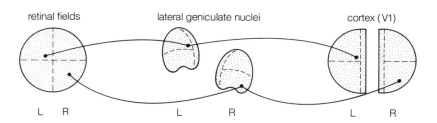

much smaller than are the receptive fields of neurons receiving input from the periphery of the visual field. This holds for retinal ganglion cells, lateral geniculate neurons, and cortical neurons.

If an electrode is passed vertically through area V1 of the cortex and the receptive fields of the encountered cells are characterized by noting their response properties while various stimuli are presented to the retina, some variation in receptive field size is observed. However, receptive field sizes are roughly similar for the neurons recorded. Furthermore, all of the receptive fields receive input from the same region of the visual field, although again some variation is noted. The total visual field area covered by the receptive fields recorded in such an experiment is called the *aggregate receptive field,* and it represents an area no more than 2 to 4 times the size of the smallest recorded receptive field.

If the electrode is inserted elsewhere in area V1, the aggregate receptive field will be smaller or larger, depending on whether one is recording from an area receiving input from a more central or more peripheral region of the visual field. Furthermore, as one moves the electrode around in the visual cortex, the area of visual field covered also changes systematically, as one would expect. A key question is, how far does one need to move laterally on the cortex to go from one aggregate receptive field to another, and does this distance vary across the cortex? That is, how much cortex is needed to look after one part of the visual field?

The answer is straightforward. Points on the cortex separated by 2 mm encompass aggregate receptive fields that cover entirely separate regions of the visual field, and this holds regardless of where in area V1 the recordings are made. Because the aggregate receptive fields are much smaller for central vision than for peripheral vision, more of the cortex is devoted to the processing of

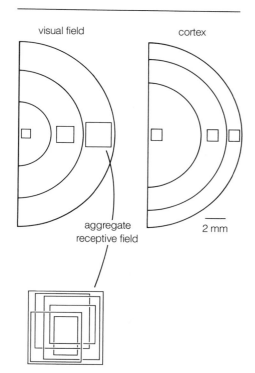

16.3 Schematic representations of the sizes of the aggregate receptive fields for neurons receiving input from the central, mid-peripheral, and far-peripheral regions of the visual field *(left)* and the relative amount of cortex devoted to analyzing information coming from these three regions *(right)*. An aggregate receptive field is the total area covered by the receptive fields of neurons receiving input from a particular part of the visual field. The aggregate receptive fields increase substantially in size from central to peripheral visual field, but all across the cortex, an area of about 2 mm² is required to analyze information coming from a separate part of the visual field. Therefore, more cortex is required to analyze the information coming from the central visual field than from the visual field periphery.

information from the central visual field as compared with the processing of information from the visual field periphery, as shown in Figure 16.3. To the left is a representation of the visual field divided into three equal zones, representing central, mid-peripheral, and far-peripheral visual space. Aggregate receptive fields are sketched for the three areas, with the smallest aggregate fields in the central area and the largest in the far-periphery. To the right is sketched the division of the cortex into the three zones. All across the cortex, an area of 2 mm² is concerned with a separate part of the visual field. But in the periphery, 2 mm² of cortex looks after a much larger part of the visual field than does a 2 mm² piece of cortex receiving central visual field input. The greater acuity of central vision is accounted for by this cortical arrangement.

Lateral Geniculate Architecture

To begin our study of the organization of the cortex, it will be instructive to consider first the architecture of the lateral geniculate nucleus. The lateral geniculate nucleus, like the retina and the cortex itself, is layered, as shown in Figure 16.4. In primates, six distinct cellular layers are seen, each 6–10 cells deep. A layer receives input from only one eye. Three layers (1, 4, and 6) receive input from the opposite (contralateral) eye and three layers (2, 3, and 5) receive input from the eye on the same side of the head (the ipsilateral eye). The nucleus illustrated in Figure 14.4 is the right lateral geniculate nucleus; the left lateral geniculate nucleus is its mirror image.

In the lateral geniculate nucleus, input from the two eyes is kept separate. Thus, binocular neurons are not encountered in the visual system until the cortex. And, as pointed out in Chapter 15, cells close to the lateral geniculate input in the cortex tend to be monocular; binocular cells are found away from the geniculate input, and in general cells farther away from the cortical input layers are more binocular. Why input from the two eyes is kept separate in the lateral geniculate nucleus is not understood, nor is it completely understood why the primate lateral geniculate nucleus contains so many layers. Close examination of Figure 16.4 reveals that the cells in layers 1 and 2 are larger than cells in the other layers and that they also stain slightly darker. Layers 1 and 2 are called the

16.4 The right lateral geniculate nucleus in the primate. Six cellular layers are evident. Layers 1 and 2 are the magnocellular layers; 3–6, the parvocellular layers. Each layer has a separate hemi-visual field map, and these maps are in register. So, for example, an electrode passing along the radial line will encounter cells receiving input from exactly the same region of the visual field.

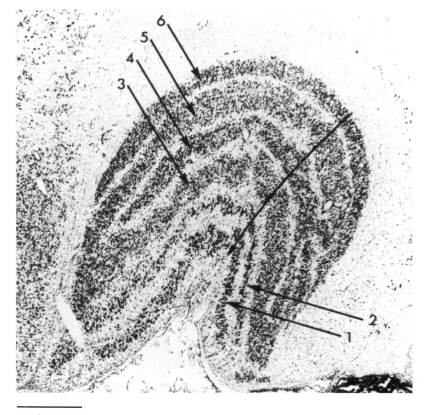

1 mm

magnocellular layers; the other four layers are called the *parvocellular layers.*

The physiological properties of the cells in the magnocellular and parvocellular layers have distinct differences. The responses to light of magnocellular neurons are more transient than are the responses of cells in the parvocellular layers, though neurons in both layers have a center-surround receptive field organization. Hence, magnocellular neurons are more sensitive to moving stimuli than are parvocellular neurons. In contrast, parvocellular neurons have more sustained responses to prolonged light stimuli and many of these neurons are color-opponent cells. The magnocellular neurons project to a slightly different region of the cortex than do the parvocellular neurons, and so part of the reason for the layering

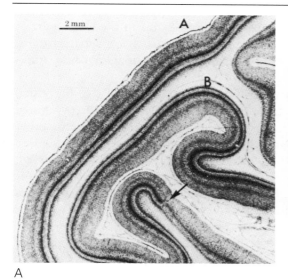

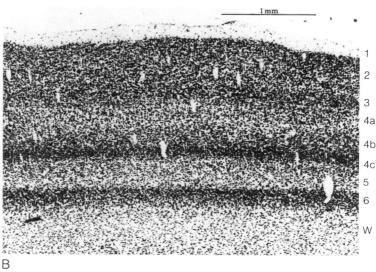

A   B

16.5 Low- **(A)** and high-magnification **(B)** light micrographs of area V1 of the visual cortex. The cortex is about 2 mm thick and is layered. The 6 major layers that are distinguished consist of cells of different packing densities. Below layer 6 is white matter *(W)*, which contains myelinated axons that carry information to and from specific cortical regions. In **(A)**, the cortex marked *B* appears to be inside the region marked *A*. This is because of the complex folding of the cortical surface in higher mammals. The arrow in **(A)** indicates the border of visual areas 1 and 2.

of the geniculate is to segregate neurons concerned with different aspects of the visual image. But why there are four parvocellular layers rather than just two (one for each eye) is not understood.

The Visual Cortex: Area V1

Figure 16.5 shows portions of visual area V1 from the monkey cut in cross section at two different magnifications. The cortex is like a plate covering the cerebral hemispheres, and it has a thickness of about 2 mm (see, for example, Figure 12.6). Layering of area V1 is obvious in both micrographs, although the cellular layers in the cortex are not nearly as distinct as they are in the retina or lateral geniculate nucleus. The cortical layers, then, consist mostly of cellular layers with different packing densities; no cell-free layers are interspersed between the cellular layers. Note also that the cellular layering is even less distinct in area V2, the first visual association area (arrow, Figure 16.5A).

Area V1 is divided into 6 layers, as indicated in Figure 16.5B. Layers 2 and 3 cannot be distinguished in this micrograph, and layer 4 is divided into three sublayers. Most input from the lateral geniculate nucleus terminates in sublayer 4c. Input from the magnocellular layers terminates in the upper half of layer 4c (termed 4cα), whereas parvocellular input terminates in the lower part of layer 4c (4cβ). Input and output fibers reach or depart from a specific region of the cortex by way of the white matter fiber tracts that run underneath layer 6.

Figure 16.6 is Ramón y Cajal's drawing of cells in the visual cortex. Two major classes of cells are distinguished: pyramidal cells, which project axons to other cortical regions or to other regions of the brain, and stellate cells, whose axons remain within this region of the cortex. Two types of stellate cells are observed, spiny and smooth cells. Spiny stellate cells have many dendritic spines, smooth stellates have none. All pyramidal cell dendrites have spines. Pyramidal cells and spiny stellate cells are excitatory neurons; smooth stellate cells are inhibitory.

The highest density of stellate cells is in and around layer 4 and pyramidal cells are generally above or below layer 4. As noted in Chapter 15, cells with a center-surround receptive field organization are close to the lateral geniculate input (that is, near layer 4c), and simple cells are also mainly in this region. These observations imply that center-surround cells and simple cells are stellate cells and that complex cells, found away from the geniculate input, are pyramidal cells. There is some evidence for this, but the data are not entirely consistent. That is, occasionally a stellate cell is found to be a complex cell and a pyramidal cell to be a simple cell.

Note in Figure 16.6 that some pyramidal cells in layers 5 and 6 extend their apical dendrites through much of the thickness of the cortex. Thus, the apical dendrites can extend well over a millimeter. The apical dendrites typically extend much farther than the more horizontally directed basal dendrites of these cells. This suggests that information flows mainly vertically within a region of the cortex. Indeed, we have good evidence that input from a certain region of the visual field comes into layer 4c and then spreads mainly up or down to cells either above or below the input layers.

As noted earlier, simple cells are located near layer 4c whereas complex cells are found above and below, in layers 2, 3, 5, and 6. There are some systematic differences among the complex cells in the various layers. In layers 2 and 3, all cells (except for end-stopped cells) respond most vigorously to long slits of light, but in layer 5 the complex cells respond as well to short slits as to

16.6 Golgi-stained neurons in the cortex drawn by Ramón y Cajal. The cells marked *H* are typical pyramidal cells; those marked *F* are typical stellate cells. The other cells are variant pyramidal and stellate cells.

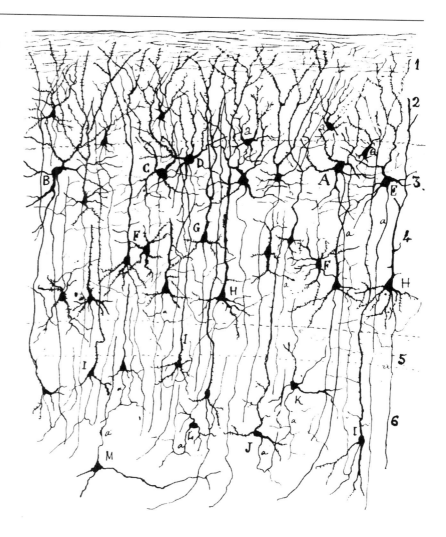

long slits. In layer 6, the receptive fields of the complex cells tend to be long and narrow; so long slits covering the entire length of the field are required to drive the cells maximally, but the width of the field (the area over which the moving slit evokes a response) is small. The conclusion drawn from these observations is that the complex cells in each layer carry out somewhat different functions. And, as would be expected, the pyramidal cells in the different layers project to different areas of the brain. Cells in layers 2 and

3 project to other cortical areas. Cells in layers 5 and 6 project to deeper brain regions, such as midbrain nuclei, lateral geniculate nuclei, or to other visual areas of the cortex.

Columnar Organization

How is the cortex organized? Are there subdivisions or groupings of cells within the cortex? How can one begin to make sense of cortical architecture? An early and important clue came from the study of binocularity of cortical neurons. As we have learned, visual information from the two eyes is kept separate until the cortex, but in the cortex most simple and complex cells receive input from both eyes. Yet most cortical cells have a preference for one eye or the other: they have the feature of ocular dominance.

Cells within the cortex having an ocular preference are not distributed randomly. Indeed, physiological experiments have shown clearly that cells with an ocular preference are grouped in columns. This was demonstrated by passing an electrode through the cortex and noting the ocular preference for each cell encountered.

Figure 16.7 depicts how such experiments were done and the results. An electrode passed vertically through the thickness of the cortex will encounter cells having the same ocular preference. If the electrode is moved to a different spot on the cortex, and again passed vertically through the cortex, the ocular preference may change, but again all the cells will have the same preference.

Oblique electrode penetrations reveal how the ocular dominance columns are arranged. As an electrode passes at an angle through the cortex, it first encounters cells that prefer one eye, then cells that prefer the other eye. In such an experiment, ocular preference usually alternates from one eye to the other in a regular and consistent fashion. The width of an ocular dominance column is approximately 0.5 mm; so as an electrode passes across 1 mm of cortex, ocular dominance columns for both eyes are encountered.

How are the columns themselves organized? Are they free-standing columns, or are they continuous strips across the cortex? An answer to this question came from anatomical studies that took advantage of the fact that the geniculate input to the cortex is monocular. Specifically, if the geniculate axon terminals are stained and flat sections of the cortex taken through layer 4c, the arrangement of the columns is revealed. One such experiment, shown in

16.7 Ocular dominance columns in the monkey. When an electrode is passed vertically through the cortex *(1 and 2)*, all of the cells encountered have the same eye preference. In layer 4c, the cells are monocular, in all other layers they are binocular. With an oblique penetration of the cortex *(3)*, an electrode first encounters cells with one eye preference, then the other. The ocular dominance columns have a width of about 0.5 mm.

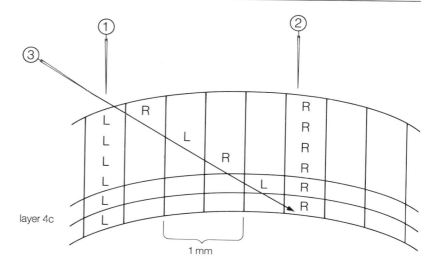

Figure 16.8, was performed as follows. A radioactive amino acid, proline, was injected into one eye of a monkey. The proline was taken up by the ganglion cells of that eye and transported via axoplasmic transport mechanisms to the lateral geniculate nucleus. There, some of it was released and taken up by the lateral geniculate neurons. From the lateral geniculate nucleus, a small amount of the injected radioactivity was transported to the cortex and detected by autoradiography. It takes about a week for the radioactivity to travel from the eye to the cortex.

Figure 16.8 illustrates that the ocular dominance columns are arranged in stripes. Furthermore, the dimensions of the stripes match closely the measurements of column thickness determined by physiological experiments; each stripe is approximately 0.5 mm wide. Figure 16.9, a reconstruction of area V1, reveals unexplained complexities of the stripe pattern across the cortex—some stripes end in blind loops and others are discontinuous or swirl in curious ways.

Above and below layer 4c a clear columnar organization cannot be demonstrated anatomically, because the cells are mainly binocular. Right and left eye inputs blur, indicating that lateral connections between columns above and below layer 4c mediate the neurons' binocularity.

16.8 Anatomical demonstration of the ocular dominance columns. Radioactive proline was injected into one eye in a monkey and after it had been transported to the cortex, the monkey was sacrificed, the cortex was sectioned transversely through layer 4c, and autoradiograms were prepared. This figure is a montage of dark-field photomicrographs. The light areas represent exposed silver grains and show the distribution of radioactivity across layer 4c.

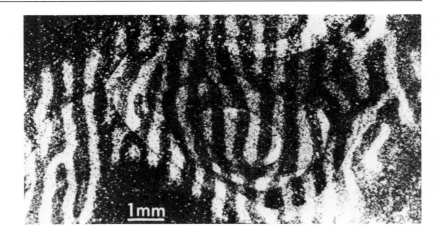

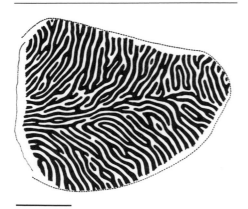

5 mm

16.9 A reconstruction of the ocular dominance columns (or stripes) across one (the right) cortical hemisphere of a monkey. The reconstruction was based on experiments similar to that shown in Figure 16.8.

Orientation Columns

What about other cells that share common features? How are they organized? The next obvious feature to examine is orientation preference, a characteristic of both simple and complex cortical cells. If an electrode is passed vertically through the cortex, all the cells encountered, whether simple or complex, have the same orientation preference, as illustrated in Figure 16.10. In and around layer 4c, some cells have concentrically organized receptive fields and show no orientation selectivity, but above and below layer 4c all the cells have the same orientation preference. If the electrode is moved elsewhere on the cortex and another vertical penetration made, we get the same result, although it is likely that the orientation preference is now different. These results indicate that cells with common orientation preference also are organized in a columnar fashion.

Oblique penetration of the cortex with a microelectrode shows that the orientation selectivity of the neurons alters in a regular and precise way. As the electrode moves from one orientation column to the next, the preferred orientation shifts by about 10° in either a clockwise or counterclockwise direction. Furthermore, such experiments reveal that the orientation columns are much narrower than are the ocular dominance columns. Indeed, a change of orientation occurs every 50 μm that the electrode advances across the cortex.

16.10 Orientation columns in the monkey. Vertical penetration of the cortex with an electrode (1 and 2) records cells with the same orientation. Oblique penetrations (3) encounter cells with systematically altered orientation preference. Cells in layer 4c usually have no orientation preference (marked by open circles). The orientation columns have a width of about 50 μm.

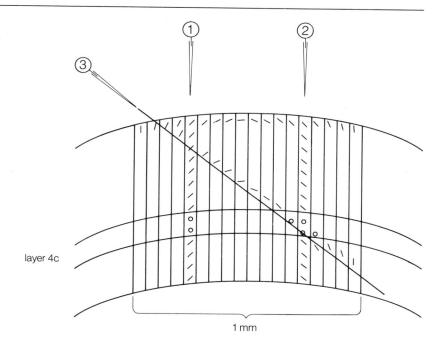

Since the shift of orientation is about 10° between columns and each column is 50 μm wide, the electrode must move laterally along the cortex about 1 mm to traverse the 18–20 columns necessary to go from one orientation back to the same orientation. This is approximately the same distance covered from one position in an ocular dominance column back to the same position in the next ocular dominance column receiving input from the same eye.

The organization of the orientation columns has been much more difficult to determine than that of the ocular dominance columns. The best data so far have come from anatomical studies using radioactive 2-deoxyglucose. This substance, like glucose, is taken up by neurons, and when the cells become active they begin to metabolize it. However, cells cannot break down 2-deoxyglucose completely; it accumulates in the cells and serves as a marker for metabolic activity. For studies of orientation columns, an anaesthetized animal was injected with 2-deoxyglucose and its retina was stimulated with slowly moving stripes of fixed orientation. After several minutes, the animal was killed, the cortex sliced horizontally, and an autoradiograph prepared. The patterns observed were complex, but they do suggest that orientation columns, like ocular dominance columns, are arranged in a stripelike fashion. Furthermore, the periodicity observed, the distance along the cortex from one orientation back to the same orientation, was 1 mm.

In some experiments ocular dominance columns were demonstrated at the same time as the orientation columns, for the purpose of determining the relation between the two kinds of columns. No correlations were obvious, probably because of the twists and turns each set of columns takes.

Hypercolumns

A reasonable way to think about how the ocular dominance columns fit together with the orientation preference columns is to assume that they are positioned at right angles to one another. An arrangement of this type, shown in Figure 16.11, suggests that a 1 mm × 1 mm × 2 mm block of cortex has all the machinery needed to analyze a bit of visual space. Such a block of cortex, called a *hypercolumn* by David Hubel and Torsten Wiesel, has within it neurons that respond to input from both eyes—there are 2 ocular dominance columns—and cells that respond to all orientations—there are 18–20 orientation columns. Furthermore, all types of simple and complex cells are represented in a hypercolumn, and the neurons have varying degrees of binocularity. The arrangement of cells with various properties in the hypercolumn is also shown in Figure 16.11.

A hypercolumn is clearly not a discrete unit but part of a continuum. Furthermore, as we learned, shifting from one part of the visual field to a completely different part of it requires a movement on the cortex of about 2 mm. This means that more than one hypercolumn contributes to the analysis of a bit of visual space. It is also the case that a hypercolumn analyzes a larger and larger area of visual space going from central to peripheral parts of the visual field.

Color Processing

So far, I have concentrated on how cells in the visual cortex respond to orientation, movement, and binocular input, and how the cells responding to these features are organized. How does

16.11 A hypercolumn: a 1 mm × 1 mm × 2 mm block of cortex containing all the cells required to analyze a bit of visual space. In a hypercolumn, input from both eyes is represented. Furthermore, all types of simple and complex cells are present, and the cells have all possible orientation preferences and varying degrees of binocularity. In addition, color-sensitive cells are found in the pegs inserted into the hypercolumn. Degree of binocularity is represented by (+) symbols; in layer 4c, where lateral geniculate input enters the cortex, the cells are monocular (represented by an open circle); away from layer 4c the cells are more and more binocular. Cell types also vary through the thickness of the cortex. Around layer 4c, the cells are mainly center-surround (CS); away from layer 4c they are first simple (S), then complex (C), and finally specialized complex (SC).

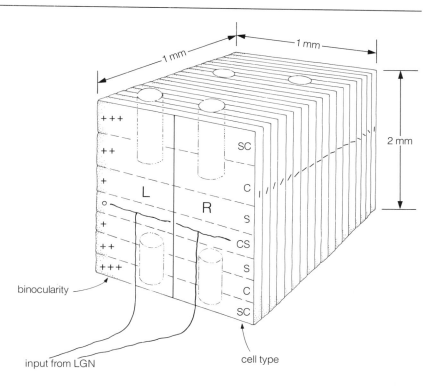

color processing fit into the picture? Information on how the cortex processes color has only recently been forthcoming, partly because early work on cortical processing was carried out in the cat, an animal that lacks good color vision. Even in the monkey, however, discovering color cells in the cortex was difficult, and only recently have color-sensitive cells been identified routinely.

The breakthrough in the identification of color cells in the monkey cortex came from an anatomical observation. When area V1 was stained for the enzyme cytochrome oxidase, discrete regions of dark staining were observed, particularly in the upper layers of the cortex (layers 2 and 3). The areas, called blobs, pegs, or patches, were about 0.15 mm across and were about 0.5 mm apart. They did not seem to correspond to ocular dominance or orientation columns, and so Margaret Livingstone and David Hubel at the Harvard Medical School decided to record from cells

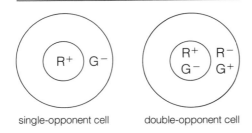

single-opponent cell   double-opponent cell

16.12   Receptive field maps for single- and double-opponent color cells. R^+ means that red light excites that region of the receptive field; G^- means that green light inhibits that region of the field.

within them. Much to their surprise, they found that within the center of a blob the cells had no orientation selectivity, their receptive fields were concentrically organized, and many cells (but not all) were color-opponent cells. Most color-opponent cells recorded in the blobs were more complicated than were the color-opponent cells found in the retina and lateral geniculate nucleus. They gave almost no response to white light stimuli of any configuration, but small colored spots presented to the retina provoked a vigorous response or strong inhibition. Furthermore, either activation or suppression of activity could occur in both the center and surround of a single cell, depending on the wavelength of light used. Hence, these cells were double-opponent cells, as compared with the single-opponent cells in the retina or lateral geniculate nucleus. The commonest type of double-opponent cell had a receptive field center that was excited by red light and inhibited by green light (R^+G^-) and a receptive field surround that was inhibited by red light and excited by green light (R^-G^+); but other combinations were observed as well. Figure 16.12 presents receptive field maps for single- and double-opponent cells.

What role do color-opponent cells play in color perception? It has been proposed that single-opponent cells underlie the phenomenon of successive color contrast. That is, following the sustained viewing of a red field, a person will perceive a blank piece of white paper not as white but as greenish in color, and vice versa. The double-opponent cells, by contrast, seem appropriate to account for simultaneous color contrast: when a gray field is surrounded by a red background, the gray field appears greenish, and vice versa.

How do the blobs fit into the structure of the hypercolumn? The assumption is that the blobs may be thought of as pegs inserted partway through the column from either side. The dimensions suggest four pegs per hypercolumn, extending through layers 1, 2, and 3 and also through layers 5 and 6. The approximate arrangement of the blobs in a hypercolumn is shown in Figure 16.11.

Livingstone and Hubel have suggested that the blobs are a relatively independent part of the primary visual cortex that is concerned with color processing. In support of this notion is evidence that the blobs appear to get an independent input from the lateral geniculate nucleus, from the interlaminar regions of the geniculate. There is good evidence that beyond area V1 color is processed separately from other aspects of the visual image, and this processing subdivision appears to begin already in area V1. Indeed, as we shall see in Chapter 18, many aspects of the visual image—such as form, movement, depth, and color—are analyzed independently in higher visual centers.

Cortical Plasticity: The Effects of Visual Deprivation

A most important question in neuroscience is, how malleable is the mammalian brain? How much of our nervous system is hard-wired or genetically determined, and how much is affected by experience and environment? Because so much is known about the organization of the visual system, it is natural that such questions have been asked about this part of the brain. Furthermore, the visual system is convenient for experimentation addressed to these issues because it is easy to restrict visual input to an animal for long periods of time or to vary the parameters of visual stimuli presented to one or both eyes.

Responses of Cells in Newborn Animals

The first question to be asked when one is investigating the effect of experience on neural processing is, how much of the cortical machinery is present at birth? How do the cells of the visual system respond in a visually inexperienced animal? Recordings from newborn monkeys show that the responses of the cells are remarkably adult-like in their behavior—with a few ifs, ands, and buts. The responses of the neurons from the newborn monkey tend to be less vigorous than they are in adults, and cells are occasionally encountered in the newborn that cannot be driven by light stimuli. Even so, it is possible to record from neurons that have simple, complex, and specialized complex receptive fields. The cells show good orientation selectivity and most are binocular with a preference for one eye or the other.

The one consistent difference that at first glance seems paradoxical is that neurons in layer 4c of the newborn and young animal are usually binocular, yet they are mainly monocular in the adult animal. I explain later the reason for this anomaly; for now, the significant conclusion is that in newborn monkeys much of the basic cortical organization is present at birth. The neurons appear to be wired correctly, and they respond physiologically much as do cells in the adult cortex. Thus, visual experience is not necessary for establishing the elementary response properties of the cortical neurons.

Visual Deprivation

If a young animal is deprived of *form* vision by lid closure, lens removal, or the application of a light-diffuser to its eye, clear and persistent visual changes will develop. Visual acuity is severely reduced, and if the deprivation is monocular, there will be striking changes in the binocularity of the visual system. Similar effects in young human beings, a condition called *amblyopia,* are seen frequently in youngsters whose eyes are misaligned or who have a congenital cataract that is not removed early in life.

An example of the effects of monocular deprivation on the responses of cells in the visual cortex of the cat is depicted in Figure 16.13. In this experiment, the eyelid of one eye was sutured shut in the first week of life. The eye remained closed for 14 weeks and, shortly after it was opened, recordings were made from neurons in the retina, the lateral geniculate nucleus, and area V1 of the cortex. No major changes in the physiological properties of the cells innervated by the deprived eye were seen up to the cortex, but dramatic differences were noted in area V1, particularly in the binocularity of the cells.

As you can see in the ocular dominance histogram of Figure 16.13, few cells in this animal were binocular. Furthermore, few cells preferred the closed eye, and the responses of the neurons that did prefer the deprived eye were often abnormal. Most of the cells that responded overwhelmingly preferred the open eye, and the great majority of these were monocular; they could be driven only by the open eye. Several nonresponsive cells were also encountered in the cortex; these cells failed to respond to stimulation of either eye.

The Critical Period

Can one induce these kinds of changes in cortical cells at any time during the life of an animal by similar kinds of visual deprivation? The answer is clearly no. In adult cats, monkeys, and humans, visual deprivation of long duration (months to years) does not

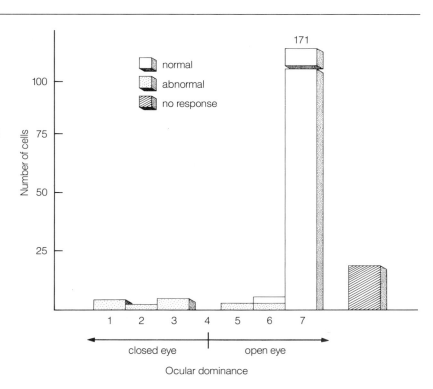

16.13 Ocular dominance histograms of cell responses recorded in the cortex of a cat that had one eyelid closed from the first to the fourteenth week of life. Most of the cells responding well to light stimuli were monocular, activated only by the control eye. The cells that were driven by the deprived eye gave mostly abnormal responses. A number of cells that could not be driven by either eye were also encountered (unresponsive cells).

have dramatic effects on visual performance or on the responses of cortical neurons. To induce changes like those noted in Figure 16.13, deprivation must occur soon after birth. The period of susceptibility is called the *critical period*. In cats, the critical period is 4 weeks to 4 months of age; deprivation during the earlier times (4–6 weeks) will cause more extensive changes. In monkeys, deprivation between birth and 6 weeks has severe effects, whereas deprivation between 6 weeks and 6–12 months induces some changes but of a lesser degree. In humans, deprivation between 6 months and 5–6 years has been reported to induce amblyopia.

Deprivation need not be extended during the early part of the critical period to create severe changes. Indeed, a few days' deprivation in the first 2 weeks of a monkey's life can result in changes about as severe as those seen in animals whose eyes are kept shut for several months early in life.

RECOVERY

Once changes are induced in the cortical neurons as a result of monocular visual deprivation, can they be reversed? The answer is complicated. If nothing is done other than to open a closed eye after a period of deprivation, little recovery is usually noted. In one experiment, for example, the eyelid of a monkey was closed for just 9 days during the first two weeks of life. The eyelid was then opened and nothing further was done to the animal. At 4 years of age, cells recorded from the animal's cortex showed changes similar to those in another monkey that had one eye closed from birth to 4 years of age!

Substantial recovery in acuity can be induced if, after a short deprivation time, the deprived eye is opened and the undeprived eye is closed for a time. So, if the animal is forced to use the deprived eye, significant recovery is often noted. Ophthalmologists long ago learned this trick to treat amblyopia. Children with amblyopia in one eye wear a patch over the normal eye for a part of every day to improve visual acuity in the amblyopic eye.

In other experiments, alternating vision between the two eyes after a period of visual deprivation to one eye has led to informative results—namely, visual acuity may recover but binocularity of the neurons usually does not. Hence, a disparity develops between visual acuity and binocularity in the visual system. These observations show that loss of binocularity in cortical neurons is not responsible for loss of visual acuity, nor do these two attributes of the visual system always run in parallel.

Strabismus

A similar separation of visual acuity and binocularity is apparent when the eyes are not aligned properly and the visual fields are not in register on the two retinas. Eye misalignment, known as *strabismus,* results when the eyes turn out (a condition referred to as wall-eyed) or when they turn in (cross-eyed).

In wall-eyed individuals, vision typically alternates: the right eye is used to view the right visual field, and the left eye is used to view the left visual field. Visual acuity of the two eyes is normal but no binocular interactions take place between the eyes; no binocularly mediated depth perception occurs in such individuals. Animals that are made wall-eyed by surgical manipulation, and

16.14 Ocular dominance histograms from animals made "wall-eyed" *(left)* or "cross-eyed" *(right)*. In wall-eyed animals, most cells are monocular; in many cross-eyed animals, one eye becomes dominant and most cells recorded are monocular, receiving input only from the dominant eye.

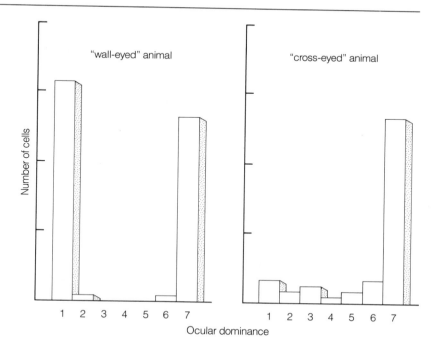

whose cortical neurons are subsequently recorded, demonstrate an interesting ocular dominance pattern (Figure 16.14). About half of the cells recorded are driven by one eye and the other half by the other eye. Virtually all of the cells are monocular, however; very few show any binocular input.

In cross-eyed individuals, by contrast, an occlusion-like deficit usually occurs. One eye, typically the straighter eye, gradually becomes dominant and visual input from the other eye is ignored. The ignored eye then develops a high degree of amblyopia. In animals made cross-eyed, the ocular dominance pattern of the cortical cells closely resembles that of an animal monocularly deprived, as shown in Figure 16.14.

When form vision is deprived equivalently in both eyes—either by covering them or because of bilateral cataracts—the visual acuity of the animal or human is severely reduced. Both eyes become amblyopic, but most of the neurons in the cortex receive

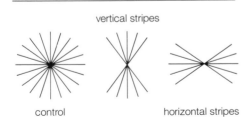

vertical stripes

control   horizontal stripes

16.15   Representations of the orientation preferences for cortical cells in a control animal and in animals exposed only to vertical or horizontal stripes. In control animals, all orientations are represented; in animals exposed only to horizontal or vertical stripes, the orientation preferences of the cells reflect the stimuli to which the eyes had been exposed.

binocular input and ocular dominance histograms constructed from cortical neuron recordings in animals are comparable to ones obtained from normal animals.

Pattern Deprivation

In the experiments described so far, all form vision was withheld from one or both eyes for a period of time. Is it possible to induce more specific deficits in cortical neurons by restricting one or another aspect of the visual world? An obvious experiment is to raise animals in environments where they are exposed to bars or stripes of only one orientation. Such experiments have been done by several groups of workers and the results are consistent. Whereas in a normal cortex, cells are found that respond to all possible orientations, in animals raised in an environment where they saw only horizontal or vertical stripes, the cells responded only to orientations to which the eye had been exposed. This is shown schematically in Figure 16.15.

Other experiments along the same general idea have been carried out with similar results. For example, if animals are restricted at an early age to environments where they see little movement of objects or movement of objects in only one direction, cells in their cortex appear to be less movement-sensitive or they are sensitive only to movement in the direction to which they have been exposed.

Mechanisms Underlying Deprivation Changes

We do not understand in detail the cellular and synaptic changes that occur in the cortex when an eye is visually deprived in the critical period of an animal's life. Some insights have come from the study of ocular dominance columns in animals that have had one eyelid closed during the first few weeks of life. Figure 16.13 suggests that the open eye takes over cells from the closed eye. Most of the recorded cells receive strong input from only one eye, the open eye. Is there evidence that the open eye's input occupies more territory than does the input from an eye in a normal animal? What do the ocular dominance columns look like in a monocularly deprived animal?

Physiological experiments offer convincing proof: the ocular dominance columns have quite disparate sizes in the cortex of a monocularly deprived animal. The column receiving input from

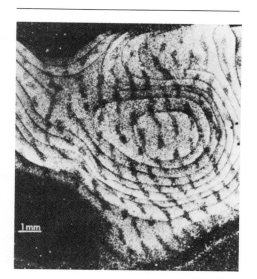

16.16   Anatomical demonstration of ocular dominance columns in a monocularly deprived monkey. The experiment was similar to the one described in Figure 16.8. Note that columns from the deprived eye *(dark stripes)* are much thinner than are the columns from the open eye, and that they are discontinuous.

the open eye is strikingly larger than is the column receiving input from the deprived eye. Anatomical studies confirm these observations, as we can see in Figure 16.16. In the experiment depicted here, radioactive material was injected into the open eye and sections were then cut through layer 4c. Not only is the amount of cortex devoted to the open eye considerably enlarged as compared with the amount devoted to the closed eye, but the columnar stripes reflecting input from the deprived eye are discontinuous.

How does this come about? How does one eye take over from the other eye? A partial answer to this question came from an attempt to examine ocular dominance columns in newborn animals. Much to everyone's surprise, no ocular dominance columns could be detected anatomically in the newborn. Why? A clue to the answer was noted above: in the newborn animal, cells in layer 4c receive input from both eyes, whereas in the adult layer 4c cells are monocular. Input from the geniculate is not segregated into columns in the young animal. Indeed, it appears that axon terminals from the lateral geniculate nucleus extend widely across layer 4c in the newborn animal. Only after several weeks do these axons begin to retract and segregate into ocular dominance columns, as shown in Figure 16.17. Since the axon terminals of geniculate fibers from the two eyes extensively overlap, it is easy to understand why no ocular dominance columns can be demonstrated anatomically in layer 4c of the newborn animal.

Why the open eye claims much more territory than does the closed eye after monocular deprivation is explained as follows. The axon terminals of the geniculate fibers coming from the open eye do not retract as they ordinarily do, yet the axon terminals from the geniculate fibers coming from the closed eye retract much more than normally. An implication of this explanation is that lateral geniculate axons compete for cortical space and synaptic connections in the young animal. So long as each eye provides the same input to the cortex, the competition is even and both eyes end up having equal cortical representation. If one eye has less input to the cortex, however, the other eye dominates the competition and ends up with more cortical area and presumably more cortical synapses.

Another possibility that may contribute to the open eye having more cortical representation than the closed eye is that its axons have more terminals. In addition to an increased retraction of

16.17 A possible explanation of how the ocular dominance columns develop in layer 4c. In the newborn, the terminals of the input axons from the lateral geniculate nucleus extend widely across layer 4c. In the normal animal, the terminal fields gradually contract and eventually extend the width of a column. In the deprived animal, the terminals from the deprived eye retract excessively, whereas the terminals from the open eye retract less than they do in the normal animal.

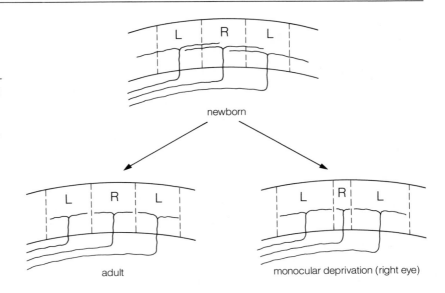

newborn

adult

monocular deprivation (right eye)

axonal processes of the geniculate fibers receiving input from the closed eye, the axon terminals from the open eye may sprout new processes.

We do not understand very well the mechanisms underlying neuronal growth and synaptic formation during development. The special field that focuses on these problems, developmental neurobiology, is an active one and will undoubtedly contribute to an understanding of these phenomena. But at the moment it is not clear why the cortex that in a newborn appears physiologically adult-like can then be modified significantly by altered visual input for the first few months of life. In the next chapter, I provide an overview of how the brain develops, focusing on the mechanisms underlying neuronal differentiation and how neurons make connections with other neurons.

Summary

Right visual field information is processed by the left half of the retina in each eye, and left visual information is processed by the right half. Thus, the left half of each retina projects to the left lateral geniculate nucleus and to area V1 in the left cortical hemisphere, and vice versa. Half of the visual field is mapped on each geniculate nucleus and on area V1 of each hemisphere, and these maps are topographic. Much more cortex is devoted to the analysis of central vision than to peripheral vision; the receptive fields of neurons receiving input from central visual field areas are much smaller in aggregate than are the receptive fields of neurons receiving input from the visual field periphery. Points on the cortex separated by 2 mm encompass receptive fields that receive input from different regions of the visual field. A bit of cortex receiving input from the visual field periphery is concerned with a much larger piece of visual space than a bit of cortex receiving input from the center of the visual field.

The lateral geniculate nucleus in higher primates consists of six cellular layers. Each layer receives input from only one eye: Layers 1, 4, and 6 receive input from the contralateral eye; layers 2, 3, and 5 from the ipsilateral eye. Layers 1 and 2 (the magnocellular layers) contain larger neurons that are more sensitive to movement than are the neurons in layers 3–6 (the parvocellular layers). Area V1 in the cortex is also differentiated into 6 layers, but the cortical layers do not correspond to the geniculate layers, nor are they as distinct. Input from the lateral geniculate nuclei terminates in layer 4c of the cortex—magnocellular input in layer 4cα, parvocellular input in 4cβ.

Two major classes of cells are found in the cortex: stellate cells, which are confined mainly to the layer 4 region; and pyramidal cells, which are found above and below layer 4 and which project axons to other regions of the brain. Pyramidal cells in layers 2 and 3 project their axons to different regions of the brain than do the pyramidal cells in layers 5 and 6. Furthermore, the pyramidal cells in the various layers have slightly different receptive field properties, although most are complex cells.

Cortical cells receiving input from only one eye or preferring that eye are grouped together in slab-like columns that run across the cortex. The ocular dominance columns are about 0.5 mm in width and alternate between the two eyes; thus, a 1 mm^2 block of cortex encompasses two ocular dominance columns. In area 4c, cortical cells are mainly monocular, but away from this layer the cells tend to be more binocular. Cells preferring a certain orientation are also grouped together in a columnar organization. Orientation columns are narrower, 0.05 mm wide; thus, across 1 mm of cortex, all orientations (18–20 columns) are included. Ocular dominance and orientation columns could be positioned at right angles to one another; if so, a 1 mm × 1 mm × 2 mm block of cortex (called a hypercolumn) includes all of the machinery necessary to analyze a bit of visual space.

Cortical cells that respond preferentially to color stimuli are present in blob-like regions found across area V1. These cells have little orientation-selectivity but instead have color-opponent receptive field properties. Most cells are double-opponent; activation and suppression of responses can be induced in both center and surround regions of the receptive fields with light spots of one wavelength or another. Blobs may represent a relatively independent part of the primary visual cortex that is concerned specifically with the analysis of color.

The neurons in area V1 of the newborn monkey have properties that are remarkably adult-like. They show orientation-specificity and other receptive field properties of cortical neurons in adult animals. If a young animal is deprived of form vision, however, significant changes in the cortical neurons are induced. Monocular deprivation leads to loss of binocularity; the cortical cells now overwhelmingly prefer the open eye. Behavioral tests reveal that restriction of form vision also leads to loss of visual acuity, thereby resulting in *amblyopia*. The period of susceptibility for inducing these changes is called the *critical* period; in cats it extends between 4 weeks and 4 months of life, in monkeys between 1 week and 1 year, and in humans between 6 months and 5–6 years.

Recovery from deprivation-induced deficits is often meager. The best result occurs if the good eye is patched and the subject is forced to use the deprived eye, although this treatment may improve acuity but not binocularity. Animals or humans that are wall-eyed tend to have good visual acuity in both eyes but little binocular interaction between the eyes. Animals or humans that are cross-eyed tend to ignore the visual input from one eye; the result is similar to that of monocular form deprivation in one eye.

After monocular form deprivation, the ocular dominance columns receiving input from the open eye become much larger, but the amount of cortex serving the deprived eye is decreased. In the newborn monkey, the geniculate axons extend widely in layer 4c and then retract with age. With monocular deprivation, geniculate axons from the open eye may not retract as in the normal cortex, whereas the geniculate axons from the closed eye may retract excessively.

Further Reading

BOOKS

Hubel, D. H. 1988. *Eye, Brain and Vision*. Scientific American Library. New York: W. H. Freeman.

ARTICLES

Hubel, D. H., and T. N. Wiesel. 1963. Receptive fields of cells in striate cortex of very young, visually inexperienced kittens. *Journal of Neurophysiology* 26:994–1002.

———— 1970. The period of susceptibility to the physiological effects of unilateral eye closure in kittens. *Journal of Physiology* 206:419–436.

———— 1977. Ferrier Lecture: Functional architecture of Macaque monkey visual cortex. *Proceedings of the Royal Society of London* B198:1–59.

Hubel, D. H., T. N. Wiesel, and M. P. Stryker. 1978. Anatomical demonstration of orientation columns in Macaque monkey. *Journal of Comparative Neurology* 177:361–380.

Le Vay, S., T. N. Wiesel, and D. H. Hubel. 1980. The development of ocular dominance columns in normal and visually deprived monkeys. *Journal of Comparative Neurology* 191:1–51.

Livingstone, M. S., and D. H. Hubel. 1984. Anatomy and physiology of a color system in the primate visual cortex. *Journal of Neuroscience* 4:309–356.

Wiesel, T. N., and D. H. Hubel. 1963. Single-cell responses in striate cortex of kittens deprived of vision in one eye. *Journal of Neurophysiology* 26:1003–1017.

SEVENTEEN

The Development of the Brain

GAINING AN understanding of how the brain works seems simple compared with the challenge of learning how the nervous system develops. From a relatively small number of undifferentiated cells, all of the neurons and glial cells form. Neurons differentiate into a variety of types; and these types extend abundant processes that find their way, often over considerable distances, to make specific synaptic junctions with other neurons. In the human, virtually all of the neurons form before birth; thus, neurons must be generated in the developing brain at an average rate of at least 250,000 per minute! But even this average figure understates the case in that cell replication is exponential, and thus many more than 250,000 cells per minute are often being generated. And this number was calculated on the assumption that the human brain contains about 100 billion cells—a conservative estimate.

I will focus on two key questions concerning brain development. First, how do neurons differentiate? What tells a cell to become a certain type of neuron? Second, how do axons find their way to the correct target in the brain? And how do the specific synaptic connections form between neurons? I begin, though, by reviewing the origin of neural cells and the development of brain structures.

The Neural Plate, Tube, and Crest

The neurons and glial cells of the nervous system all derive from a group of cells on the dorsal surface of the embryo. These cells, about 125,000 in number, form a flat sheet known as the *neural plate*. They come from ectodermal cells; that is, cells of the embryo's outer layer. In humans, the neural plate forms about 3 weeks after conception.

During the third and fourth week of human development, the neural plate invaginates from anterior to posterior and forms a groove that eventually closes into a long tube, the *neural tube*.

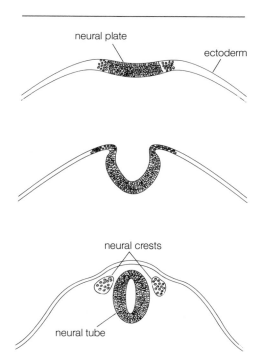

neural plate

ectoderm

neural crests

neural tube

17.1 Formation of the neural tube and crest cells from the neural plate. During the third and fourth week of development in the human embryo, the neural plate invaginates, forming the neural tube. The neural crest cells, which come to lie on either side of the neural tube, derive from cells that initially lie laterally along the neural plate.

The central nervous system develops from the neural tube; the anterior end becomes the brain; the posterior part develops into the spinal cord. Three swellings become apparent along the anterior part of the neural tube shortly after it forms. These structures are precursors to the brain's three major regions; forebrain, midbrain, and hindbrain.

Some cells that initially lie along each side of the neural plate remain separate from the neural tube and form the *neural crests*. Much of the peripheral nervous system develops from neural crest cells. These include sensory neurons of the dorsal root ganglia, cells of the sympathetic and parasympathetic ganglia, Schwann cells, and numerous neurons whose axons contribute to several cranial nerves and even nonneuronal tissues. Figure 17.1 shows in a schematic way the formation of the neural tube and crest from the neural plate.

What causes the formation of the neural plate? Can any ectodermal cells become neural plate cells? Experiments by Hans Spemann in Germany in the early part of this century provided important answers and insights to these questions. Spemann and his colleagues showed that cells of the middle layer of the salamander embryo (the mesoderm) induce the formation of the neural plate cells. They proved this by transplanting mesoderm from one part of the embryo to another. When mesoderm underlies ectoderm, it induces the ectodermal cells to become neural plate cells, even if the ectodermal cells would normally have become ordinary skin cells. Conversely, if mesodermal cells are prevented from migrating beneath ectodermal cells in the early stages of embryonic development, no neural plate develops and the embryo fails to form a nervous system.

What is the nature of the interaction between mesodermal and ectodermal cells that causes the ectodermal cells to form neural plate cells? Much evidence indicates that a chemical signal from mesodermal cells mediates this induction. For example, pieces of embryonic ectoderm will become neural tissue when cultured in the presence of embryonic mesoderm but not in the absence of mesodermal cells. Furthermore, by placing porous filters between ectodermal and mesodermal layers in the intact embryo, one can control neural plate induction. One can also define the size of the signal molecules that cause induction if one uses filters of different bore diameters. These experiments suggest that the inducers are small peptides with a molecular weight that is less than 1,000. As we shall see, chemical signaling between cells is crucial not only for the induction of the neural plate but also for the differentiation of specific neuronal types and the formation of synaptic contacts.

Another pertinent question concerns the ability of neural plate cells to form the various parts of the nervous system. Initially, as one might expect, the cells are not highly specified, but specification progresses rapidly. That is, if a small piece of the neural plate is removed early enough in development, neighboring cells will proliferate and fill in the defect, and neural development proceeds normally. A short time later this does not happen; a permanent defect occurs in the nervous system when a small piece of the neural plate is removed. In other words, the fate of the cells quickly becomes sealed; the cells can differentiate into certain kinds of neurons but not all kinds. The factors that give rise to this specification are not known.

The Proliferation and Migration of Neural Cells

I noted above that nerve cells proliferate abundantly and rapidly during neural development. Proliferation begins upon closure of the neural tube and takes place mainly along the inner surface of the neural tube—an area called the *germinal zone*. Initially, the neural tube consists of a single layer of cells, but it rapidly becomes many cells thick, growing from the inside out. Along the inner germinal zone, dividing cells undergo characteristic movements, as shown in Figure 17.2. The dividing cell is bipolar in shape and initially its processes reach both surfaces of the neural tube. DNA is synthesized while the nucleus of the cell is deep in the tube,

17.2   The proliferation of cells in the germinal zone of the neural tube. The nucleus of the dividing cell migrates up and down during this process. DNA synthesis occurs when the nucleus is deep in the zone. The nucleus migrates to the inner surface of the neural tube and the cell divides. The nuclei of the two daughter cells then migrate away from the inner margin of the germinal zone and the process repeats itself.

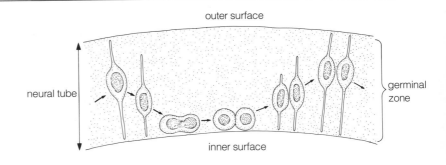

away from the inner surface. The nucleus then migrates to the tube's inner surface, the distal process of the cell withdraws, and the cell rounds up and divides. After cell division the two daughter cells extend new processes distally, the nuclei migrate deeper into the tube, and the process repeats itself.

After the cells divide several times, most of them appear to lose the ability to replicate their DNA. At this point, they migrate away from the germinal zone and form a distinct cellular layer distally. The cells in this intermediate layer are mainly young neurons that never will divide again. Furthermore, most of them are now specified. That is, where they will reside in the brain is now established—and sometimes even their synaptic connections have been determined. Some cells that migrate from the germinal zone in certain brain regions do retain the ability to divide, and they form important brain structures, including the basal ganglia and other deep nuclei of the forebrain. Certain cerebellar cells also proliferate after they migrate away from the germinal zone, and neural crest cells divide mainly after they have migrated to their final destination. In a few special situations, germinal cells remain in the adult brain and continue to divide. An example is the retina of some fishes, which adds neurons during the animal's life. But this is quite exceptional; in most species, especially mammals, no new neurons are produced in the adult.

From the intermediate zone the young neurons undergo a further migration to take up their final positions. How this happens varies somewhat from brain region to brain region. In many parts of the brain, the cell bodies of specialized glial cells lie in the germinal zone but their distal processes extend to the brain surface (see Figure 17.3). These cells, called *radial glial cells*, are present early in development and persist until after neuronal migration is completed. It is believed that the neurons migrate along the glial cells, and substantial evidence in favor of this proposal has been presented. In electron micrographs, the migrating neurons are inevitably found associated with these glial cells, and in tissue culture neurons migrate along radial glial cells. In addition, Richard Sidman and Pasko Rakič of the Harvard Medical School have shown that in mice with a genetic defect affecting the cerebellum, the radial glial cells degenerate early, and consequently certain cerebellar cells undergo disruptions in their migration.

The migration of cells in the developing brain differs among

17.3 The arrangement of radial glial cells in the developing cerebral hemispheres of a monkey *(left)* and the migration of a neuron along a radial glial process *(right)*. The radial glial cells extend across the neural tube and migrating neurons are observed typically in close association with radial glial processes.

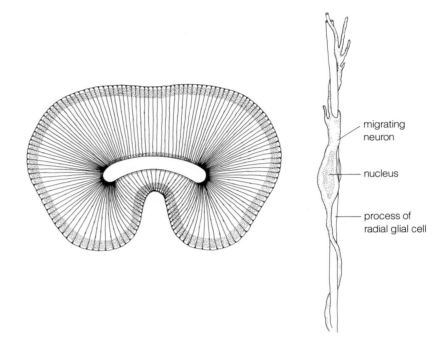

migrating neuron

nucleus

process of radial glial cell

brain structures. In the cortex the first neurons to complete cell division and to migrate form the deepest cortical cell layer (layer 6), whereas cells that proliferate and migrate later form the more superficial cortical cell layers (layers 2–5). In the retina the opposite is observed. The first cells generated (the ganglion cells) migrate to the opposite side of the retina, while cells that are generated later form the retina's middle layers.

Figure 17.4 shows schematically the development of the cortex. Initially the neural tube giving rise to the cortex is a few cells thick and consists mainly of a germinal zone (A). As the cells proliferate, some stop dividing and migrate to the intermediate zone (B). From there they migrate radially to the more superficial layers (C). As noted, however, the first cells to migrate and differentiate form the deepest cortical layer, and cells migrating later form the more superficial layers. The younger cells therefore migrate past the older cells to assume their final positions (D). Given this sequence of events, it is evident that the prominent feature of cortical struc-

17.4 The formation of the cerebral cortex. **(A)** Initially the neural tube is just a few cells thick and consists mainly of the germinal zone *(GZ)*. **(B)** As cells proliferate, some stop dividing and move to the intermediate zone *(IZ)*. **(C)** The intermediate zone cells then migrate distally to take up their final positions; the first cells to migrate form the deepest cortical layers (layer 6). **(D)** Cells migrating later form the more superficial layers (here, layers 4 and 5).

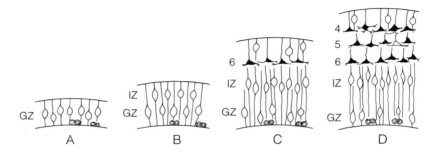

ture (and that of other brain regions)—namely, a radial or columnar organization superimposed on horizontally arranged cellular layers—can be explained by developmental mechanisms.

The Differentiation of Neurons

The specific types of neurons found in a brain region are typically generated only at one time during development and virtually simultaneously. In the retina, for example, the ganglion cells differentiate first, followed by horizontal, amacrine, and photoreceptor cells. The bipolar cells are the last to differentiate. When does commitment to a specific type neuron occur and what mechanisms carry out the specification? Two theories have been proposed, and there is evidence for both. The first theory states that cells inherit developmental directives and are not strongly influenced by their environment. Thus, a cell's lineage determines its fate. The second theory states that cells differentiate on the basis of their position in the embryo and the clues they receive from the environment. In this scenario, signals from nearby cells induce changes that determine a cell's fate. The lineage theory says that precursor stem cells and their daughters are already committed to specific fates; whereas the induction theory predicts that precursor cells, even after final cell division, can become one of a number of cell types, depending on position-dependent signals that they encounter during or after migration.

The best evidence for the lineage theory comes from invertebrate studies, particularly from small and simpler animals such as the nematode worm, *Caenorhabditis elegans*. But even in inverte-

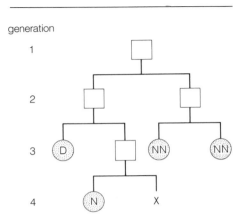

generation

17.5   A cell lineage scheme for two neurons (dopaminergic and nondopaminergic) in *C. elegans*. The precursor cell gives rise to two daughter cells. When the daughter cell on the left divides, one of its offspring differentiates into a dopaminergic neuron *(D)*; the other divides once again, producing a nondopaminergic neuron *(N)* and another cell that dies *(X)*. The second daughter of the precursor cell *(right)* divides to yield two nonneural cells *(NN)*. Differentiated cells are indicated by shading.

brates, good evidence for positional signals driving cell differentiation has been provided, and the developing eye of the fruitfly presents an elegant example. It is widely held that development in more complex animals, including all vertebrates, is generally indeterminant and hence induction of cell type is the rule. In this view, development is quite flexible.

Cell Lineage in the Nematode

The nematode worm, *C. elegans*, has 302 neurons and about 7,000 synapses. (This animal is so small that it has been possible to count the number of neurons and synapses in serial sections of the whole worm examined by light and electron microscopy.) Sydney Brenner and his colleagues in Cambridge, England, showed that it is also possible to observe cell divisions in the animal and to determine the fate of each of the daughter cells after fertilization because *C. elegans* contains so few cells and because the embryo is translucent. The cells that give rise to the nerve cells are found along the embryo in two rows, one on each side of the body. During development, these cells divide in a highly stereotyped way and therefore give rise to various types of neurons and other cell types. Neurons that have the same function in different regions of the animal are almost always descendants of different precursor cells, but they have a similar lineage.

Figure 17.5 shows a cell lineage scheme for two neurons, one that is dopaminergic and one that is not. The precursor cell, the topmost square, gives rise to two daughter cells. When one of the daughter cells divides (left), it produces one offspring that differentiates into a dopaminergic neuron (*D*) and another that divides once more. In this fourth generation, one of the daughters becomes a nondopaminergic neuron (*N*) and the other daughter cell invariably dies (*X*). The second daughter of the precursor cell (right) divides to yield two nonneural cells (*NN*). Following the final cell divisions, then, two of the cells differentiate into distinct types of neurons, two become nonneural cells, and one cell is programmed to die. Cell death in the developing nervous system is a phenomenon seen commonly in all species and will be discussed further later on.

One genetic mutant of *C. elegans* is known to alter this lineage scheme. In the mutant, the cell that normally dies does not; rather, it becomes a dopaminergic neuron. The mutants, therefore, have extra neurons of this type. Ordinarily *C. elegans* has 8 dopaminergic neurons; mutant animals have twice this number of dopaminergic cells.

A test of the invariability of preprogrammed development of the type displayed by *C. elegans* is to destroy a cell during development (by a laser beam, for example) and to observe the results. Usually deletion of a cell has little effect on the development of nearby cells—the animal now has a permanent deficit. Sometimes, though, when one cell is deleted another cell assumes the role of the deleted cell or nearby cells proliferate to try to replace the missing cells. Thus, even in *C. elegans,* some cell-cell interactions can occur that will affect cell fate.

Cell Interactions in the Fruitfly Eye

The developing eye of the fruitfly is an advantageous system in which to study how cell interactions govern cell fate. The eye of the fruitfly, *Drosophila melanogaster,* like the eye of the horseshoe crab (see Chapter 9), is a compound eye, made up of hundreds of identical units, the ommatidia. Each ommatidium in the fruitfly consists of eight distinguishable photoreceptor cells that develop in a strict sequence. If this developmental sequence is disturbed, the ommatidium does not form properly; that is, the earlier cells to develop induce the differentiation of the later cells. Photoreceptor cell 8, called R8, is the first cell to differentiate, followed by a pair of cells (R2 and R6). Two further pairs of cells (R3 and R4, and R1 and R5) develop in sequence, and the last cell to differentiate is R7. This sequence is shown schematically in Figure 17.6.

17.6   The sequence of photoreceptor cell differentiation in an ommatidium of the fruitfly eye. R8 is the first cell to differentiate, followed by cells R2 and R6. R3 and R4 differentiate next and then R5 and R1. R7 is the last cell to differentiate. The spatial arrangement of the cells in an ommatidium, approximated in this drawing, is consistent from ommatidium to ommatidium. The *sevenless* mutation blocks the differentiation of R7. Differentiating cells are shaded.

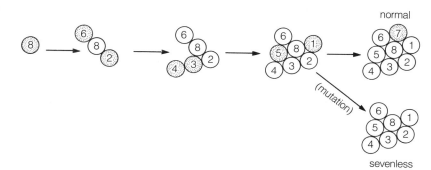

Several mutations of the fruitfly affect eye development, many of which have provided clues as to the nature of the cell-cell interactions and signals that give rise to this developmental sequence. In *sevenless* mutants, discovered in Seymour Benzer's laboratory at the California Institute of Technology, every ommatidium lacks the R7 photoreceptor. Early ommatidial development is identical in these animals to normal, wild-type flies, but the cell in the ommatidium that normally differentiates into R7 becomes instead a nonneural cone cell, a cell involved in forming the lens of the ommatidium. That the mutation affects specifically the R7 precursor cell was shown in flies that were genetic mosaics; that is, flies produced in the laboratory that contain both normal and mutant cells. Ommatidia that had normal R7 precursor cells, but other ommatidial cells that were mutant, formed normal R7 cells. Yet, R7 cells never formed in ommatidia that had mutant R7 cells even when all the other cells in the structure were normal, which means that normal photoreceptor cells in an ommatidium cannot rescue a mutant cell.

The gene that, if it is defective, gives rise to the *sevenless* mutation has been isolated. It codes for a protein similar to a number of membrane receptors. It is a transmembrane protein with a large extracellular component and an associated intracellular kinase. The receptor-like structure of the protein has led to the proposal that it participates in transducing extracellular signals. The binding of a molecule to its extracellular parts leads to activation of the kinase. Subsequent intracellular changes induced by the activated kinase lead to differentiation of the cell.

Support for the view that differentiation of R7 requires a signal from another photoreceptor cell has come from the analysis of another mutant, called *bride of sevenless* or *boss*. This mutation also results in an ommatidium lacking the R7 photoreceptor cells, but in this case the mutation affects the R8 cell. The guess is that the *boss* gene product is the molecule that activates the receptor protein affected in the sevenless mutant. If a genetic mosaic fly is made of a mix of normal and *boss* mutant cells, R7 cells form in those ommatidia that possess normal R8 cells but not in ommatidia that contain mutant R8 cells; the *boss* gene product is necessary for an ommatidium to form an R7 cell. There is evidence that the *boss* gene product, the presumed signal molecule in the R8 cells, is a membrane protein. If so, direct cell-cell contact between R8 and the precursor cell for R7 would be required for induction.

A host of interesting questions arise from these observations, but they cannot be answered at the present time. For example, since R8 is also crucial for the development of the other ommatidial photoreceptor cells, are similar signaling mechanisms likewise at work? Why does R7 differentiate into a cone cell if it does not receive the appropriate signal from R8? Is this the result of another signal to the cell or is this fate genetically determined? And how, finally, does activation of the *sevenless* protein kinase lead to the differentiation of the precursor cell into a photoreceptor cell? Even though the experiments are as yet incomplete, they do emphasize the power of a genetic approach for the analysis of development.

Axon Projection and Synapse Formation

One of the most challenging questions to be answered about brain development is how axons find their way to appropriate targets. As was the case with cell differentiation, chemical signals have long been implicated. A chemoaffinity hypothesis was first suggested by John Newport Langley at Cambridge University as long ago as 1895, but the theory was formulated more forcefully by Roger Sperry of the University of Chicago in the early 1940s. This hypothesis states that during development neurons become chemically specified as a result of factors such as their position after migration. They have chemical markers, in other words, that enable them to recognize complementary markers on appropriate target cells. Stated in more modern terms, cell surface molecules mediate recognition between neurons.

Much of Sperry's evidence for the chemoaffinity hypothesis came from his study of the projections of retinal ganglion cell axons to the midbrain or tectum in cold-blooded vertebrates, such as fish and frogs. Adult fish and frogs have a precise map of the retina on the tectum that is easy to determine. A spot of light projected onto the retina will stimulate a small number of ganglion cells whose axons project to a defined region of the tectum. An extracellular electrode readily records the incoming activity, telling the investigator which part of the retina activates which part of the tectum. Information from the left eye projects to the right

17.7 Projections from the retina to the tectum in the goldfish. The drawing on the left shows that the right retina projects to the left tectum and vice versa; on the right are maps of the right visual field and left tectum. Corresponding areas between the retina and tectum are indicated by numbers, showing an arrangement known as a topographic projection.

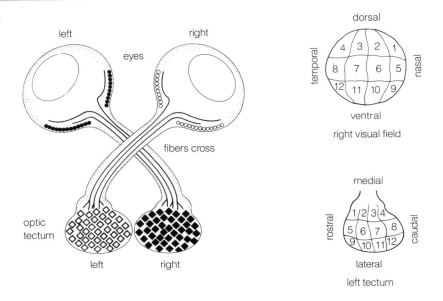

tectum, and vice versa. Figure 17.7 shows how the right retina projects on the left tectum in a goldfish; corresponding areas between the retina and tectum are indicated by numbers.

An early and influential experiment, carried out by Sperry in the late 1930s, was to cut the optic nerve in newts (an amphibian) and frogs, rotate the eye by 180°, and then fix it back in place. Severed optic nerves regenerate in these animals (and in fish), and investigators had previously shown that after optic nerve regeneration, vision is restored. The astounding result was that after the optic nerve had regenerated, Sperry's animals could see but their visual world was inverted and shifted from left to right! They consistently misdirected attacks by about 180°; if a fly was up and to the right, they attacked down and to the left (Figure 17.8). No amount of training reversed this deficit.

The conclusion drawn from these experiments was that the axons had grown back to their original locations; they could recognize the tectal cells they had been connected to originally. The logical explanation for this observation was that the axons and tectal cells had complementary markers. Subsequently, it was shown more directly, by electrophysiological experiments, that the

17.8 The effect of eye rotation on a frog's behavior. The optic nerve was severed and the eye rotated in its socket by 180° *(top)*. After regeneration of the optic nerves, the frog was unable to respond appropriately to a visual stimulus. A fly in the upper right field caused the frog to direct its attack toward the lower left field.

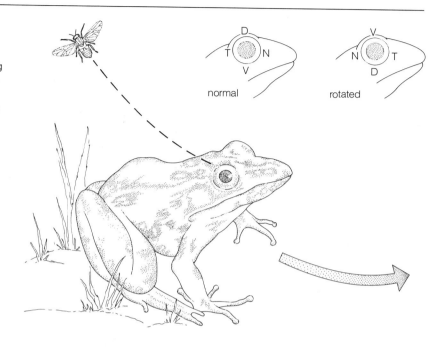

normal   rotated

regenerating axons in similar experiments innervate their original target areas.

In another series of experiments carried out in the late 1950s, Sperry and his colleagues, now at California Institute of Technology, showed that if the optic nerve in fish were cut and then part of the retina were removed, the regenerating axons would grow back to that part of the tectum from which they had originated. They did so even when they had to cross large stretches of the tectum that had been innervated by those parts of the retina that had been removed. It became evident that the regrowing axons initially bypassed populations of neurons that had no retinal innervation to reach their previously assigned targets.

When and how does this specification occur? If an eye is rotated in a young embryo, the subsequent projection of the ganglion cell axons to the tectum does not cause an inverted map; somewhat later the maps are inverted, however. In this regard it is interesting that the dorsal-ventral axis is specified slightly before the temporal-nasal axis, so at one stage of development inversion of the eye results in inversion of the dorsal-ventral axis but not the temporal-nasal axis. Ganglion cells appear to be specified just after they stop dividing and have migrated to the retina's inner surface.

It is also possible in developing amphibians to excise and rotate the tectum, and in these experiments, too, vision is eventually restored. If the tectum is inverted early in development, before the ganglion cell axons arrive, vision develops normally and the maps are not inverted. If the experiment is done later, after the ganglion cell axons have innervated the tectum, the resulting visual behavior is inverted and the retino-tectal maps are abnormal. These experiments appear to suggest that positional clues for the tectal cells are provided by the innervating ganglion cell axons, but an alternate view is that surrounding brain structures supply the positional information. Once specified, however, the ganglion cell axons and tectal cells readily recognize each other. They do so even if pieces of tectum are excised and moved to different and abnormal tectal locations.

Cell Adhesion Experiments and Molecules

Direct evidence for intercellular recognition between retinal and tectal cells has been obtained from experimentals on dissociated retinal ganglion and tectal cells maintained in culture (see Figure 17.9). Cells dissociated from the retina's ventral part adhere to dorsal tectal cells, whereas dorsal retinal cells prefer ventral tectal cells. A suspension of retinal cells from the dorsal or ventral half of the retina is labeled with a radioactive isotope and is then placed in a culture dish containing a piece of dorsal or ventral tectum or onto a monolayer of cells from dorsal or ventral tectum. Following an incubation period, measurements of the radioactivity bound to the tectal cells give an indication of the adhesivity between the cells. The result: specificity of adhesion between retinal and tectal cells and between specific cell populations.

The molecules that aid in this process of recognition and adhesion of retinal to tectal cells are not known. Several cell adhesion molecules have been tentatively identified, the best-known of which is a molecule called N-cam (neural cell adhesion molecule) found originally in the chicken retina by Gerald Edelman and his colleagues at Rockefeller University. Subsequently N-cam was identified at various times of development on all nerve cells, and

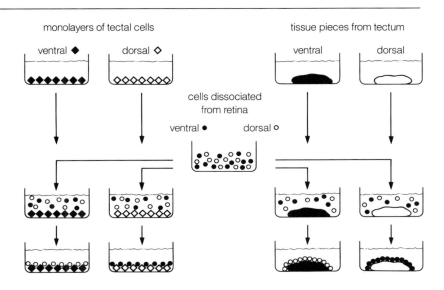

17.9 Dissociated retinal cells adhere preferentially to tectal cells maintained in culture. Furthermore, ventral retinal cells adhere preferentially to dorsal tectal cells and vice versa, as shown here. On the left, suspensions of tectal cells from the dorsal and ventral retinas were placed in culture dishes containing a monolayer of cells from either the dorsal or ventral tectum. On the right, suspensions of cells were placed in culture dishes containing a piece of dorsal or ventral tectum. In both kinds of experiments, specificity of adhesion between retinal and tectal cells is observed.

anti-N-cam antibodies injected into the developing eye of the chick were shown to alter retinal axon outgrowth.

N-cam is a protein of about 120,000 molecular weight and it has a high content of sialic acid, a carbohydrate. Because the sialic acid part of the molecule changes significantly in amount during development, it would appear that this molecule is important in neural development and in neuronal recognition. But the precise role of N-cam is still not understood.

The Chemoaffinity Theory Revisited

In the 1960s, 1970s, and 1980s, investigators carried out numerous experiments on the retino-tectal system that led to modifications in thinking about the chemoaffinity hypothesis. Whereas many interpreted the early experiments by Sperry and others to mean that connections in the brain initially form and are maintained in a rigid and inflexible way, the new experiments demonstrated much more flexibility and plasticity in development and regeneration. Indeed, some experiments indicated that even in the adult brain synaptic connections are not strictly fixed.

The first experiments of this sort were similar to Sperry's earlier

17.10 Reorganization of retino-tectal maps over time. When a portion of the retina is ablated following optic nerve section, the area of the tectum that received input from the excised retinal region is unresponsive to light stimuli for 1–3 months. By 7–9 months later, the entire tectum responds to light stimuli, and mapping of the retino-tectal projections shows that the visual field map has expanded to fill the tectum.

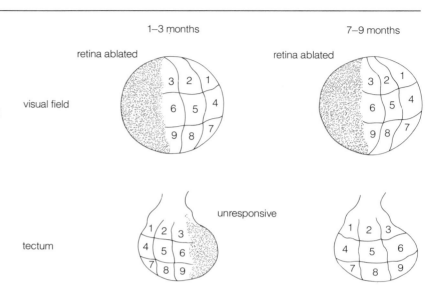

ones, in which half the retina or tectum was removed following optic nerve section. While the regenerating axons initially projected to the appropriate tectal locations as a consequence of their original connections (as Sperry had found), rearrangements took place over time: the maps expanded or compressed depending on whether part of the retina or tectum was excised. Typical results, presented in Figure 17.10, indicated that retinal projections adjust eventually to fit the available space. In addition, histological studies of regenerating axons growing back to the tectum indicated that their terminal branches covered a larger area than they did initially or after regeneration was completed. The axons were not reinnervating the tectum in a strictly fixed or stereotyped fashion.

Studies of normal retino-tectal development also indicated plasticity of connections. During the development and lifetime of many fish, the retina and tectum add more cells. In goldfish, for example, about 50 new ganglion cells are added to the retina each day of the animal's life. The retina grows like a tree trunk, adding cells in a ringlike fashion, but the tectum adds cells on just one side. Projection maps determined in young and old animals are similar, however, so connections in the tectum must continually rearrange themselves over the life of the animal to keep the topographical relations constant.

The conclusions drawn from these experiments are that the retinal axons are not strictly wired to specific tectal cells. Rather, the axons have a strong affinity for tectal cells in certain regions and they have positional sense. They know where they are relative to the other ganglion cell axons, and they form and re-form connections in the appropriate tectal areas to yield orderly maps. Furthermore, the axonal fields of the ganglion cells, during both development and regeneration, are flexible. The terminals fill the available tectal space and appear to compete for space.

Observations made in many parts of the developing nervous system are in agreement with these general ideas. An example, described in the Chapter 16, is the innervation of the primary visual cortex by lateral geniculate neurons in mammals. These axons project to the cortex in a topographic fashion in the developing brain, but initially they innervate a much larger area in layer 4c than they do in the adult; for this reason a columnar organization cannot be demonstrated anatomically in layer 4c in the newborn cat or monkey cortex. The axon terminals in the young animal then compete for cortical space. If equivalent input comes from the two eyes, the axon terminals retract equally; if one eye is visually deprived, it does not compete successfully and ends up with much less cortical space (see Figures 16.16 and 16.17).

Axonal Guidance

The experiments described so far indicate that axons can recognize potential target cells in specific brain regions. The next question is how the axons are guided to the appropriate brain region containing the target cells. Little of substance is known about the mechanisms underlying this process, yet many suggestions have been made. It may well be that several mechanisms come into play in axonal path finding. Sperry suggested that diffusible chemical gradients could guide axons to their target areas and that synaptic contacts are then made depending on cell-cell recognition factors. He pointed out that two or more gradients would give growing axons positional information that would allow them to make proper directional decisions as they advance toward their eventual targets.

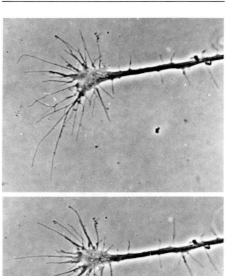

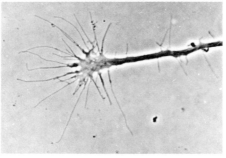

17.11 Light micrographs of a growth cone at the end of an axon in culture. The three photographs, taken at 20-second intervals, show that the fine processes of the growth cone change shape very rapidly as they probe the surface of the culture dish.

A complementary view has come from experiments on grasshoppers and other invertebrates; this research has shed some light on how axons find their way over long distances in the developing nervous system. Axons growing into the legs of these animals are attracted by specific cells along the way. These neurons, termed *guidepost neurons,* do not stop the growing axons; rather, the axons grow to them and make contacts with them, and then they continue on their way toward the next guidepost cell. If the guidepost cells are destroyed, axonal growth may be disrupted. An obvious suggestion is that the guidepost neurons secrete an attractant substance toward which the axons grow.

Another idea to help explain how axons find their way over long distances is that some neurons serve as "pioneers." Early in development, distances between structures in the brain are much shorter than they are later on or in the adult. It may be that a few connections between brain structures are formed early (by *pioneer neurons*) and that as the brain grows the axons simply stretch. Axons from neurons that develop later then follow along the pioneer axon pathways, or perhaps along the pioneer axons themselves.

Axons grow via growth cones, prominent expansions of the tips of axons discovered by Ramón y Cajal in 1890. Growth cones are visualized very well in tissue culture, as Figure 17.11 shows. While the axon is growing in culture, the growth cone is in constant motion, extending and retracting fine processes that explore the surrounding area. As the growth cone moves ahead, new membrane is added to it, thereby resulting in axonal growth.

Investigators have shown that growth cones in culture respond to different substrates and that axonal growth can be dictated by the substrates. Adhesiveness of the substrate is one significant variable. Specific recognition molecules may also be important in encouraging and directing axonal growth. Thus, growth cones and axonal growth respond to a variety of clues of both a neuronal and nonneuronal nature. For instance, the texture of the substrate on which the axons are located may be as important for growth as are the specific molecules in the substrate. Also, the growth cones may sense the presence of diffusible substances in the environment and move in accord with the concentrations of these substances. But growing axons respond to a variety of stimuli, including electrical fields, so no one mechanism provides a completely satisfactory explanation for axonal growth.

Neuronal and Synapse Maturation

Several themes with regard to neuronal and synapse maturation have been noted already. During development, axons typically have larger terminal fields than they do in the mature nervous system. During maturation of the nervous system, these fields are typically rearranged, restricted, and stabilized. Furthermore, many neurons die during development. Initially, neurons are overproduced in many parts of the brain, and during normal development these cells are eliminated. We noted earlier that cell death is programmed in *C. elegans*. In vertebrates, the death of neurons during development is not predetermined in the same way but appears to involve competition for synaptic sites; "winners" form synapses and "losers" that do not, die. Removal of input to a neural center during development often causes an excessive loss of neurons in that nucleus. Yet, if a structure receives extra innervation, more neurons may survive in it than is ordinarily the case. The extent of cell death varies in the different brain structures. Some regions, including the cerebral cortex, appear to undergo only a modest amount of neuronal degeneration during development. In other structures (including the spinal cord, brain stem nuclei, and autonomic ganglia) 30–75 percent of the neurons originally present may die during development.

Cell death and terminal field retraction are not the only important features of the developing and maturing nervous system; another is the rearrangement of synaptic connections. Neurons initially establish qualitatively appropriate connections during development, but then during final development and maturation the connections are rearranged to provide the precise quantitative relations found in the adult brain. Well-studied examples include the early innervation of muscles and autonomic ganglia in mammals. At birth, axons typically innervate several muscle fibers or autonomic ganglion neurons. Over the first few weeks of life, the innervation pattern in these structures alters such that in the mature animal one axon innervates one muscle fiber or autonomic neuron. Hence, axonal branches become eliminated. The surviving axons tend to elaborate more terminal branches, however, which

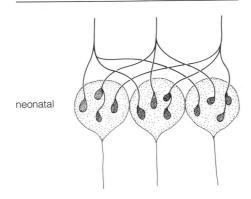

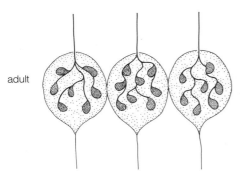

17.12 Scheme showing synapse rearrangement and maturation on autonomic ganglion cells. In the newborn animal, one axon innervates several ganglion cells, but an axon makes only one or a few synapses on each ganglion cell *(top)*. In the adult, only one axon innervates each ganglion cell, but each axon makes many contacts with that ganglion cell.

means that the total innervation of a muscle fiber or autonomic neuron often increases during maturation while the innervation becomes more specific. A synaptic rearrangement during maturation is displayed schematically in Figure 17.12.

Trophic Substances and Nerve Growth Factor

What mechanisms might explain such phenomena as the death of cells, the retraction of processes, and the rearrangement of synapses observed during nervous system development and maturation? All suggest that an exchange of information takes place between neurons and their target cells and that this exchange regulates neuronal shape, growth connectivity, and even cell survival. In Chapter 5, a specific example of this kind of interaction was presented: denervation of muscle fibers leads to the synthesis of new acetylcholine channels in the muscle cell. These channels become inserted over the muscle membrane and the muscle then becomes sensitive to acetylcholine over its entire surface. Reinnervation leads to less channel synthesis and a restriction of the channels to the neuromuscular junction region. This pattern has led to the conclusion that trophic substances are released by neurons and that they regulate many aspects of neuronal structure and function; it appears that trophic substances are released not only from axonal terminals, which regulate the target cell, but also from target cells, which regulate the innervating neurons.

Several neuroactive substances released at synapses may act as trophic substances. For example, many of the neuromodulatory agents mentioned in Chapter 6 cause the outgrowth or retraction of processes in cultured neurons. Dopamine, for instance, causes a 30–40 percent decrease of process length in cultured retinal horizontal cells, but the decrease is reversed when the dopamine is removed from the culture medium. This action of dopamine is mediated via activation of phospholipase C, formation of diacylglycerol, and stimulation of protein kinase C (see Figure 6.6).

In addition to neuroactive substances released at synapses, it is believed that other trophic factors are crucial for the regulation of neuronal form and connectivity. The best-characterized of these substances is nerve growth factor (NGF), discovered in the early 1950s by Rita Levi-Montalcini, a young postdoctoral fellow from Italy working in Viktor Hamburger's laboratory at Washington

University in St. Louis. These investigators were studying why an excessive number of neurons die in the spinal cord of the chick embryo following the removal of a nearby limb bud. Their studies indicated that, although some cell death was normal during development, an excessive number of cells died when a limb bud was destroyed. They proposed that the target cells in the limb send signals to the spinal cord neurons that allow the neurons to survive. That signal is limited, they reasoned, and therefore some of the neurons will routinely die. Following ablation of the limb, though, much less of the signal is available and excessive cell death occurs.

What was the nature of the signal? Following up on an earlier unexplained observation that implantation of pieces of a specific mouse tumor at the site of the ablated limb decreases neuronal cell death in the chick embryo, Levi-Montalcini and Hamburger demonstrated that a soluble factor released from the tumor stimulated the survival and growth of neurons in the spinal cord and nearby sympathetic ganglia. In the mid 1950s, this factor, termed nerve growth factor, was isolated by Stanley Cohen, a colleague at Washington University, and found to be a protein of 130,000 molecular weight, a relatively large molecule consisting of 5 subunits. In 1986, Levi-Montalcini and Cohen received the Nobel Prize for their studies of nerve growth factor.

Nerve growth factor, or NGF, does not act on all neurons. In the chick, it acts mainly on the dorsal root sensory neurons and on sympathetic ganglion neurons. During normal development, about one-third of these neurons die, but when NGF is supplied in sufficient amounts, many of the cells are rescued. Conversely, when little NGF is available, or when an antibody against NGF is chronically administered to an animal, virtually all of the dorsal root sensory neurons and sympathetic ganglion cells die.

In addition to ensuring the survival of neurons, NGF also promotes the growth of processes and the formation of synapses by sensory and sympathetic neurons. Figure 17.13 shows the effects of NGF on the growth of dendrites of sympathetic ganglion cells. Newborn animals (rats) were given NGF daily for 2 weeks, and their neurons were injected with horseradish peroxidase so they could be examined histologically. The dendritic arbors of the cells from the treated animals were larger and more complex than those

17.13 Effects of NGF on dendritic growth of sympathetic ganglion cells of the newborn rat. The presence of NGF increased significantly the dendritic branches of the ganglion cells.

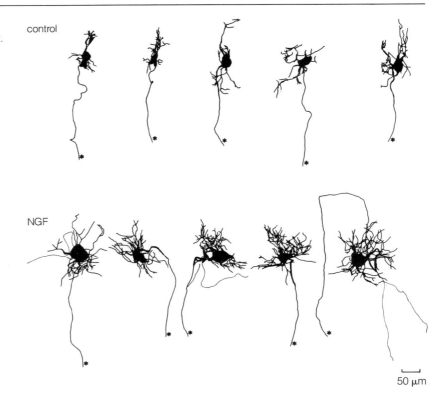

of the control animals. The size of the dendritic trees in newborn animals after 2 weeks of NGF treatment was about the same as in an untreated animal 3 months of age.

Nerve growth factor also stimulates the growth of axons. Both in culture and in an intact animal, axons will turn toward a source of NGF. For this reason NGF seems capable of guiding axons to target cells. An experiment of this sort is shown in Figure 17.14. In response to a source of NGF (the growth factor diffused from a micropipette that was gradually moved around in a culture dish), the distal tip of an axon of a dorsal root ganglion cell was induced to elongate and turn in the opposite direction from which it had been originally growing. The significance of these experiments has been questioned, however, because the concentration of NGF required to induce the effect is very high.

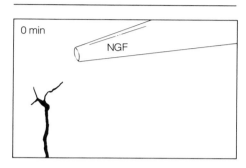

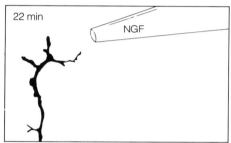

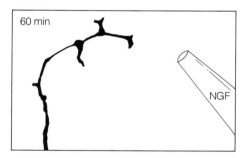

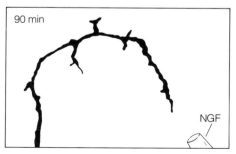

17.14 Effect of NGF on the growth of an axon of a dorsal root sensory cell in culture.

Many questions remain about NGF and its mode of action. Furthermore, other trophic factors must exist that affect neurons that do not respond to NGF. At present, though, little is known about them.

Summary

The nervous system's neurons and glial cells all derive from neural plate cells, a group of cells found on the embryo's dorsal surface. Most neural plate cells invaginate during early development to form the neural tube, from which arises the central nervous system. Some neural plate cells form the neural crest—the precursor to much of the peripheral nervous system. Underlying mesodermal cells induce ectodermal cells to become neural plate cells. This induction is mediated by a chemical signal, probably a small protein.

Nerve cells proliferate along the inner surface of the neural tube. The nucleus of a proliferating cell moves back and forth during DNA replication and cell division until it finishes dividing. The daughter cells then move to an intermediate zone in the neural tube and shortly thereafter begin migrating to their final positions. Migration often occurs along specified glial cells that extend radially across the neural tube. Some neurons divide after migration from the germinal zone and later form certain brain structures.

Two theories have been proposed to explain cell differentiation in the nervous system. One, the lineage theory, states that cells inherit development directives; the other states that cells differentiate as a consequence of environment clues—chemical signals. In simpler animals, such as the nematode *C. elegans,* cell lineage explains neural differentiation. In more complex animals, including all vertebrates, cell-cell interactions appear responsible for differentiation. In the fruitfly eye, interactions between developing photoreceptor cells are crucial for normal development. Mutants that lack certain membrane receptors or that fail to produce specific proteins show abnormal eye development. This indicates that direct cell-cell interactions are the key to normal cell differentiation.

How axons find their way in the developing brain remains a fundamental question. Chemical signals are thought to be an important part of the answer. The chemoaffinity hypothesis suggests that cells have specific chemical markers that allow cells to recognize each other during development and regeneration. In amphibians and fish, for example, retinal ganglion cell axons grow to specific regions of the tectum during development. If the optic nerve is severed in the adult, the axons grow back to their original locations. Ganglion cells become specified early in development, shortly after they stop dividing and migrate to the retina's inner surface. It has been shown that specific populations of retinal and tectal cells preferentially adhere to one another in culture experiments, but the molecules that mediate their recognition are not known. One molecule, neural cell adhesion molecule (N-cam), has been identified as a candidate.

Other experiments indicate flexibility in the formation of connections during development and regeneration. If part of the retina or tectum is removed prior to optic nerve section, connections gradually re-form during regeneration such that axon terminals fill the available space. They do so in an orderly way, however, such that axons and tectal cells always retain positional sense. We also have evidence of competition for space during normal brain development. If axons do not compete successfully, their terminal arbors are excessively restricted and they end up with fewer synaptic connections.

Various mechanisms may be involved in axonal pathfinding during development. Chemical gradients are one likely mechanism, and certain cells, guidepost neurons, may release chemical attractants to guide axonal growth. Axons may also follow after pioneer neurons, cells that form connections early in development when distances between brain structures are short. Axons grow via growth cones, the expanded tips of axons. These structures respond to the substrate on which they are growing, as well as to the presence of diffusible substances in the vicinity.

During maturation of the nervous system, the axonal fields and synaptic connections are subject to rearrangements and restrictions. In addition, many neurons are lost in most brain regions. Neuronal cell death during development appears to be brought about by competition for synaptic sites or a lack of innervation. It is believed that trophic substances are released by pre- and postsynaptic neurons that regulate many aspects of neuronal structure and function and even cell survival. The best-known agent of this sort, nerve growth factor (NGF), promotes the survival of certain neurons during development, promotes the growth of processes from specific neurons, and guides axonal growth of some neurons in culture.

Further Reading

BOOKS

Purves, D. 1988. *Body and Brain: A Trophic Theory of Neural Connections.* Cambridge, MA: Harvard University Press. (Describes the evidence that synaptic connections are flexible in both the developing and adult nervous systems.)

Purves, D., and J. W. Lichtman. 1985. *Principles of Neural Development.* Sunderland, MA: Sinauer.

ARTICLES

Cowan, W. M. 1979. The development of the brain. *Scientific American* 241(3):112–133.

Dodd, J., and T. M. Jessell. 1988. Axon guidance and the patterning of neuronal projections in vertebrates. *Science* 242:692–699.

Easter, S. S., Jr., D. Purves, P. Rakic, and N. C. Spitzer. 1985. The changing view of neural specificity. *Science* 230:507–511.

Goodman, C. S., and M. Bate. 1981. Neuronal development in the grasshopper. *Trends in Neuroscience* 4:163–169.

Hamburger, V. 1988. Ontogeny of neuroembryology. *Journal of Neuroscience* 8:3535–3540.

Hatten, M. E. 1990. Riding the glial monorail: A common mechanism for glial-guide neuronal migration in different regions of the developing mammalian brain. *Trends in Neuroscience* 13:179–184.

Horvitz, H. R., P. W. Sternberg, I. S. Greenwald, W. Fixsen, and H. M. Ellis. 1983. Mutations that affect neural cell lineages and cell fates during the development of the nematode *Caenorhabditis elegans. Cold Spring Harbor Symposium on Quantitative Biology* 68:453–463.

Levi-Montalcini, R. 1982. Developmental neurobiology and the natural history of nerve growth factor. *Annual Review of Neuroscience* 5:341–362.

Rakič, P., and R. L. Sidman. 1973. Sequence of developmental abnormalities leading to granule cell deficit in cerebellar cortex of weaver mutant mice. *Journal of Comparative Neurology* 152:103–132.

Rutishauser, U., A. Acheson, A. K. Hall, D. M. Mann, and J. Sunshine. 1988. The neural cell adhesion molecule (NCAM) as a regulator of cell-cell interactions. *Science* 240:53–57.

Rutishauser, U., M. Grumet, and G. M. Edelman. 1983. Neural cell adhesion molecule mediates initial interactions between spinal cord neurons and muscle cells in culture. *Journal of Cell Biology* 97:145–152.

Schmidt, J. T., C. M. Cicerone, and S. S. Easter. 1978. Expansion of the half retinal projection to the tectum in goldfish: An electrophysiological and anatomical study. *Journal of Comparative Neurology* 177:257–278.

Sidman, R. L., and P. Rakič. 1973. Neuronal migration with special reference to developing human brain: A review. *Brain Research* 62:1–35.

Sperry, R. W. 1963. Chemoaffinity in the orderly growth of nerve fiber patterns and connections. *Proceedings of the National Academy of Science* 50:703–710. (One of the key papers providing evidence for the chemo-affinity hypothesis.)

Tomlinson, A. 1988. Cellular interactions in the developing *Drosophila* eye. *Development* 104:183–193.

Walter, J., B. Kern-Veitz, J. Huf, B. Stolze, and F. Bonhoeffer. 1987. Recognition of position-specific properties of tectal cell membranes by retinal axons *in vitro*. *Development* 101:685–696.

EIGHTEEN

Toward an Understanding of Higher Brain Function

IN THIS FINAL chapter, I discuss aspects of higher brain function. We know little of the mechanisms underlying these phenomena, but a variety of approaches are being used to understand them. The emphasis here is on the human brain and its special attributes of perception, speech, face recognition, memory, and personality. I begin, though, by returning to sensory cortical mechanisms.

Higher Cortical Processing

So far I have discussed the primary visual cortex, area V1. What about the rest of the cortex? How is it organized? Present evidence suggests that the other primary sensory areas, serving the somatosensory, auditory, and olfactory systems, are organized similarly to the primary visual cortex. For example, neuroscientists have observed a columnar organization in the primary somatosensory and auditory cortex. In fact, the first columns found in the cortex were discovered in the somatosensory cortex by Vernon Mountcastle at Johns Hopkins University in the 1950s.

The columns in the primary somatosensory cortex encompass cells that respond to different types of stimuli. Cells in one column are responsive to light touch; cells in other columns respond to deep pressure, to the movement of hairs, or to joint position. The neurons in a column all receive their input from the same areas of the skin or limb, and they have receptive field areas that overlap. Stimulation within the receptive field either excites or inhibits the firing of the neurons. The sizes of the receptive fields vary, depending on what part of the body surface the neurons receive input from: neurons that receive input from areas of the skin that are most sensitive to sensory stimuli or that have the best tactile discrimination, such as the fingers and lips, have the smallest receptive fields.

The receptive fields of many neurons in the somatosensory cor-

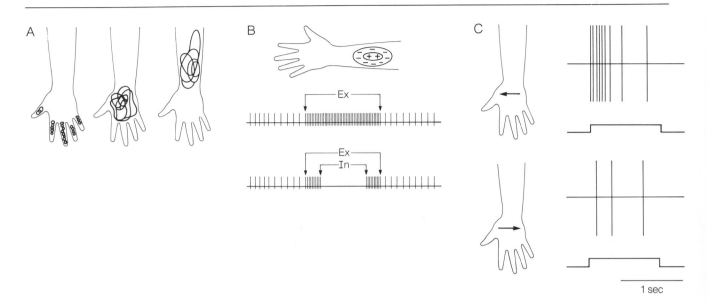

18.1 Receptive fields of tactile neurons recorded in the somatosensory cortex. **(A)** Receptive fields of neurons receiving input from the fingers, where tactile discrimination is best, are much smaller than the fields of neurons receiving input from the palm or forearm. **(B)** Many of the neurons have an antagonistic center-surround receptive field organization. Stimulation of the center of the field *(Ex)* excites the cell; surround stimulation *(In)* inhibits firing of the cell. **(C)** Some neurons are movement- and even direction-sensitive. Movement of a tactile stimulus in one direction vigorously activates the cell; movement in the other direction only weakly excites the cell.

tex are organized in ways reminiscent of the visual system. The receptive fields of tactile neurons may have an antagonistic center-surround organization. The center of the cell's receptive field is inhibited by simultaneous stimulation in the periphery of the receptive field. In the region of the somatic sensory cortex receiving input from the hands, a few neurons respond much better to moving tactile stimuli than to static touch, and some of these neurons show a preference for direction of movement; they are direction-sensitive! Figure 18.1 illustrates some aspects of receptive field size and organization for tactile neurons recorded in the somatosensory cortex.

In the primary auditory cortex, the neurons in one column respond to narrowly tuned stimuli; that is, they respond best to a specific tone or frequency. The cells in other columns respond well to multiple sound frequencies and complex sounds like a click. Another interesting feature of the auditory cortex's organization concerns the input from the two ears. In some columns, called *summation columns,* cells respond better to stimulation of both ears. In other columns, one ear is dominant and the cortical cells

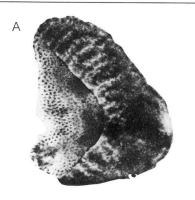

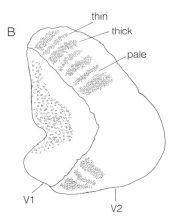

18.2   Cytochrome oxidase staining of the monkey cortex (**A**) and a partial reconstruction of the staining pattern (**B**). In area V1, a polka-dot pattern is observed. The dots are the color pegs in V1. In area V2, a stripe pattern is observed. Thick and thin dark stripes are separated by pale stripes.

respond better to stimuli reaching one ear alone than to stimuli reaching both ears.

Spreading beyond the primary projection areas are the secondary association areas of the cortex. As noted in Chapter 12, these areas are concerned with more complex processing, with integration of sensory input, and ultimately with recognition, understanding, and memory. How are these areas organized? Let us focus once again on the visual system, where we have more information, and discuss visual processing beyond area V1.

Area V2 and Higher Visual Areas

The first visual association area is area V2, the main target for the axons leaving area V1. Clues to the organization of V2 have come from anatomical studies that stained for the enzyme cytochrome oxidase. In area V1, such staining revealed the blob structure that led to the discovery of the color pegs (see Chapter 16). In area V2, tangential sections of area V2 stained for cytochrome oxidase display a pattern of dark and light staining stripes. Furthermore, the dark stripes are of two types, thick and thin (see Figure 18.2). Thus, across area V2, a regular and repeating pattern of thick and thin dark stripes is separated by pale stripes.

The obvious question raised by these observations is whether cells in the three subdivisions of V2 have different properties. This question was examined by Hubel and Livingstone, who found striking differences among the neurons in the three stripes. Cells in the thin dark stripes had no orientation-selectivity and over half of the neurons were color-coded. Most of the color-coded cells were double-opponent cells, but they possessed properties not seen in area V1. For example, the centers of their receptive fields were larger than the color-opponent cells in area V1, but they did not exhibit summation effects. Small spots elicited vigorous responses over a wide central area, yet larger spots induced smaller responses or sometimes no response at all. Thus, thin dark stripes appeared to receive input exclusively from the color pegs in area V1 and were concerned primarily with color processing.

Cells in the pale stripes were complex, orientation-selective cells. They were not direction-selective, though many (over half) were end-stopped. The cells in the thin pale stripes appeared to be concerned mainly with object shape (see Chapter 15). In the thick dark stripes, the great majority of cells were also complex cells and orientation-selective. But, in addition, most cells were direction-selective and highly binocular. They responded poorly to monocular stimulation but vigorously when both eyes were stimulated together. Further, many of the cells were disparity-tuned neurons. The cells in the thick stripes, therefore, appeared to be concerned primarily with movement and with depth.

The conclusion drawn from these findings is that further segregation of visual image processing exists in area V2. The processing subdivisions remain interdigitated in area V2, but explorations of visual areas beyond V2 indicate that the segregation becomes even more pronounced; in other words, that separate regions of the cortex become specialized for one or another kind of processing. Area V4, for example, appears to be concerned mainly with color analysis, as was first shown by Semir Zeki of University College London; it receives substantial input from the thin stripes of V2, and most of the neurons in V4 are color-selective cells. Area V5 or MT (middle temporal) seems to be specialized for the analysis of movement and perhaps depth. For example, the responses of neurons in V5 correlate closely with the ability of monkeys to discriminate direction of movement. V5 receives input from the thick stripes of V2 as well as from layer 4b in area V1, which contains many direction-selective and disparity-tuned cells.

From examining areas V4 and V5 we know that each has a representation of the retina; the retina is topographically mapped onto these areas. Indeed, many areas, perhaps as many as 15, have now been found beyond area V2 that have either complete or partial representations of the retina, suggesting that many distinct areas analyze one or another aspect of the visual image. A diagram of the various visual areas is shown in Figure 18.3.

There are consequences to this way of analyzing a visual image. For example, because form and color are processed separately, it is the case for us that form is difficult to see if an image is made from two colors that have the same brightness. To see form well, we need light-dark contrast; color contrast by itself will not do.

Another consequence of this arrangement—and one which also provides additional evidence for it—is that certain people can have brain lesions that affect their ability to see color but no other aspect of the visual image. They see shapes perfectly well, have high visual acuity, respond to moving stimuli, and have good depth perception, yet they see no colors. It is likely that they have a

18.3 A reconstruction of the visual areas in the monkey. The cortical surface of the right hemisphere is unfolded and flattened for the purposes of the figure. Fourteen areas are indicated, but more may be present. Area abbreviations refer to the location on a lobe of the area: *AIT*, anterior inferotemporal; *LIP*, lateral intraparietal; *MST*, medial superior temporal; *MT*, middle temporal (or V5); *PIT*, posterior inferotemporal; *PO*, parietal-occipital; *VIP*, ventral intraparietal; *VP*, ventral posterior.

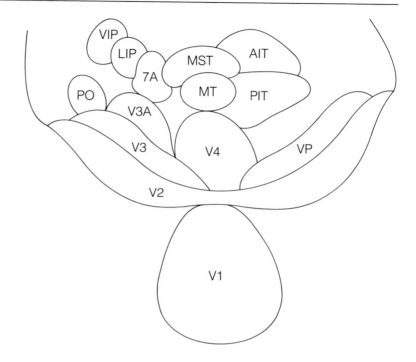

lesion confined to area V4, caused perhaps by a small stroke. Others have been known to lose the ability to see movement, although other aspects of vision—such as perception of colors, static objects, and depth—are preserved. This condition may result from a lesion in area V5.

Visual Perception

Two general themes have emerged from the study of visual processing at all levels of the visual system, from the retina to visual areas 4, 5, and beyond. These themes offer insights into the problems of visual perception.

The first is that cells and pathways are concerned with the processing of one or another aspect of the visual image. Already at the level of the photoreceptor synapse, we find that on- and off-information is segregated into two classes of bipolar cells; this segregation is maintained throughout the visual system. Furthermore, the retina's outer plexiform layer is concerned primarily

18.4 Examples of illusionary borders. Although the shapes are not drawn as such, a well-defined triangle and circle appear in the top figures. In the lower figure, a curved line separating the two halves of the figure appears to be visible.

with spatial aspects of a visual image, while the inner plexiform layer emphasizes temporal aspects of a visual signal. Two classes of ganglion cells providing information about these two fundamental aspects of the visual scene project to higher visual centers. Again, segregation of these aspects of the visual image is maintained in the lateral geniculate nucleus and throughout the cortex. Beyond visual areas V1 and V2, more specific components of the visual image—including color, form, movement, and depth—are dealt with separately and in separate cortical areas. There exist, then, parallel processing streams in the cortex that give rise to the relatively unified visual world we perceive.

The second common theme is that the visual system is not designed to make absolute judgments but rather to make comparisons. This characteristic also is seen first in the retina and is reflected in the receptive field organization of bipolar and ganglion cells. That is, the receptive fields of bipolar and many ganglion cells consist of antagonistic center-surround or color-opponent regions. And, as noted in Chapter 14, it is not the absolute intensity of light that comes from an object that makes it appear light or dark; it is the intensity of light coming from the object *relative* to the intensity of light coming from surrounding objects. The same holds true for color perception. The perceived color of an object depends very much on the colors of surrounding objects. This pattern of analyzing the visual world by making comparisons is seen throughout the visual system. It characterizes our perception of many aspects of the visual scene, including motion and even depth.

The retinal image is two-dimensional, yet we live in a three-dimensional world and perceive it so. Our visual system uses many pieces of information to reconstruct an image. If all the pieces are not consistent, or if they are not all there, the system nevertheless attempts to provide a complete and usually coherent picture. The system can be confounded, however, as anyone who has recognized a visual illusion is aware. Figure 18.4 displays examples of illusionary borders. Although no lines define the shapes, most people will easily make out a well-demarcated triangle and circle at the top and a contoured line running through the lower figure. Furthermore, the triangle and circle seem brighter than the surrounding white areas. Figure 18.5 is a marvelous example of how we judge size by comparing objects to one another and to their

18.5 Context affects the perception of size: the scene was photographed twice; once with both women present, once with only the woman in the foreground. The image of the other woman was cut out from a print of the first picture and pasted onto the second picture. She now appears much smaller than in the first picture, but she is exactly the same size in both photographs (measure for yourself!).

surround. The two women seem to be of comparable size in the left picture but radically different in size in the right picture, but the only difference between the two scenes is the inappropriate position of one of the figures.

Visual perception is, therefore, reconstructive and creative. This was appreciated in the beginning of this century by the Gestalt psychologists, who put forward the idea that the brain creates an image both by taking in information concerning the components of a visual scene and by making certain assumptions about the shapes, colors, and movements within the scene—both are needed to create a coherent visual picture. If something is missing from the scene, it is filled in; if the information is ambiguous, we perceive one thing or another but not a mixture of two percepts. An example of the latter phenomenon is the famous vase-face illusion shown in Figure 18.6. At any moment we may see two black faces in the picture or a white vase, and it is fairly easy to switch back

18.6 The face-vase illusion. At any one time two black faces or a white vase are perceived. Never are the two percepts seen simultaneously.

and forth between the two percepts. We see only one of the two percepts at a time, however.

Where in the cortex images are created, if indeed they are created, is not known. Interactions between the processing streams must occur, and perhaps at several levels. Psychological experiments have emphasized the role of attention in extracting information from the processing streams and in visual perception. To pick out an object from the surround in a scene requires that specific attention be paid to it, as exemplified by the effort needed to switch from one percept to another in the face-vase illusion.

Speech

Humans are distinguished from animals by their ability to communicate by speech and the written word. And speech, reading, and writing are controlled in the cortex. Two areas of the cortex have been identified as playing key roles in speaking and understanding the spoken or written word. Curiously, both are located in the left hemisphere. Broca's area, named after Pierre Paul Broca, a nineteenth-century French neurologist who discovered the area, is adjacent to the face region of the primary motor projection area. Wernicke's area, named after Carl Wernicke, a German psychiatrist who identified it, is in the left temporal lobe between the primary auditory and visual areas. (See Figure 18.7.)

Broca's area is involved mainly with the motor aspects of speech.

18.7 The left hemisphere of the human cortex. Broca's area is near the primary motor area, whereas Wernicke's area sits between the auditory and visual cortical areas. The face recognition area is on the underside of the occipital lobes and extends forward to the underside and inner surface of the temporal lobes.

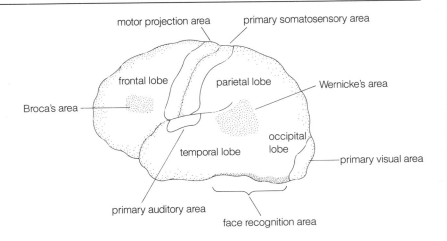

Lesions in this area cause impairments of speech articulation and language production, but word selection by the patient is usually appropriate. Interestingly, singing is not appreciably affected by lesions in Broca's area. Singing and the ability to make music appear to be controlled mainly in the right hemisphere, by an area comparable to Broca's area. Wernicke's area is concerned primarily with the comprehension of speech and with reading and writing. Patients with lesions in Wernicke's area can articulate words perfectly well, but word choice is often inappropriate. If visual input into Wernicke's area is disrupted, a person will have difficulty reading; lesions closer to the primary auditory cortex make it difficult to understand speech.

Ordinarily it makes no difference that the speech areas are confined to the left side of the brain. But if the corpus callosum, the large tract of axons that connects the two hemispheres, is cut, clear deficits can be demonstrated (see Figures 12.3 and 12.6). For example, a patient viewing an object in the left visual field, or feeling an object in the left hand, cannot describe or name the object. This is because the sensory information coming from the left is processed by the right side of the brain; since it cannot cross to the left hemisphere, where the speech centers lie, the patient cannot verbalize what is seen or felt. If the object is moved to the right visual field or to the right hand, the patient can instantly describe and name the object!

Are other mental abilities localized in one hemisphere or another? This is a question that has aroused much interest in neuroscience, and some evidence in support of additional cortical lateralization has been provided. So, for example, in addition to speech, it has been proposed that reading, writing, language acquisition, arithmetic calculations, and complex voluntary movement are localized primarily in the left hemisphere. The right hemisphere, it has been suggested, is concerned more with complex pattern recognition in vision, audition, and touch, spatial sense, intuition, and the already mentioned capacity for singing and music. The implication here is that individuals are better in one or another of these attributes because one hemisphere is more developed than the other. Although this hypothesis is intriguing, firm evidence in support of the notion that one hemisphere or the other is dominant in an individual, and governs the abilities of that individual, is not strong.

Face Recognition

Another fascinating ability that humans and perhaps some animals have is to recognize faces rapidly and reliably. This ability has been localized to an area that lies on the underside of both occipital lobes and extends forward to the underside and inner surface of the temporal lobes (see Figure 18.7). Someone with a lesion in this area no longer can identify people on the basis of appearance, something that normal individuals do easily. Other than the failure to recognize people visually, these people are usually remarkably free of other visual deficits. They can read, write, name objects, and so forth, but they cannot visually recognize even a very familiar person, including a spouse. They can recognize a familiar person by voice and immediately name them, and describe a face in detail, but they cannot associate a face with a person.

Exactly what is involved in face recognition is not well understood. For example, it is difficult to recognize a face that is upside down. Furthermore, faces that have been distorted do not appear nearly as upsetting when viewed upside down as when viewed right-side up. When viewed upside down the two faces in Figure 18.8 appear more or less equivalent. When viewed the other way around, the two faces are dramatically different.

Recordings from neurons in the monkey's cortical area comparable to the face recognition area in humans have yielded in-

---

18.8 Upside-down pictures of Mrs. Thatcher. The two pictures appear approximately equivalent when viewed upside down, but when turned right-side up they appear very different. In one of the photographs, the eyes and mouth have been inverted.

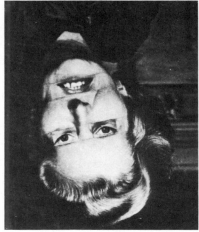

triguing results. The neurons in this region of the cortex have large receptive fields and most of them include the fovea, which means that the cells are concerned virtually exclusively with the central visual field. Most interesting is the finding that about 10 percent of the cells respond best to complex images including faces and hands. If the image is simplified by blanking out the eyes in a picture of a face, or by filling in the space between the fingers, the picture elicits a weaker response from the neuron than does the original picture.

Studying the Human Brain

The experiment cited above—concerning neurons in the monkey cortex and the monkey's response to facial images—raises the important question of how we can gain information on the human brain. Clearly we cannot record from neurons in the human cortex as we can in animals, but neither can we know if animals such as monkeys perceive stimuli in the same way we humans do. It is possible, however, to study patients who have neurological lesions or deficits, and I have cited several examples in this book. But is this the only way we can learn about cortical mechanisms in the human brain?

The answer is no. It is possible on occasion to gather more direct information on the living human brain during surgery. This approach was pioneered particularly by Wilder Penfield, a Canadian neurosurgeon, who developed a procedure in the late 1930s to treat epilepsy. Epileptic seizures occur in a brain when a population of neurons discharge abnormally and synchronously, thereby causing an involuntary and usually fixed alteration in behavior. Epilepsy can frequency be localized to a specific cortical region containing a group of abnormal cells. If those cells can be surgically removed, the epilepsy can be cured.*

Identifying the exact location of the abnormal cells and removing as little of the cortex as possible were Penfield's goals. The cortex itself has no pain fibers, so once the skull is opened and the brain exposed, it is possible to manipulate the cortex (or excise

---

*Today epilepsy often can be controlled with drugs, and so less surgery is carried out. Nevertheless, some patients do not respond to drug treatment, and surgery remains the preferred course of treatment for these individuals.

tissue) without the patient feeling any pain. Furthermore, it is possible during the operation to converse with the patient and to record sensations, feelings, and emotions. To identify the abnormal cells, Penfield stimulated the cortex electrically with a small probe. By carefully examining the effects of the electrical stimulation on the responses of the patient, he could define quite precisely the limits of the abnormal tissue. In the course of the operation, however, Penfield often stimulated normal cortex, in sensory, motor, and association areas, and he made some startling observations.

Penfield found that electrical stimulation of the primary sensory and motor areas caused two kinds of effects. In the primary sensory areas, electrical stimulation alone caused a sensory effect. For example, following electrical stimulation of the primary visual area, a flash of light might be seen; or a stimulus to the auditory area might induce the patient to hear a sound. If an electrical stimulus was applied to the primary visual cortex at the same time that a flash of light was presented to the appropriate part of the visual field, the light sensation was usually blocked. The same result could be demonstrated in the other primary sensory areas; electrical stimulation blocked the effects of a natural stimulus.

In the primary motor area, similar kinds of effects were found. Stimulation alone caused a movement of a limb, digit, or other part of the body. Stimulation of the motor cortex while that part of the body was being consciously moved stopped the movement. In the sensory and motor areas the results were highly consistent. Stimulation of part of the visual cortex always elicited a visual sensation from the same part of the visual field. Stimulation of a locus on the motor cortex always evoked a similar movement in the same part of the body. From hundreds of operations involving all parts of the cortex, Penfield and others were able to match functions to various parts of the brain and to work out the topographic representations on the cortical areas. The representations of the human body on the primary somatosensory and motor projection areas (shown in Figure 12.8) were determined in this way by Penfield and his colleagues.

Temporal Lobe Stimulation and Memory

The most surprising findings from these experiments occurred when Penfield stimulated association areas in the temporal lobes of patients. On many occasions, stimulation evoked memories so vivid that the patients believed they were reliving an experience. Related details were recalled that often were amazingly extensive, and the memory could usually be evoked again and again by repeated stimulation of the same cortical locus. In one case, a young woman could hear a song being played by instruments whenever the stimulating electrode was applied to a point on her temporal cortex. The memory always started at the same place in the song and she could hum along with the music. The music was so clear that the patient thought that there was a phonograph in the operating room! Later, when reflecting on her experience, she said: "There were instruments. It was as though it were being played by an orchestra. Definitely, it was not as though I were imagining the tune to myself. I actually heard it." She added, "I could remember much more of it right in the operating room after hearing it than I could three or four days later. The song is not as real to me now, as I recall it to memory, as it was in the operating room."

In another instance a mother felt she was in her kitchen listening to her small son, who was outside. She was even aware of neighborhood noises, including passing cars. But, at the same time, she recognized that she was in the operating room! In most instances, as these examples illustrate, the event recalled was not a major episode in the life of the patient.

What are we to make of these observations? They certainly indicate that specific memories can be evoked by stimulating a certain cortical locus, and that experiences are recorded in the brain in much more detail than we can ordinarily recall or make use of. Furthermore, the patients felt emotions during the flashback that the remembered event originally evoked; thus, stimulation of the temporal lobes brought back to consciousness an experience that happened long ago. Finally, the experiences were evoked only by temporal lobe stimulation, which suggests that they and other memories may be stored in this region of the brain.

Memory Storage: The Hippocampus
A region of the subcortex, called the hippocampus, which is tucked in behind the temporal lobes, has been implicated in the formation of long-term memories in humans and other mammals (see Figure 18.9). This was first demonstrated dramatically when the hippo-

18.9 A vertical section through the middle of the brain (see insert and compare with Figure 12.3). The corpus callosum joins the 2 hemispheres of the brain, and the hippocampi are located under the temporal lobes.

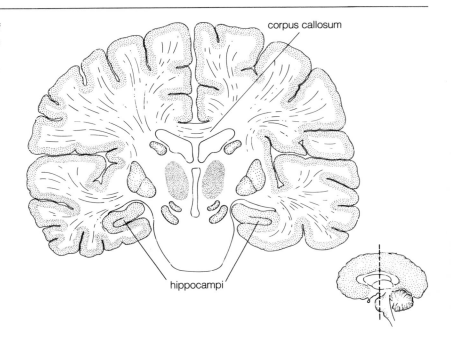

corpus callosum

hippocampi

campi on both sides of the brain were removed from a patient being treated for severe epilepsy. Much to everyone's surprise and shock, the patient no longer could store memories. His long-term memory mechanisms were permanently disrupted. Short-term memory was unaffected, but within a few minutes all recent experiences and newly learned information were lost by this unfortunate individual. Old memories were retained, but no memories of experiences after the operation were stored.*

This patient was studied for over 25 years by Brenda Milner in Canada, and virtually no changes occurred over this time in the patient's ability to retain declarative memories. Bilateral surgery

*Psychologists often distinguish two types of long-term memories: declarative memories are recollections of specific facts or events (such as the ability to name a person one has not seen for a while), and nondeclarative memories are retained motor skills or routines (such as the ability to ride a bicycle). Patients with hippocampal lesions no longer store declarative memories, but they can retain certain nondeclarative ones. There is evidence that the cerebellum is involved in learning and retaining some nondeclarative motor skills.

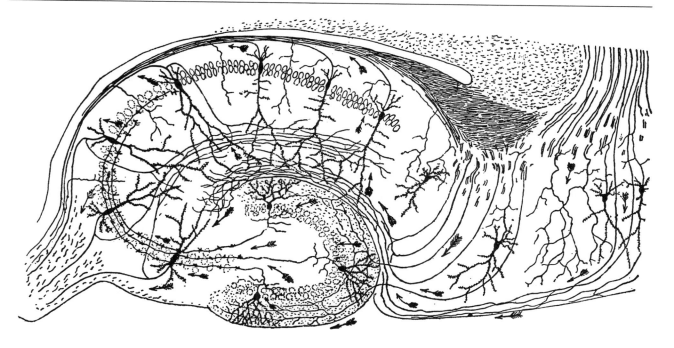

18.10   A drawing by Ramón y Cajal of the hippocampus; the arrows indicate his view of the flow of information through the hippocampus. Like the retina and cerebellum, the hippocampus has a well-defined, layered structure. Because of its role in long-term memory formation, it is under intense study.

on this region of the brain is no longer contemplated, but there are several cases reported in the literature of bilateral lesions in the hippocampi that resulted in the loss of long-term memory formation. The loss of one hippocampus, interestingly enough, does not appear to impede long-term memory formation.

Needless to say, neuroscientists have great interest in the hippocampus. Like the cerebellum and the retina, the hippocampus has a highly distinctive and relatively simple cellular organization, as illustrated in Figure 18.10, a drawing by Ramón y Cajal.

A striking physiological observation regarding the hippocampus was made in 1973 by Timothy Bliss and Terje Lømo in England. Following high-frequency stimulation of pathways leading into the hippocampus, subsequent responses recorded from hippocampal neurons are highly potentiated, which means that the postsynaptic response to a test stimulus is increased significantly. With repeated high-frequency stimuli applied to the hippocampus, the potentiation can be made to last for hours, days, and even weeks. The phenomenon has been called *long-term potentiation,* or LTP.

It is not yet clear how LTP works, but both pre- and postsynaptic mechanisms have been postulated. For example, LTP is accompanied by enhanced release of glutamate from the hippocampus, suggesting increased release of transmitter from synaptic sites. An increased number of glutamate receptors have also been detected during LTP. In addition, we have evidence that phosphorylation of ion channels and other proteins occurs during LTP. Furthermore, LTP can be blocked by the intracellular injection of a Ca^{2+} binding agent into neurons, indicating that Ca^{2+} influx is important and that Ca^{2+} serves as a second-messenger mediating LTP. The Ca^{2+} influx into hippocampal neurons is facilitated by a special glutamate channel that opens when the neurons are depolarized. This glutamate-sensitive channel is voltage-dependent and admits Ca^{2+} into a cell only when the neuron has been strongly activated by other excitatory synapses. It is called the NMDA channel because it is selectively activated by the glutamate analogue N-methyl D-aspartate. Thus, present evidence suggests that the mechanisms underlying LTP are similar in principle to the mechanisms responsible for sensitization and habituation in *Aplysia*. These include altered release of transmitter, involvement of second-messengers, special membrane channels, protein phosphorylation, and perhaps protein synthesis.

Imaging Brain Function

Over the past decade new imaging techniques have been developed that promise to tell us a great deal about human brain function. These are noninvasive techniques that can be used in both normal and diseased subjects. One of these techniques, positron emission tomography or PET scanning, is particularly useful because it allows us to observe activity in specific brain regions when a subject is asked to perform a specific neural task. In Chapter 13, I described the use of PET scanning for dissecting motor control processes. In that example, a subject was asked to move his fingers in a complex sequence, and increased activity was observed both in the primary motor area of the cortex and in premotor areas. The activity in the primary motor area could be related to the movements themselves, whereas the activity in the premotor area appeared to be directed to motor programming.

The PET scanning technique uses short-lived isotopes that contain an extra proton. The elements employed are nitrogen, oxygen, carbon, or fluorine, all of which ordinarily have stable nuclei. The extra protons in the nuclei of the isotopes rapidly break down to unstable positrons and stable neutrons. The positrons leave the nucleus and quickly collide with electrons. The particles are annihilated by the collision, but two gamma rays are produced which travel away from the site of collision at exactly 180° from one another. Detectors placed all around the subject's head respond to gamma rays, but they record an event only when two detectors 180° apart are simultaneously activated. In this way the site of gamma ray emission can be pinpointed to within a few millimeters.

The experiments are done in the following way: A biologically relevant molecule, such as CO_2, O_2, 2-deoxyglucose, or an amino acid, is labeled with an isotope and given to a subject. The isotopes last only a few minutes, which means that fairly high concentrations can be used safely. The various molecules that are administered give measures of blood flow (CO_2), oxygen consumption (O_2), glucose metabolism (2-deoxyglucose), or protein synthesis (amino acids). Increased activity in a particular brain region correlates with increased cerebral blood flow, O_2 consumption, and glucose metabolism, so areas in the brain related to a specific neural function or mental task can be identified. Simple visual stimuli activate area V1 predominantly, whereas complex visual scenes activate areas V1, V2, and other visual areas. When a subject is asked to perform a mental task involving language, activity is increased predominantly in the left hemisphere, whereas musical tasks usually increase activity in the right hemisphere. When a subject is asked to remember something, increased activity is observed in the hippocampus.

The use of this method is still in its infancy. The resolution of the images is still relatively poor, and interpretation is difficult. Nevertheless, improved PET scanning techniques should enable neuroscientists to learn much about the locus of neural activity in awake and behaving human subjects. In addition, other imaging techniques are being developed. Neuromagnetic imaging records the weak magnetic signals produced by neuronal electrical activity,

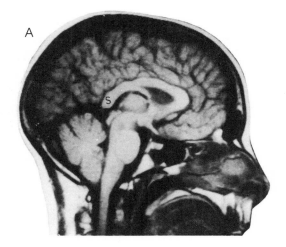

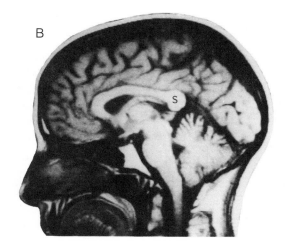

18.11 Magnetic resonance images obtained from a living human male (**A**) and female (**B**). Compare the images of the brain with the drawing of the primate brain shown in Figure 12.3. *S* marks the posterior part of the corpus callosum, which is more bulbous in the female.

producing maps of the sites of electrical activity within the brain. Magnetic resonance imaging uses imposed magnetic fields to visualize brain anatomy and provides impressive structural detail of neural structures in living subjects.

Figure 18.11, for example, shows magnetic resonance images obtained from two living human subjects, a male *(left)* and a female *(right)*. The images show that the posterior part of the corpus callosum (marked with an *S*) is more bulbous in the female brain. The significance of this difference is poorly understood; one idea is that it could relate to the extent of connectivity between the two hemispheres.

Personality and the Frontal Lobes

We still know very little about much of the cortex, and especially about most of the frontal lobes. We do know that lesions of the frontal lobes can cause significant changes in personality. This first became evident from a famous neurological case involving a railroad construction worker in New Hampshire who was struck by a metal bar propelled by an excavating explosion. The metal bar penetrated his skull and damaged his frontal lobes. The man, Phinneas Gage, survived but his personality was altered dramatically. The responsible foreman changed into an irascible and unpleasant individual.

Subsequent animal experiments confirmed that changes in behavior resulted from lesions in the frontal lobes. When it was discovered that lesions occasionally decreased levels of aggressiveness, a surgical procedure called the frontal lobotomy was introduced for the treatment of mentally ill patients. Beginning in the 1930s and extending through the 1940s and 1950s, thousands of these so-called psychosurgical operations were performed around the world, with mixed results. Although it was claimed that some patients were helped, others clearly were not. Furthermore, careful postoperative examinations of the patients indicated a significant loss of mental ability. Frontal lobotomies to treat mental illness are no longer performed. Today mental illness is usually treated with drugs, and quite successfully so.

Bits and pieces of information about the mechanisms underlying higher cortical and brain function are slowly being uncovered. At a cellular level the picture is becoming coherent—we can describe in considerable detail how individual neurons receive, process, carry, and transmit information. We know much less about the interactions of neurons that lead to perception, learning, memory, and consciousness. Some hints have emerged, and these are based mainly on a cellular analysis of neuronal behavior. But it is clear that there is much to be learned if we are to understand how we behave.

Summary

Other primary sensory areas have a cortical organization similar to the one found in the primary visual cortex. A columnar organization has been described in both the somatosensory and the auditory cortex, and many neurons in the somatosensory cortex have receptive field properties like the neurons in the visual system. For example, certain tactile neurons have a center-surround receptive field organization; others are movement-sensitive and even direction-sensitive.

Area V2, a visual association area, has a characteristic cytochrome oxidase staining pattern consisting of thick and thin dark stripes separated by pale stripes. These stripes appear to represent areas that process different aspects of the visual image. The thin dark stripes contain many color-opponent cells; the thick dark stripes have many direction-selective and disparity-tuned cells; and the pale stripes contain many end-stopped cells. It is believed that the cells in the thin dark stripes are processing color information, cells in the thick dark stripes process movement and depth information, and cells in the pale stripes process form and shape information.

Beyond area V2 are about 15 separate visual areas that process one or another aspect of the visual image. Area V4 is primarily concerned with color; area V5 (or MT) is concerned with movement and perhaps depth. Lesions in these areas lead to a specific visual deficits, such as a loss of color vision or movement detection. We can draw two conclusions from observations made at all levels in the visual system: (1) cells and pathways are specialized to process one or another aspect of the visual image; (2) the system operates by providing comparative information about the visual world, not absolute information. For example, the brightness and color of an object depend on the brightness and color of the surround. Furthermore, visual processing of illusions suggests that visual perception is reconstructive and creative. The brain constructs a coherent and logical picture by making certain assumptions as well as by utilizing incoming visual information.

Two cortical areas concerned with speech have been identified. Broca's area, adjacent to the face region of the primary motor projection area, is concerned with speech articulation; Wernicke's area, in the temporal lobe between the auditory and visual areas, is concerned with the comprehension of speech and with reading and writing. Both areas are found in the left hemisphere only, indicating a lateralization of function in the two cortical hemispheres. Face recognition is localized to an area of the cortex under the occipital and temporal lobes. Certain neurons recorded from this area in animals have receptive fields that respond best to complex images such as faces or hands.

During surgery for epilepsy, electrical stimulation of the brain is used to identify diseased cells. Information gathered during such procedures has elucidated which parts of the brain are responsible for particular functions. Stimulation of the temporal lobes often results in evoking vivid memories, implying that this area plays a role in memory storage. Lesions of the hippocampi, which are tucked under the temporal lobes, prevent long-term memory storage. Cellular studies of the hippocampal neurons have demonstrated a long-term potentiation (LTP) of synaptic potentials that may help explain how memories are stored. The functions of many areas in the cortex remain unknown, including much of the frontal lobes. Lesions in the frontal lobes often cause personality changes and other effects, including loss of intelligence.

New imaging techniques, such as position emission tomography or PET scanning, promise to reveal much about higher brain function. These noninvasive techniques, which can locate activity in specific brain regions, can be used with human subjects. Thus, not only can areas concerned with a specific neural function be identified, but also those parts of the brain concerned with cognitive mental tasks.

Further Reading

BOOKS

Penfield, W., and J. Jasper. 1954. *Epilepsy and the Functional Anatomy of the Human Brain*. Boston: Little, Brown. (A detailed review of Penfield's brain stimulation experiments.)

Springer, S. P., and G. Deutsch. 1989. *Left Brain, Right Brain,* 3d ed. New York: W. H. Freeman. (A readable discussion of the lateralization of function in the brain.)

Valenstein, E. S. 1986. *Great and Desperate Cures*. New York: Basic Books. (A fascinating account of the rise and fall of psychosurgery.)

ARTICLES

Bliss, T. V. P., and T. Lømo. 1973. Long-lasting potentiation of synaptic transmission in the dentate area of the anaesthetized rabbit following stimulation of the perforant path. *Journal of Physiology* (London) 232:331–356. (The first description of long-term potentiation.)

DeYoe, E. A., and D. C. Van Essen. 1988. Parallel processing streams in monkey visual cortex. *Trends in Neuroscience* 11:219–226.

Gazzaniga, M. S. 1967. The split brain in man. *Scientific American* 217:24–29.

Geschwind, N. 1979. Specializations of the human brain. *Scientific American* 241(3):108–117.

Goldman-Rakič, P. 1984. The frontal lobes: Uncharted provinces of the brain. *Trends in Neuroscience* 7:425. (Introduction to a special issue devoted to the frontal lobes.)

Livingstone, M., and D. Hubel. 1988. Segregation of form, color, movement, and depth: Anatomy, physiology, and perception. *Science* 240:740–749.

Madison, D. V., R. C. Malenka, and R. A. Nicoll. 1991. Mechanisms underlying long-term potentiation of synaptic transmission. *Annual Review of Neuroscience* 14:379–397.

Mountcastle, V. B. 1957. Modality and topographic properties of single neurons of cat's somatic sensory cortex. *Journal of Neurophysiology* 20:408–434. (The paper that first described cortical columns.)

——— 1986. The neural mechanisms of cognitive functions can now be studied directly. *Trends in Neuroscience* 9:505–508.

Posner, M. I., S. E. Petersen, P. T. Fox, and M. E. Raichle. 1988. Localization of cognitive operations in the human brain. *Science* 240:1627–1631.

Raichle, M. E. 1983. Positron emission tomography. *Annual Review of Neuroscience* 6:249–267.

Squire, L. R., and S. Zola-Morgan. 1991. The medial temporal lobe memory system. *Science* 253:1380–1386.

Zeki, S. 1990. Colour vision and functional specialisation in the visual cortex. *Discussions in Neuroscience* 6:11–64.

APPENDIX
CREDITS
INDEX

Electrical Charge and Circuits

ELECTRICAL CHARGE is a fundamental property of matter, and electrical forces determine the chemical properties of matter. Charge comes in two *polarities*, negative and positive. Charges of the same polarity repel; charges of the opposite polarity attract.

ATOMS AND IONS

Electrons possess negative charge, whereas protons possess positive charge. In atoms, there are equal numbers of electrons and protons and the charges exactly cancel one another—atoms are electrically neutral. Put in other words, the negative charge on an electron is exactly the same magnitude but of opposite polarity to the charge on a proton. Atoms can gain or lose electrons and *become* charged. Charged atoms are termed *ions*. Ions with more electrons than protons are negative ions or *anions* (an example is Cl^-); ions with fewer electrons than protons are positive ions or *cations* (such as Na^+).

POTENTIAL AND VOLTAGE

Whenever charges are separated, a force exists between them. Stationary electrical charges attract (if they are of opposite polarity) or repel (if of like polarity) one another with a force proportional to the product of the magnitude of the charges and inversely proportional to the square of the distance between the charges. *Voltage (V)* is a measure of the work required to move a unit of charge from one point to another; also called *potential difference*, it is measured in *volts*.

CURRENT

In a region of charge separation, charges will move if they can. Positive charges move toward regions of negativity, whereas negative charges move toward positivity. The movement of charge is *current (I)*. Current is the rate of movement of charge past a point;

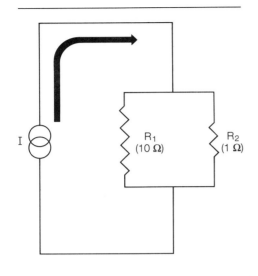

it is measured in *coulombs* per second or *amperes (A)*. (A coulomb is equivalent to the charge on 6×10^{18} electrons.) In wires, current is carried by free electrons, which move through the metal. In cells, current is carried by positive and negative ions that move in aqueous solutions.

RESISTANCE

Neither electrons through metal wires nor ions through solutions in cells move unimpeded. The difficulty that charged particles encounter when moving through a medium is termed *resistance (R)*. The unit of electrical resistance is the *ohm (Ω)*. *Conductance (g)*, the reciprocal of resistance, measures the ease with which current flows through a medium. Its units are *siemens (s)* and 1 siemen = 1/ohm.

ELECTRICAL CIRCUITS AND OHM'S LAW

Current flows through *conductors*—materials that let charged particles (electrons or ions) move. *Insulators,* on the other hand, are very resistant to current flow. The path that a current flows along defines an *electrical circuit. Ohm's law* is the fundamental law that all electrical circuits obey:

$$V = IR$$

If voltage is constant, the amount of current flowing depends inversely on the resistance of the circuit. The higher the resistance, the lower the current, and vice versa. Conversely, if the current flowing in a circuit is kept constant, the voltage varies directly with the resistance. It takes a higher potential (voltage) to drive the same amount of current through a high resistance.

No current is lost in a circuit; the current flowing into a circuit must be equal to the current flowing out of it. I illustrate this with the simple circuit shown in Figure A.1, which connects a current generator and two resistors in parallel, the one on the left having ten times the resistance of the one on the right. (Symbols that are used to indicate various electrical components are shown in Figure A.2.) Current flows through the path of least resistance; that means, in the circuit shown in Figure A.1, most of the current (about 90 percent) flows through the right-hand resistor. But the current returning to the generator is equal to the current leaving it: none is lost traveling around the circuit.

A.2 Symbols used in electrical circuit diagrams.

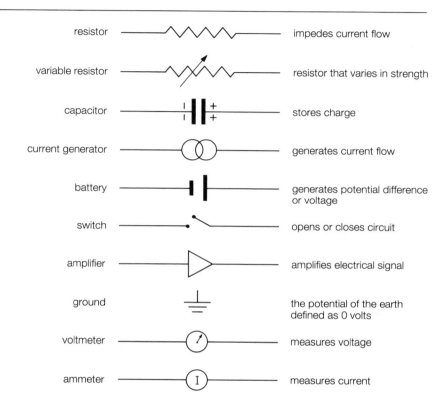

resistor		impedes current flow
variable resistor		resistor that varies in strength
capacitor		stores charge
current generator		generates current flow
battery		generates potential difference or voltage
switch		opens or closes circuit
amplifier		amplifies electrical signal
ground		the potential of the earth defined as 0 volts
voltmeter		measures voltage
ammeter		measures current

Circuits can be very complex, especially in biological systems. Current can take many paths, but the important point to remember is that an electrical circuit is always complete. That is, the current flowing back is identical in magnitude to the current leaving the point of current origin, although it may follow different paths in its return.

CAPACITANCE

Capacitance introduces a time element into the flow of current around a circuit and the changes in voltage that occur, and thus it is an important concept for our understanding of electrical signaling by cells. A capacitor stores charge and thus prevents it from flowing through the circuit. The units of capacitance are *farads*.

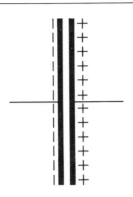

A.3 In a charged capacitor, excess negative charge on one plate results in excess positive charge on the other.

Capacitors consist of two conducting regions separated by an insulator. In electronic devices they are made up of two metal plates separated by air or another nonconductive medium. When current begins to flow in an electronic circuit that contains a capacitor, electrons flow into one plate of the capacitor but they cannot cross to the other plate because of the intervening insulator. As negative charge builds up on one plate, however, it drives negative charge from the other plate. The other plate then becomes positive and partially neutralizes the effect of the charge on the first plate. This is illustrated in Figure A.3. Larger plates store more charge and plates closer together have greater capacitance.

A cell membrane acts as a capacitor; the membrane is a thin insulator separating two conductive solutions (the intracellular and extracellular fluids). Charges on one side of the membrane affect the distribution of charge on the other side of the membrane. As charge builds up on one side of the membrane, interactions with charges on the other side of the membrane occur, initially buffering the buildup of voltage or potential difference across the membrane.

A membrane's capacitance can be illustrated by a circuit with a resistor and capacitor in parallel (see Chapter 3 and Figure A.4). When current begins to flow in the circuit, most of the current initially flows into the capacitor, little through the resistor. Thus, in accord with Ohm's law, little voltage initially occurs across the resistor. As charge builds up on the capacitor, more current flows through the resistor and voltage increases, but relatively slowly. Once the capacitor is fully charged, all of the current flows through the resistor and the voltage is maximal. The change in voltage is shown in graphic form in Figure A.4: the current is turned on and off instantly by the switch, but voltage builds up and dissipates slowly as the capacitor charges up and discharges.

By moving the plates of a capacitor apart, the charge interactions decrease and the capacitance decreases. The same effect occurs in neural tissue by *myelination* of nerve cell axons (see Chapter 2). Membrane is wrapped around and around a myelinated axon, essentially increasing the thickness of the insulator separating the conducting solutions.

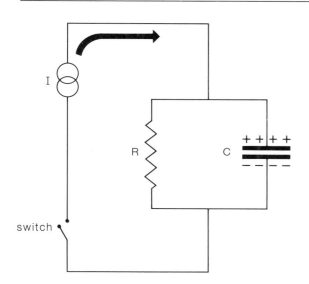

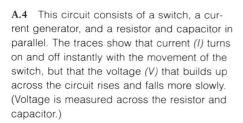

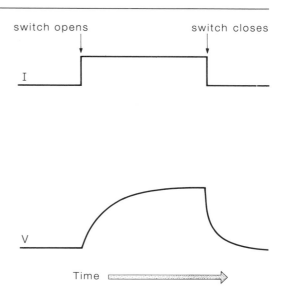

A.4 This circuit consists of a switch, a current generator, and a resistor and capacitor in parallel. The traces show that current *(I)* turns on and off instantly with the movement of the switch, but that the voltage *(V)* that builds up across the circuit rises and falls more slowly. (Voltage is measured across the resistor and capacitor.)

Credits

Figure 1.4 J. E. Dowling, "Synaptic organization of the frog retina: An electron microscopic analysis comparing the retinas of frogs and primates," *Proceedings of the Royal Society, B* 170(1968):205–228. Used by permission of The Royal Society, London.

Figure 1.5 O. F. K. Deiters, *Untersuchungen über Gehirn und Ruckenmack des Menschen und der Saugethiere* (Braunschweig: Vieweg und Sohn, 1865). An original copy of this publication was kindly provided by Sanford L. Palay, M.D.

Figure 1.6 E. Crosby, T. Humphrey, and E. Lauer, *Correlative Anatomy of the Nervous System* (New York: Macmillan, 1962). Used by permission of Macmillan Publishing Co.

Figure 1.7 S. Ramón y Cajal, *Histologie du Système Nerveux de l'Homme et des Vertébrés*, 2 vols. (Paris: A. Maloine, 1911; reprint, Madrid: Consejo Superior de Investigaciones Científicas, Instituto Ramón y Cajal, 1955). Used by permission of Instituto Ramón y Cajal.

Figure 1.8 Photograph kindly provided by Antonio Gallego, Instituto Ramón y Cajal, Madrid.

Figure 1.9A B. Ehinger, B. Falck, and A. M. Laties, "Adrenergic neurons in teleost retina," *Zeitschrift für Zellforschung und Mikroskopische Anatomie* 97(1969):285–297. Used by permission of the authors and Springer-Verlag.

Figure 1.9B Photomicrograph courtesy of Charles L. Zucker, Eye Research Institute, Boston.

Figure 1.10 C. L. Zucker and J. E. Dowling, "Centrifugal fibres synapse on dopaminergic interplexiform cells in the teleost retina," *Nature* 330(1987):166–168. Copyright 1987 Macmillan Magazines Ltd.; reprinted by permission.

Figure 2.1 S. Ramón y Cajal, *Histologie du Système Nerveux de l'Homme et des Vertébrés*, 2 vols. (Paris: A. Maloine, 1911; reprint, Madrid: Consejo Superior de Investigaciones Científicas, Instituto Ramón y Cajal, 1955). Used by permission of Instituto Ramón y Cajal.

Figure 2.4B S. L. Nam, K. J. Kim, J. W. Leem, K. S. Chung, and J. M. Chung, "Fiber counts at multiple sites along the rat ventral root after neonatal peripheral neurectomy or dorsal rhizotomy," *Journal of Comparative Neurology* 290(1989):336–342. Copyright © 1989; reprinted by permission of Wiley-Liss, a Division of John Wiley and Sons, Inc., New York, and the authors.

Figure 2.5A,C Photomicrographs courtesy of Cedric S. Raine, Albert Einstein College of Medicine, New York.

Figure 2.7 S. K. Fisher and B. B. Boycott, "Synaptic connexions made by horizontal cells within the outer plexiform layer of the retina of the cat and the rabbit." *Proceed-*

Index